18th Russian Youth Conference on Physics of Semiconductors and Nanostructures, Opto- and Nanoelectronics (RYCPS 2016)

Journal of Physics: Conference Series Volume 816

St. Petersburg, Russia
28 November – 2 December 2016

Editors:

R.A. Suris
L.E. Vorobjev
D.A. Firsov
V.A. Shalygin

ISBN: 978-1-5108-3913-7
ISSN: 1742-6588

Printed from e-media with permission by:

Curran Associates, Inc.
57 Morehouse Lane
Red Hook, NY 12571

CURRAN ASSOCIATES INC. proceedings.com

Some format issues inherent in the e-media version may also appear in this print version.

Printed by Curran Associates, Inc. (2017)

For permission requests, please contact the Institute of Physics
at the address below.

Institute of Physics
Dirac House, Temple Back
Bristol BS1 6BE UK

Phone: 44 1 17 929 7481
Fax: 44 1 17 920 0979

techtracking@iop.org

Additional copies of this publication are available from:

Curran Associates, Inc.
57 Morehouse Lane
Red Hook, NY 12571 USA
Phone: 845-758-0400
Fax: 845-758-2633
Email: curran@proceedings.com
Web: www.proceedings.com

Table of contents

Volume 816

18th Russian Youth Conference on Physics of Semiconductors and Nanostructures, Opto- and Nanoelectronics

28 November to 2 December 2016, St Petersburg, Russian Federation

Accepted papers received: 23 February 2017
Published online: 11 April 2017

Preface

Papers

Bulk properties of semiconductors

Structure growth, surface and interfaces

Heterostructures, superlattices and quantum wells

Quantum dots, quantum wires and other low-dimensional systems

Editorial Preface

The 18th Russian Young Conference on Physics of Semiconductors and Nanostructures, Opto- and Nanoelectronics was held on November 28 – December 2, 2016 at Peter the Great St. Petersburg Polytechnic University.

The program of the Conference included semiconductor technology, heterostructures with quantum wells and quantum dots, opto- and nanoelectronic devices and new materials. A large number of participants with more than 200 attendees from many regions of Russia provided a perfect platform for the valuable discussions between students and experienced scientists.

Conference included two invited talks given by Doctor of Science I.S. Burmistrov ("Topology in condensed matter physics") and Dr. N.V. Kryzhanovskaya ("Micro-lasers with whispering-gallery modes"). Students, graduate and postgraduate students presented their results on plenary and poster sessions. The total number of accepted papers published in Russian (the official conference language) was 138. Here we publish 42 of them in English.

Like previous years, the participants were involved in the competition for the best report. Certificates and cash prizes were awarded to a number of participants for the presentations selected by the Program Committee. Two special E.F. Gross Prize were given for the best presentation in semiconductor optics. Works with potential applications were recommended for participation in the following competition for support from the Russian Foundation for Assistance to Small Innovative Enterprises in Science and Technology.

The conference was supported by the Russian Foundation for Basic Research and the innovation company "ATC – Semiconductor Devices", St. Petersburg.

The official conference website is http://www.semicond.spbstu.ru/conf2016

Editors

R.A. Suris
L.E. Vorobjev
D.A. Firsov
V.A. Shalygin

IOP Conf. Series: Journal of Physics: Conf. Series **816** (2017) 011001 doi:10.1088/1742-6596/816/1/011001

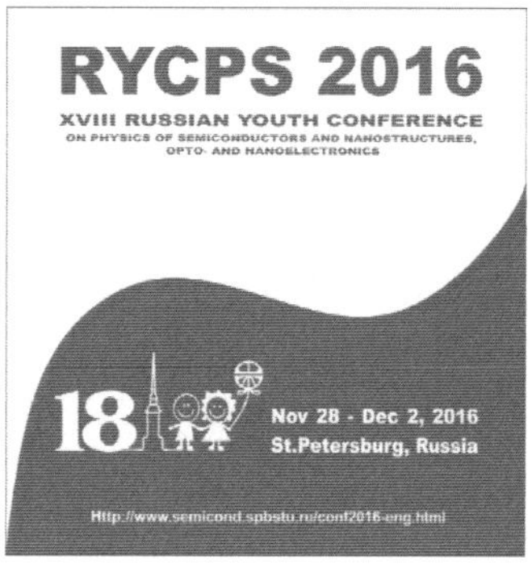

Winners of the E.F. Gross Prize for the best presentation in semiconductor optics

Michael Roiz

A postgraduate student at Ioffe Institute
(Saint Petersburg)

Presentation:

"Semiconductor disc whispering-gallery-mode lasers for mid IR range"

Eldar Stepanets-Hussein

A student at Moscow Institute of Physics and Technology
(Moscow)

Presentation:

"Coherent spin dynamics in 2D electron system near odd filling factors "

IOP Conf. Series: Journal of Physics: Conf. Series **816** (2017) 011001 doi:10.1088/1742-6596/816/1/011001

Winners of the competition for the best report

First Prize

Dmitry Veselov

a postgraduate student at Ioffe Institute
(Saint Petersburg)

Presentation:

"*Internal optical loss and current leakage in high-power laser diodes for the 1400–1600 nm spectral range* "

First Prize

Mikhail Ivanov

a student at Peter the Great St. Petersburg Polytechnic University

Presentation:

" *Quasi-streamer mode of delayed avalanche breakdown initiated by technological imperfections* "

Peer review statement

All papers published in this volume of *Journal of Physics: Conference Series* have been peer reviewed through processes administered by the proceedings Editors. Reviews were conducted by expert referees to the professional and scientific standards expected of a proceedings journal published by IOP Publishing.

IOP Conf. Series: Journal of Physics: Conf. Series **816** (2017) 012001 doi:10.1088/1742-6596/816/1/012001

The spin-flip scattering effect in the spin transport in silicon doped with bismuth

A A Ezhevskii[1], A P Detochenko[1], A V Soukhorukov[1], D V Guseinov[1], A V Kudrin[1], N V Abrosimov[2] and H Riemann[2]

[1]Lobachevsky State University of Nizhni Novgorod, 23 Gagarin Avenue, Nizhni Novgorod 603950, Russia
[2]Leibniz Institute for Crystal Growth, 2 Max-Born str., Berlin D-12489, Germany

E-mail: ezhevski@phys.unn.ru

Abstract. Spin transport of conduction electrons in silicon samples doped with bismuth in the $1.1 \cdot 10^{13}$ - $7.7 \cdot 10^{15}$ cm^{-3} concentration range was studied by the Hall effect measurements. The dependence of the Hall voltage magnitude on the magnetic field is the sum of the normal and spin Hall effects. The electrons are partially polarized by an external magnetic field and are scattered by the bismuth spin-orbit potential. Spin-flip scattering results in the additional electromotive force which compensates the normal Hall effect in strong magnetic fields.

1. Introduction

Silicon doped with donors having a large spin-orbit potential (such as Bi or Sb) is an attractive material for semiconductor spintronics due to a possibility of realization of the spin transport in these structures. Furthermore, the weaker spin-orbit interaction in silicon provides a longer lifetime of the excited spin state in comparison with A_3B_5 compounds.

In a recent paper [1], we have shown that a deviation from the linear dependence of the Hall resistance in the magnetic field in the temperature range 10 - 80 K is the manifestation of the spin Hall effect. The deviation decreases with rise of bismuth concentration as well as with temperature increasing. We have previously reported that the Elliott-Yafet mechanism [2,3] prevails in the spin-lattice relaxation of conduction electrons in silicon doped with bismuth. So, these results were explained by spin-dependent and spin flip scattering induced by heavy bismuth impurity centres.

In this work, we experimentally demonstrate how doping with bismuth affects the galvanomagnetic properties of silicon and propose the theoretical model of spin-flip scattering which explains the experimental results.

2. Samples and experimental technique

Natural silicon samples doped with bismuth in the $1.1 \cdot 10^{13}$ - $7.7 \cdot 10^{15}$ cm^{-3} concentration range were prepared by the floating-zone growth process in the Leibniz Institute for Crystal Growth (Berlin, Germany).

The contribution of spin-dependent scattering to the Hall electromotive force (EMF) was studied with the Hall technique using a measuring system with a Janis Research CCS-300S/202 Cryostat.

IOP Conf. Series: Journal of Physics: Conf. Series **816** (2017) 012001 doi:10.1088/1742-6596/816/1/

3. Experimental results and discussion

The experimental dependencies of the Hall voltage on the magnetic field represent the sum of the normal and the spin Hall effects at $T = 30$ K (figure 1). Furthermore, the spin Hall effect compensates the normal one in the strong external magnetic field. It is easy to see that increasing the concentration of impurities leads to increasing the contribution of the spin Hall effect.

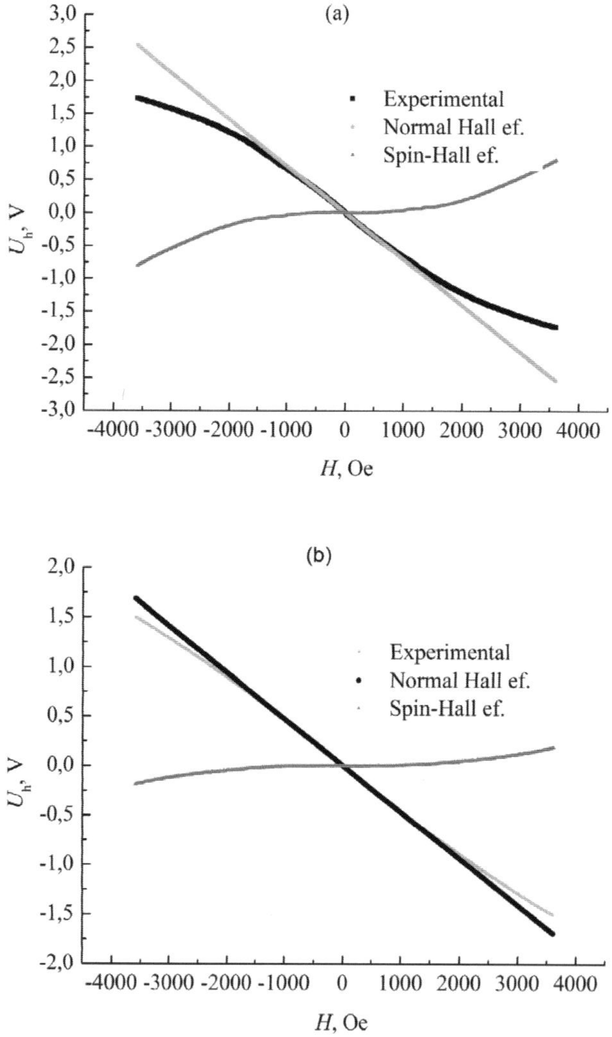

Figure 1. The dependencies of the Hall voltage on the magnetic field at $T = 30$ K. (a) - $N_d = 7.7 \cdot 10^{14}$ cm^{-3}, (b) - $N_d = 2.4 \cdot 10^{13}$ cm^{-3}.

Polarization of conduction electrons is necessary for an appearance of the electromotive force due to the spin Hall effect. Our samples are not ferromagnetic, therefore, only the mechanism of spin polarization by the magnetic field is obvious. We calculated the contribution of the spin Hall effect, taking into account only the polarization of carriers by the magnetic field and using the formulas presented in [4]. The calculation results are presented in figure 2. It is easy to see the linear dependence of the Hall voltage on the magnetic field which is quite a good description of the experimental dependence in the weak magnetic field. However, the experimental dependence is growing faster than the theoretical one with increasing magnetic field. Thus, there is another mechanism of carriers polarization which is associated with spin-dependent scattering induced by heavy bismuth impurity centres.

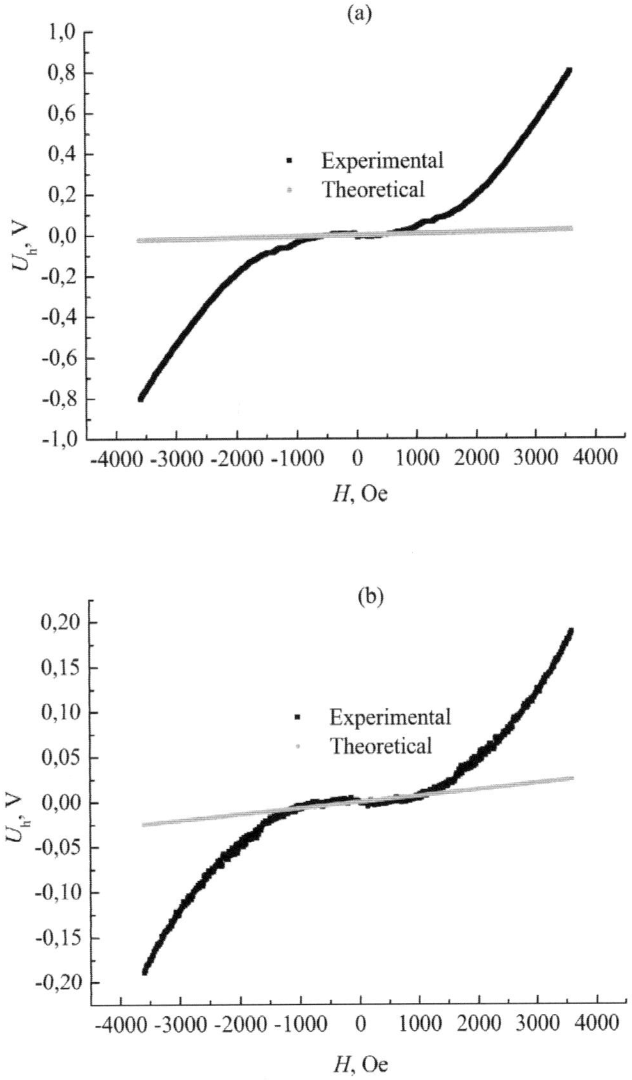

Figure 2. The theoretical and experimental dependencies of the Hall EMF on the magnetic field at $T = 30$ K. (a) - $N_d = 7.7 \cdot 10^{14}$ cm^{-3}, (b) - $N_d = 2.4 \cdot 10^{14}$ cm^{-3}.

We offer the following model of the spin-flip scattering induced by heavy bismuth impurity centres: electrons are scattered through the polarized Bi donor and change the spin orientation in accordance with the direction of the donor polarization (figure 3).

The time variation of the changing of the spin-up carrier density polarized due to spin-flip scattering $\frac{dn_{\uparrow}^{\text{flip}}}{dt}$ has the form:

$$\frac{dn_{\uparrow}^{\text{flip}}}{dt} = CN_{\uparrow}n_{\downarrow}W_{\downarrow\uparrow}, \tag{1}$$

where C is constant, $N_{\uparrow}$ is the spin-up donor concentration, $n_{\downarrow}$ is the spin-down electron concentration, $W_{\downarrow\uparrow}$ is the probability of electron spin flip from the position $\downarrow$ to $\uparrow$. The similar expression can be written for the opposite direction of the spin of the free carriers and donors.

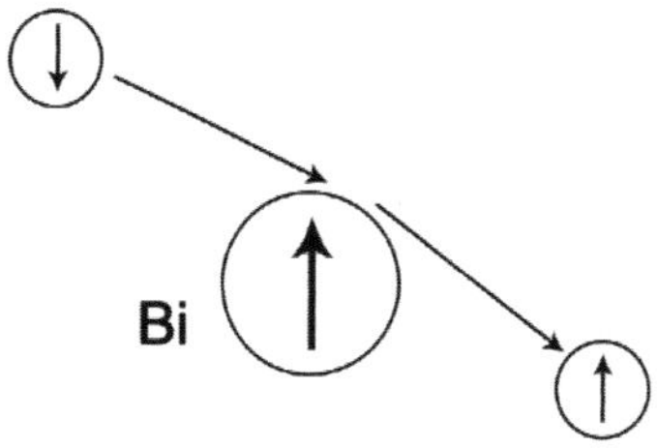

Figure 3. The schematic representation of the spin-flip mechanism of polarization of the conduction electrons.

Taking into account the coupling of the probability of electron spin flip with the relaxation time, the change of the concentration of free carriers with spin up due to spin-flip scattering can be written as:

$$\Delta n_{\uparrow}^{\text{flip}} = CNn\left(\frac{1}{4} - \left(\frac{e^{\frac{g\mu_B H}{kT}} - 1}{e^{\frac{g\mu_B H}{kT}} + 1}\right)^2\right)\left(\frac{1}{\tau_{\uparrow}} - \frac{1}{\tau_{\downarrow}}\right)\Delta\tau, \tag{2}$$

where $\Delta\tau$ is the integration time, N is the donor concentration, n is the full concentration of the free carriers, g is the g-factor, H is the magnetic field strength, μ_B is the Bohr magneton, k is the Boltzmann constant, T is temperature, $\tau_{\uparrow}$ and $\tau_{\downarrow}$ are the life times of the spin-up and spin-down electrons, respectively.

The electromotive force according to [4] has the form:

$$V_{\text{SH}} = 4\pi R_{\text{SH}} L j_x \mu_B (n_{\uparrow} - n_{\downarrow}) \tag{3}$$

where R_{SH} is the spin Hall constant, L is the sample length, j_x is the current density which has three terms in our structures:

$$j = j_{\text{dr}}^{0} + \Delta j + \Delta j^{\text{flip}}, \tag{4}$$

where the first term is the drift current density, the second one is the change of current density due to the magnetic field, the third one is the change of current density due to the spin-flip scattering. The concentrations of electrons with specific spin orientations also have three components:

$$\begin{aligned} n_{\uparrow} &= n_{\uparrow}^{0} + \Delta n_{\uparrow} + \Delta n_{\uparrow}^{\text{flip}} \\ n_{\downarrow} &= n_{\downarrow}^{0} - \Delta n_{\downarrow} - \Delta n_{\downarrow}^{\text{flip}} \end{aligned} \tag{5}$$

IOP Conf. Series: Journal of Physics: Conf. Series **816** (2017) 012001 doi:10.1088/1742-6596/816/1/012001

where the first terms are the initial concentrations of the carriers with a certain spin, the second ones are the changes of the electron concentrations due to the magnetic field, the third ones are the changes of the electron concentrations due to the spin-flip scattering

Then, the expression for the EMF spin Hall effect can be represented as:

$$V_{\mathrm{SH}} = 8\pi R_{\mathrm{SH}} L e \upsilon_x \mu_{\mathrm{B}} \left(2n^2 \frac{e^{\frac{g\mu_{\mathrm{B}}H}{kT}} - 1}{e^{\frac{g\mu_{\mathrm{B}}H}{kT}} + 1} + n^2 N \frac{1}{\tau_{\mathrm{s}}} C(2\Delta\tau\upsilon_x) \frac{e^{\frac{g\mu_{\mathrm{B}}H}{kT}} - 1}{e^{\frac{g\mu_{\mathrm{B}}H}{kT}} + 1} \left[\frac{1}{4} - \left(\frac{e^{\frac{g\mu_{\mathrm{B}}H}{kT}} - 1}{e^{\frac{g\mu_{\mathrm{B}}H}{kT}} + 1} \right)^2 \right] \right) \tag{6}$$

where υ_x is the velocity of the free carriers.

Then, the total Hall electromotive force appearing in our samples has the form:

$$V_{\mathrm{H}} = R_{\mathrm{H}} L H e n \upsilon_x - V_{\mathrm{SH}}, \tag{7}$$

where the first term is the normal Hall effect, R_{H} is the constant of the normal Hall effect. Figure 4 shows the theoretical dependencies of the Hall voltage on the magnetic field at $T = 30$ K.

It is easy to see from the equation (6) and figure 4 that the spin-flip process (gray) gives a cubic contribution and a nonlinear dependence of the Hall EMF on the magnetic field. As a result, the form of the theoretical dependence repeats the experimental one.

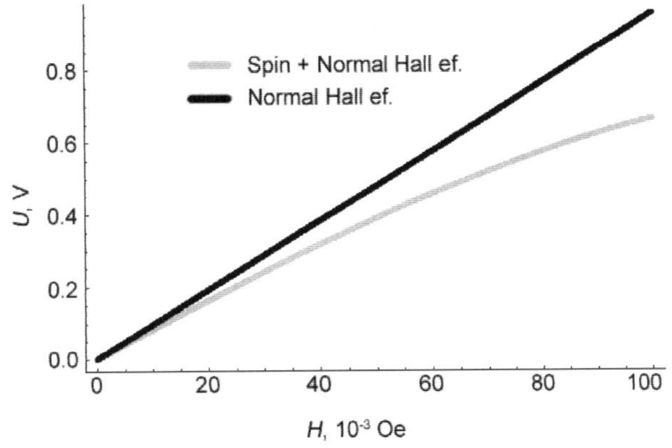

Figure 4. The theoretical dependencies of the Hall voltage on the magnetic field at $T = 30$ K.

4. Conclusion

The dependence of the Hall voltage magnitude on the magnetic field is the sum of the normal and spin Hall effects. The electrons are partially polarized by an external magnetic field and are scattered by the bismuth spin-orbit potential. Spin-dependent scattering results in the additional electromotive force which leads to a nonlinear dependence of the Hall EMF on the magnetic field. The spin-flip scattering of electrons in bismuth makes a significant contribution to the electromotive force spin Hall effect.

5. References

[1] Soukhorukov A V, Guseinov D V, Kudrin A V, Popkov S A, Detochenko A P, Koroleva A V, Ezhevskii A A, Konakov A A, Abrosimov N V and Riemann H 2016 *Sol. State Phenom.* **242** 327

[2] Elliott R J 1954 *Phys. Rev.* **96** 266

[3] Yafet Y 1963 *Sol. State Phys.* edited by Seitz F and Turnbull D **14** 1

[4] Hirsch J E 1999 *Phys. Rev. Lett.* **83** 1834

Nature of intensive defect-related broadband luminescence of heavily doped $Al_xGa_{1-x}N$:Si layers

I V Osinnykh[1,3], T V Malin[1], V F Plyusnin[2,3], K S Zhuravlev[1,3], B Ya Ber[4] and D Yu Kazantsev[4]

[1]Rzhanov Institute of Semiconductor Physics of the Siberian Branch of the Russian Academy of Sciences, 13 Lavrentieva av., Novosibirsk 630090, Russia

[2]Voevodsky Institute of Chemical Kinetics and Combustion of the Siberian Branch of the Russian Academy of Sciences, 3 Institutskaya, Novosibirsk, 630090, Russia

[3]Novosibirsk State University, Pirogova 2, Novosibirsk 630090, Russia

[4]Ioffe Physical Technical Institute of the Russian Academy of Sciences, 26 Polytekhnicheskaya str., St. Petersburg 194021, Russia

igor-osinnykh@isp.nsc.ru

Abstract. We report photoluminescence investigations of heavily doped $Al_xGa_{1-x}N$:Si films grown by molecular beam epitaxy on sapphire substrates. The wide intensive defect-related band dominates in the photoluminescence spectra of $Al_xGa_{1-x}N$:Si films with the Al content higher than 0.46 covering the whole visible spectral range. This band is attributed to donor-acceptor transitions. The acceptor ionization energy of about 1.87 eV for heavily doped AlN:Si was obtained, decrease of Al content leads to decrease of the acceptor ionization energy. The donor was assigned to the Si atom on the Ga/Al site; the acceptor might be the (2-/3-) transition level of the V_{Al}.

1. Introduction

GaN and its alloys with AlN have emerged as promising semiconductor materials with applications to laser sources for the blue-green spectral range. Bokhan et al. resently reported the finding of. a blue-green superluminescence related to optical transitions through levels of defects in heavily doped $Al_xGa_{1-x}N$:Si layers with high Al content covering the whole visible and near-infrared region of the spectrum [1]. This luminescence band was attributed to superposition of donor-acceptor and free electron-acceptor transitions involving the same acceptor [2]. Super-radiance within the broad band was obtained by pumping these samples up with an electron beam with power more than 200 kW. Overall, however, the study of luminescent properties and obtaining stimulated emission in the optical transitions via defect levels in wide band-gap solid alloys $Al_xGa_{1-x}N$:Si have not been given attention.
The aim of this work was further study of the characteristics of broadband emission from heavily doped $Al_xGa_{1-x}N$:Si layers by steady-state and time-resolved PL spectroscopy, as well as photoluminescence excitation (PLE) spectroscopy.

IOP Conf. Series: Journal of Physics: Conf. Series **816** (2017) 012002 doi:10.1088/1742-6596/816/1/012002

2. Experimental details

We investigated 1.2 μm-thick $Al_xGa_{1-x}N$:Si layers grown by molecular beam epitaxy techniques using ammonia as a source of active nitrogen on nitridated (0001)-oriented 400-μm-thick sapphire substrates with an about 300-nm-thick AlN nucleation layer. The Al content (x_{Al}) in the layers was set by the Al-to-Ga flux ratio during the growth. The mix of 0.7% silane (SiH_4) with nitrogen (N_2) was used as a source of Si. The first series of studied layers was grown at the constant silane flux $F_{SiH4} = 3$ sccm, x_{Al} was varied from 0 to 1. The second series of studied layers was grown at the constant $x_{Al} = 0.67$, F_{SiH4} was varied from 0 to 6 sccm. The silicon concentration in the layers was estimated to be about $(1.5\text{-}2.0)\times10^{20}$ cm^{-3} at $F_{SiH4} = 3$ sccm by secondary-ion mass spectrometry (SIMS) using an IMS7f (CAMECA) setup with primary Cs^+ ions. The real x_{Al} in $Al_xGa_{1-x}N$ layers was also obtained from SIMS data. The sample charge was neutralized by the EB irradiation of the samples. The concentration of free electrons in the samples under study was determined by measuring the Hall effect in the Van der Pauw geometry in a constant magnetic field of 0.5 T at room temperature. Measurements of photoluminescence (PL) of the studied samples were conducted in a wide temperature range of 5-1200 K using a continuous He-Cd laser with a photon energy $E_{ph} = 3.81$ eV and a pulsed 4th harmonic of Nd:YLF -laser ($E_{ph} = 4.71$ eV) for excitation. The studied samples were mounted in a helium closed-circuit refrigerator for measurements below 300 K and in a custom-made vacuum camera for measurements above 300 K. Measurements of photoluminescence excitation (PLE) spectra were carried out at room temperature on a FLS920 spectrofluorimeter with a 450 W xenon lamp as the excitation source. This setup allowed the measurement of both PLE and PL spectra.

3. Results

The PL spectra of heavily silicon doped $Al_xGa_{1-x}N$ layers for the first series at room temperature are shown in figure 1. PL spectra of the heavily doped GaN structure ($x_{Al} = 0$) comprise only the band-edge ultraviolet band ($E_{em} \approx 3.6$ eV) related to band-to-band transitions, the defect-related "yellow" band typical for undoped GaN is reduced likely because of Si incorporation in Ga vacancies which are considered as the origin of the "yellow" band. The free-electron concentration in the GaN structure was about 1.1×10^{20} cm^{-3}. The rise of x_{Al} leads to decrease of both the intensity of band-edge PL and the free-electron concentration (see figure 2) whereas the very weak defect-related PL appears in the visible spectral range. The wide defect-related PL band dominates in PL spectra of the $Al_xGa_{1-x}N$ layers with $x_{Al} > 0.46$. The increase of Al content from $x_{Al} = 0.5$ to $x_{Al} = 1$ leads to the blue shift of the energy position of the PL band from the red-orange spectral range ($E_{em} \approx 2.0$ eV) to the violet spectral range ($E_{em} \approx 3.1$ eV).

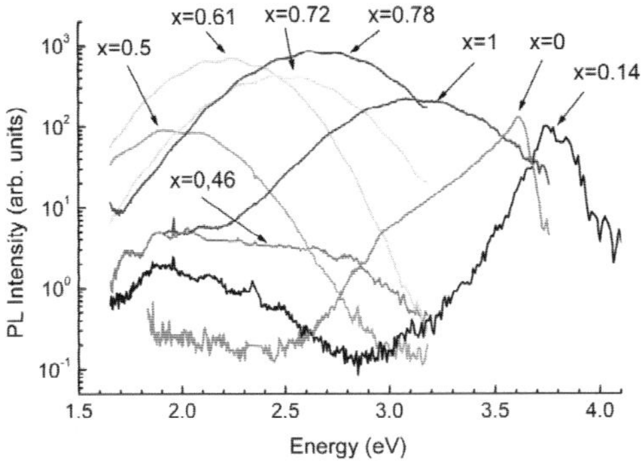

Figure 1. PL spectra of heavily doped $Al_xGa_{1-x}N$:Si layers grown at the constant silane flux $F_{SiH4} = 3$ sccm measured at room temperature.

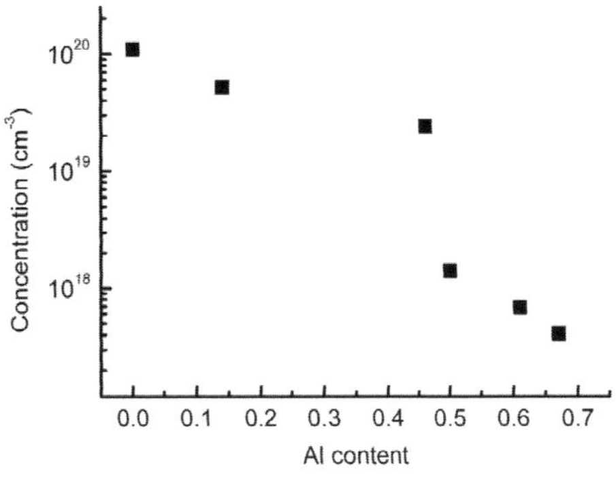

Figure 2. Dependence of the electron concentration on Al content.

Figure 3 shows the energy position of the defect PL band (E_{em}) and the band gap of the $Al_xGa_{1-x}N$ layers (E_g) estimated using the Vegard law:

$$E_{gAlGaN} = x_{Al} \cdot E_{gAlN} + (1-x_{Al}) \cdot E_{gGaN} - x_{Al} \cdot (1-x_{Al}) \cdot b = 1 \cdot x^2 + 1.59 \cdot x + 3.43 \quad (1)$$

where $E_{gGaN} = 3.43$ eV [3] and $E_{gAlN} = 6.015$ eV [4] are the GaN and AlN band gaps, respectively, and $b = 1.0$ eV [5] is the bowing parameter. E_{em} rises weaker with x_{Al} increase than E_g. The intensity of the green band in PL spectra of the second series of samples ($E_m \approx 2.4$ eV) increases with the rise of the silicon concentration up to $F_{SiH4} = 4.5$ sccm then the PL intensity decreases with further rise of the silicon concentration (see figure 4). The electron concentration was about 1.6×10^{19} cm^{-3} and 4.1×10^{17} cm^{-3} for the samples grown at $F_{SiH4} = 1.5$ sccm and $F_{SiH4} = 3.0$ sccm, respectively, the undoped sample and samples grown in the highest silane flux were very low conductive. That is in agreement with our previous results [6].

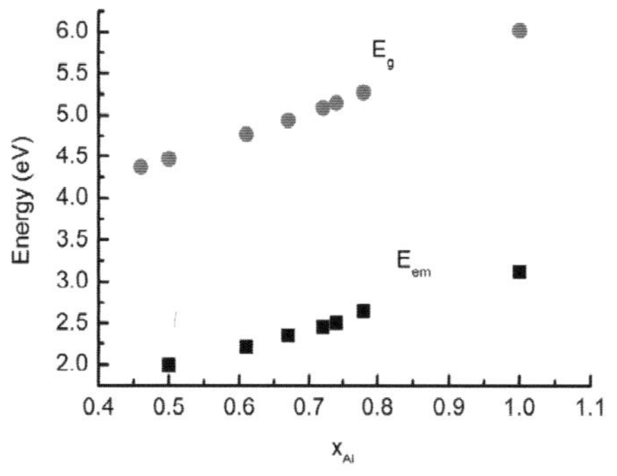

Figure 3. Dependences of the band-gap (E_g) and the defect PL band energy position (E_m) on Al content.

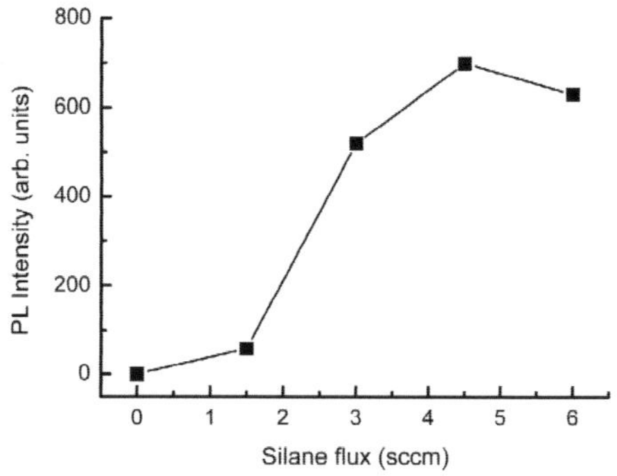

Figure 4. Dependences of the intensity of the defect PL band in $Al_{0.67}Ga_{0.33}N$ layers on silane flux.

Figure 5 shows the PL and PLE spectra of the $Al_{0.78}Ga_{0.22}N$ and AlN ($x_{Al} = 1$) samples at room temperature. The peak in the PLE spectra marked E_1 at 5.2 eV and 6.0 eV for the $Al_{0.78}Ga_{0.22}N$ and AlN layers, respectively, corresponds to band-to-band transitions in heavily doped $Al_xGa_{1-x}N$ layers because its energy position matches with the E_g from equation (1). The second peak E_2 at 4.7 eV and 4.94 eV for the $Al_{0.78}Ga_{0.22}N$ and AlN layers, respectively, was attributed to acceptor-to-band transitions in the heavily doped AlN layer. The large energy difference between the PL peak and the PLE peak E_2 (around 2.04 eV and 1.86 eV for the $Al_{0.78}Ga_{0.22}N$ and AlN layers, respectively) indicates a large Stock shift due to strong electron-phonon coupling in a deep center. These results are in a good agreement with our previous PL and PLE investigations of heavily doped $Al_xGa_{1-x}N$ layers with high x_{Al} [2]. Temperature-dependent PL spectra of the $Al_{0.78}Ga_{0.22}N$ layer demonstrate unusual behavior of the PL intensity, i.e., the PL intensity increases with increasing temperature from 5 to 500 K, while the further rise in temperature causes the PL quenching (see figure 6). The enhancement of the PL intensity is explained by the emission of carriers from states arising due to the band edge perturbations caused by the Al content and Si concentration fluctuation. The quenching of the PL intensity with temperature is described by the expression:

$$I(T) = A / (1 + Be^{-\frac{E_0}{kT}}), \quad (2)$$

where E_0 is the activation energy, and A and B are constants. Fitting of the PL quenching curve (see figure 7) by equation (2) gives $E_0 = 1.57$ eV.

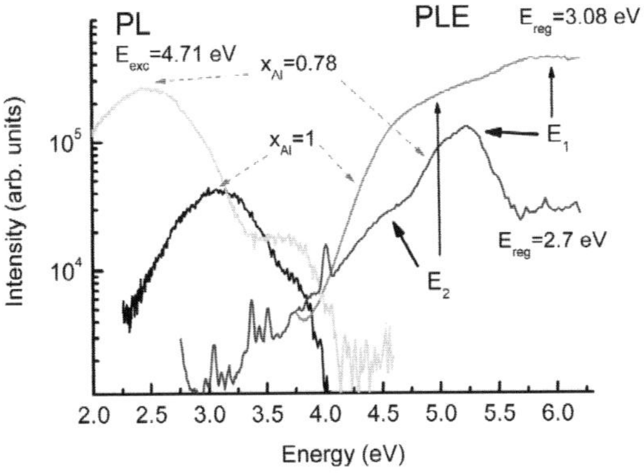

Figure 5. PLE and PL spectra of the $Al_{0.78}Ga_{0.22}N$ and AlN ($x_{Al} = 1$) samples at a detection energy of 3.08 eV and 2.7 eV, respectively, and PL spectra measured at the excitation energy of 4.71 eV at room temperature.

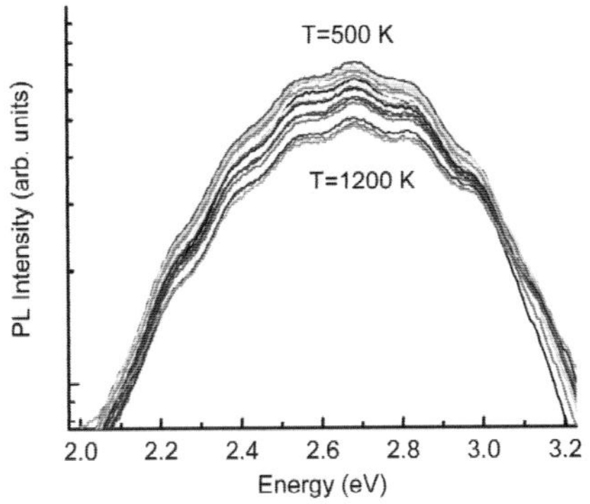

Figure 6. PL spectra of the $Al_{0.78}Ga_{0.22}N$ sample at different temperatures

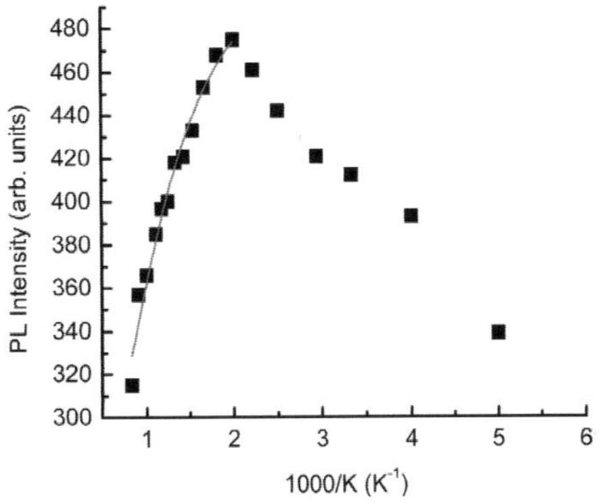

Figure 7. Temperature dependence of the PL intensity. Solid line depicts the approximation curve (see the text for details)

4. Discussion

The large Stock shift is typical for the centers with strong electron-phonon coupling. The configuration coordinate model describes such systems well [7, 8]. The configuration coordinate diagram describing optical absorption and emission at defects with strong electron-phonon coupling is shown in figure 8. S_{em} and S_{ab} are the numbers of photon with energies $h\nu_g$ and $h\nu_e$ involved in the transitions, E_m and E_{ab} are energies of the transitions. We assume that $S_{em} = S_{ab} = S$ and $h\nu_g = h\nu_e = h\nu_0$. The emission peak energy is given by:

$$E_{em} = E_{gAlGaN} - E_D - E_A - S_{em}h\upsilon_0 \, , \tag{3}$$

where S_{em} is the number of photons with the energy $h\nu_g$ involved in the transitions. The Stock shift in this model should be:

$$E_2 - E_{em} = 2S_{em}h\upsilon_0 + E_D \, . \tag{4}$$

Since the intensity of green PL of the samples with $x_{Al} = 0.67$ increases with increasing Si concentration, the observed PL band is related to the silicon impurity in $Al_xGa_{1-x}N$. The Si atom on the aluminum site Si_{Al} is considered as a deep donor with a binding energy of 250 meV in AlN [9-11]. This allows estimating the $S_{em}h\nu_o$ value:

$$S_{em}h\upsilon_0 = 0.5(E_2 - E_{em} - E_D) \approx 0.8eV. \qquad (5)$$

After that we obtain the acceptor ionization energy in AlN

$$E_A = E_{gAlN} - E_{em} - E_D - S_{em}h\upsilon_0 \approx 1.87eV. \qquad (6)$$

Because of its ionization energy, possible candidates for a deep acceptor are the (0/-) transition level of the C_N situated at about 1.88 eV [12] or the (2-/3-) transition level of the cation vacancy V_{III} [13] above the valence-band maximum. It could be possible that Si_N acts as an acceptor, however, the formation energy of Si_N is so high that its formation is highly unlikely [14]. The carbon concentration in studied $Al_xGa_{1-x}N$ layers was about $(1.0\text{-}5.0)\times10^{18}$ cm^{-3} and reached a significant value of about 8×10^{19} cm^{-3} only in the AlN layer. The triple acceptor V_{III} is known to be an effective electron compensator [13, 15], it's formation energy decreases with the increase of the Fermi level, referenced to the valence-band maximum [13]. The intensity of the broadband PL rises with decrease of electron concentration due to compensation. Consequently, the strongest candidate for an acceptor is V_{III}.

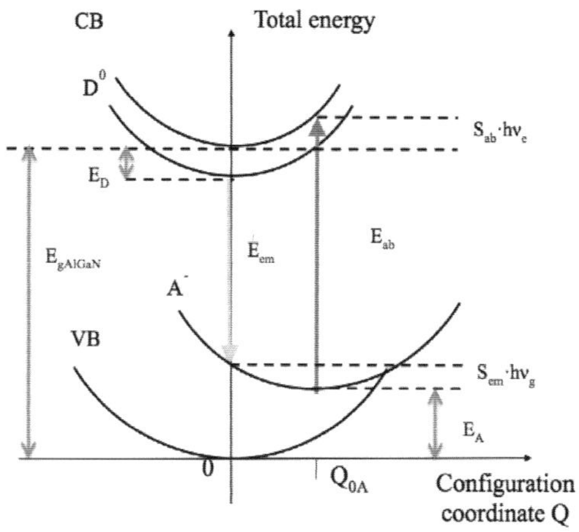

Figure 8. Configuration coordinate diagram describing optical band-to-band absorption and donor-acceptor pair emission at acceptor-like defects with strong electron-phonon coupling. The A^- term corresponds to an occupied state of the defect, and the VB, CB, and D^0 terms correspond to unfilled excited states of the defect where an electron is in the valence band, the conductivity band, and occupies the shallow donor state, respectively. The blue and green arrows indicate optical transitions during excitation and recombination, respectively.

The decrease of x_{Al} from 1 to 0.78 leads to decreasing difference in the energy position of E_1 and E_2 peaks from 1.06 eV to 0.5 eV. It's known that Si on a cation site is considered as a shallow donor with a binding energy of 28-30 meV in GaN [16-18], and remains a shallow donor with a similar energy up to $x_{Al} \sim 0.85$ with further deepening to 250 meV in AlN [10]. Assuming $E_D \approx 0.03$ eV, we obtain $S_{em}h\nu_0 \approx 1.1$ eV and $E_A \approx 1.6$ eV for $Al_{0.78}Ga_{0.22}N$. The last value is close to the value obtained from the temperature dependence of the PL intensity. So, the decrease of Al content leads to decrease of E_A and increase of S_{em} or $h\nu_0$. From expression (3) and a weaker dependence of E_{em} on x_{Al} than the one of E_g (see figure 3), we obtain:

$$\Delta(E_A + S_{em}h\upsilon_0 + E_D) = \Delta(E_{gAlGaN} - E_{em}) \approx 1.1\cdot\Delta x_{Al} \qquad (7)$$

5. Conclusions

The photoluminescence properties of the defect related emission attributed to the donor-acceptor transitions in heavily doped $Al_xGa_{1-x}N$:Si layers with $x_{Al} > 0.46$ have been investigated. The increase of Al content from $x_{Al} = 0.5$ to $x_{Al} = 1$ leads to the blue shift of the energy position of the PL band from the red-orange spectral range to the violet spectral range. The acceptor ionization energy was estimated to be 1.87 eV at $x_{Al} = 1$, decrease of x_{Al} leads to decrease of this value. The donor is supposed to be Si_{III}, a possible candidate for an acceptor is the (2-/3-) transition level of the V_{III}.

Acknowledgments
This work was supported by The Ministry of Education and Science of the Russian Federation (unique identification number of this work is RFMEFI61315X0039). SIMS measurements were done at the Center of the Multy-User Equipment "Material Science and Diagnostics for Advanced Technologies" (Ioffe Institute), that is supported by the Russian Ministry of the Education and Science.

References

[1] Bokhan P A, Gugin P P, Zakrevsky Dm E, Zhuravlev K S, Malin T V, Osinnykh I V, Solomonov V I and Spirina A V 2014 *J. Appl. Phys.* **116** 113103.
[2] Osinnykh I V, Malin T V, Plyusnin V F, Suranov A S, Gilinsky A M, and Zhuravlev K S 2016 *Jpn. J. Appl. Phys.* **55** 05FG09.
[3] Muth J F, Brown J D, Johnson M A L, Yu Zhonghai, Kolbas R M, Cook Jr J W and Schetzin J F 1999 *MRS Internet J. Nitride Semicond. Res.* 4S1 G5.2.
[4] Feneberg M, Leute R A R, Neusch B, Thonke K, and Bickermann M 2010 *Phys. Rev. B* **82** 075208.
[5] Koide Y, Itoh H, Khan M R H, Hiramatsu K, Sawaki N, and Akasaki I 1987 *J. Appl. Phys.* **61** 4540.
[6] Zhuravlev K S , Osinnykh I V, Protasov D Yu, Malin T V, Davydov V Yu, Smirnov A N, Kyutt R N, Spirina A V and Solomonov V I 2013 *Phys. Status Solidi C* **10** 315.
[7] Huang K and Rhys A 1950 *Proc. R. Soc. A* **204** 406.
[8] Henry C H and Lang D V 1977 *Phys. Rev. B* **15** 989.
[9] Silvestri L, Dunn K, Prawer S, and Ladouceur F 2011 *Appl. Phys. Lett.* **99** 122109.
[10] Borisov B, Kuryatkov V, Kudryavtsev Yu, Asomoza R, Nikishin S, Song D Y, Holtz M, and Temkin H 2005 *Appl. Phys. Lett.* **87** 132106.
[11] Taniyasu Y, Kasu M, and Makimoto T 2004 *Appl. Phys. Lett.* **85** 4672.
[12] Limpijumnong S and Van de Walle C G 2004 *Phys. Rev. B* 69 035207.
[13] Zhang Y, Liu W and Niu H 2008 *Phys. Rev. B* **77** 035201.
[14] Bogusławski P and Bernholc J 1997 *Phys. Rev. B* **56** 9496.
[15] Wagener M C, James G R, and Omnes F 2003 *Appl. Phys. Lett.* **83** 4193.
[16] Wang H and Chen A-B 2000 *J. Appl. Phys.* **87** 7859.
[17] Martı ́nez-Criado G, Miskys C R, Cros A, Ambacher O, Cantarero A and Stutzmann M 2001 *J. Appl. Phys.* **90** 5627.
[18] Moore W J, Freitas J A, Lee S K, Park S S and Han J Y 2002 *Phys. Rev. B* **65** R 081201.

IOP Conf. Series: Journal of Physics: Conf. Series **816** (2017) 012003 doi:10.1088/1742-6596/816/1/012003

Infrared magnetoreflection in $CoFe_2O_4$ single crystals

A A Buchkevich, Yu P Sukhorukov, A V Telegin, A P Nosov and V D Bessonov

M.N. Miheev Institute of Metal Physics UB RAS, 620990, Ekaterinburg, Russia

E-mail: buchkevich@imp.uran.ru

Abstract. Magnetoreflection of unpolarized infrared radiation for single crystals of magnetostrictive cobalt ferrite spinel was studied. The correlation between magnetoreflection and magnetoelastic properties of that type of spinel was observed. It is shown that the magnetoreflection is the most pronounced in the region of a middle-infrared impurity band and reflection minima near phonon bands.

1. Introduction

Currently, a new branch of spintronics – straintronics is intensively studied. Straintronics studies changes of the elastic properties of spintronic materials under deformation due to application of a magnetic or an electric field [1-2]. There are a huge number of magneto-optical effects related to magnetoelastic properties of magnetic materials, preferably in polarized light [3-5]. However, research of magneto-optical effects in unpolarized light can also be of high importance. For example, huge magnetoreflection and magnetotransmission of natural IR radiation up to few tens of percent for various magnetic semiconductors possessing the magnetoresistance effect were obtained [6]. The four different physical mechanisms except for a deformation one were defined to contribute to these effects depending on the spectral region of interest [6]. Meanwhile, mechanic deformations could strongly influence the absorption of ferromagnetic semiconductors of the spinel type [7]. However, no experimental studies have been performed in this field yet.

In this work, we report about the discovery of huge magnetoreflection of unpolarized light and its correlation with magnetoeleastic properties of ferrimagnetic spinel $CoFe_2O_4$ with a high magnetostriction effect.

2. Experimental setup and samples

The $CoFe_2O_4$ single crystals ($a_0 = 8.38$ Å) were grown by the floating zone method [8]. In the experiments the samples in the form of plates (surface (001)) with a geometrical size of 4x4 mm^2 and a thickness $d = 290$ μm were used. According to the XRD and EDXMA data, the samples under study are single-phase and correspond to nominal compositions. The samples were polished before optical measurements with an average roughness better than 1 μm. Reflectivity and magnetoreflection of samples have been explored in natural (unpolarized) light in the spectral range of 1 - 30 μm at room temperature with the help of a custom created cryomagnetic device based on a prism spectrometer. The magnetic field H up to 3.6 kOe was directed along [100] axis of the sample (in-plane geometry) and perpendicularly to the incident light (Voigt geometry). The reflection coefficient was defined as $R = I_s/I_{Al}$, where I_s and I_{Al} are, respectively, intensities of unpolarized light reflected from the sample and the aluminium mirror used as a reference. Magnetoreflection was calculated as a relative change in reflection in the presence (R_H) and absence (R_0) of a magnetic field: $\Delta R/R_0 = (R_H-R_0)/R_0 \cdot 100\%$.

Magnetostriction measurements were performed by the standard tensometric method in the same geometry of experiment as the magnetoreflection ones. The relative accuracy of the experiments was about 0.2%.

3. Result and discussion

The spectral dependence of reflectivity $R(\lambda)$ and optical conductivity $\sigma(\lambda)$ of the $CoFe_2O_4$ single crystal (insert in figure 1 (a)) are typical for band-gap semiconductors. They are formed by the fundamental absorption at $\lambda < 1.5$ μm, a frequency-independent part up to $\lambda = 10$ μm (with $R \sim 15$ %) and phonon bands E_1 at $\lambda = 16.4$ μm and E_2 at $\lambda = 24.2$ μm [9]. The optical conductivity was calculated from the reflectivity data for the $CoFe_2O_4$ crystal via Kramers-Kronig calculations (figure 1 (a)). As one can see, the results are in good accordance with the reflection spectrum.

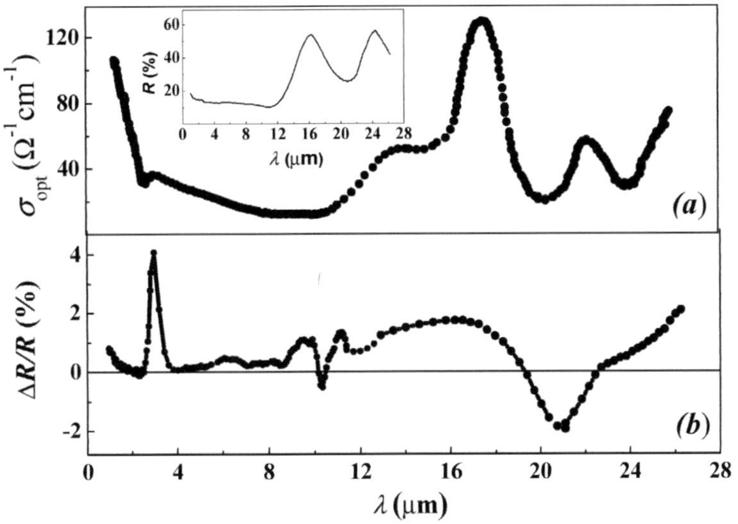

Figure 1. Spectral dependences of optical conductivity $\sigma(\lambda)$ (a) and magnetoreflection $\Delta R/R(\lambda)$ (b) for $CoFe_2O_4$ single crystals at $T = 295$ K in a magnetic field $H = 3.5$ kOe, $H \parallel [100]$. The insert shows the spectra of reflectivity $R(\lambda)$ of the sample.

The application of an external static magnetic field significantly changes the reflectivity (not shown) of $CoFe_2O_4$ and leads to appearance of the magnetoreflection effect $\Delta R/R$. It is worth to notice that this is the first experimental observation of the strong influence of a relatively weak magnetic field on the reflection coefficient of a magnetic semiconductor in the infrared range. In our case, the value of magnetoreflection for $CoFe_2O_4$ single crystals reaches about 4% at $H = 3.5$ kOe (figure 1 (b)), which is close to the values of $\Delta R/R$ for non-magnetostrictive ferromagnetic $Hg(Cd)Cr_2Se_4$ single crystals with the spinel structure [6].

The growth of $\Delta R/R(\lambda)$ at $\lambda < 1.5$ μm is associated with influence of magnetic field on the absorption edge ($E_g = 1.18$ eV for $CoFe_2O_4$), well-known for magnetic dielectrics as a "blue shift" [11]. A so-called MIR absorption band appears in magnetoreflection at $\lambda = 3$ μm. According to [12], this MIR-band is associated with the elastic modes of the crystal. We suppose that the observed peculiarities and a peak of $\Delta R/R(\lambda)$ in the region $1.5 < \lambda < 4.5$ μm are due to a change of the intensity and position of the MIR-band on temperature and magnetic field. Moreover, a small contribution of the tails of the absorption edge has to be taken into account. At $\lambda > 8$ μm, spectrum of $\Delta R/R(\lambda)$ is formed by the shift of the reflection minima near the phonon bands in the same way as it occurs in ferromagnetic semiconductors with strong electron-phonon interaction [6].

Figure 2 presents field dependences of magnetostriction and magnetoreflection at $T = 295$ K.

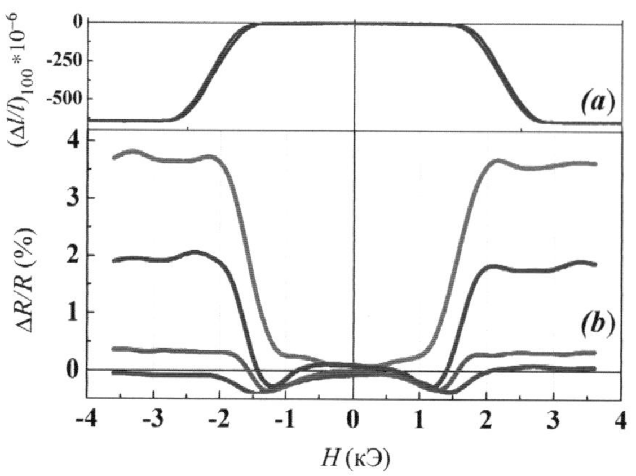

Figure 2. Field dependences of magnetostriction $(\Delta l/l)_{100}$ (a) and magnetoreflection $\Delta R/R$ (b) for $\lambda = 2.3$ μm (blue line), $\lambda = 2.9$ μm (green line), $\lambda = 3.2$ μm (red line) and $\lambda = 8$ μm (violet line) for $CoFe_2O_4$ single crystals at $T = 295$ K, $H \parallel [100]$.

Magnetoreflection is an even effect and is saturated at $H_s > 2.5$ kOe. The complex form of magnetoreflection between 1 kOe and 2 kOe can be explained by the different contributions of opposite sign associated with the fundamental absorption edge (positive $\Delta R/R$) and MIR-band (negative $\Delta R/R$). The behavior of $(\Delta l/l)_{100}$ is in good correlation with the published data for $CoFe_2O_4$ crystals [13]. One can notice a close correlation between magnetostriction and magnetoreflection field dependences. For example, magnetostriction appears at $H > 1.5$ kOe and shows a saturation at $H > 2.5$ kOe as well as magnetoreflection. The small changes of $(\Delta l/l)_{100}(H)$ and $\Delta R/R(H)$ at weak magnetic fields can be explained by the crystallographic anisotropy as well as by the crystal domain structure and the demagnetization factor. From the similarity of $\Delta R/R(H)$ and $(\Delta l/l)_{100}(H)$ dependences, we can conclude that the magnetoreflection in the region $1.5 < \lambda < 4.5$ μm may be connected with the occurrence of magnetoelastic strains in the $CoFe_2O_4$ crystals, so a new mechanism of magnetooptical effects in magnetostrictive semiconducting materials is observed.

4. Conclusion

In the wide infrared range, the giant magnetoreflection effect up to 4% was observed in the ferrimagnetic single crystal $CoFe_2O_4$ with a giant magnetostriction at room temperature. The effect was explained by the shift of the absorption edge at short wavelengths, by the change of the intensity and a position of the MIR-band and a shift of the reflection minima under the magnetic field applied. The close correlation of the magnetoreflection with magnetoelastic deformations in $CoFe_2O_4$ may pave the way for discovering a new mechanism of magnetoreflection of light in ferrimagnetic materials. The effect can be promising for creation of new magnetic-field-driven optical materials.

Asknowledgments

The work was supported by program of FASO Russia (theme «Spin» № 0120146330) and UB of RAS № 15-9-2-4.

References

[1] Roy K 2014 *Proc. SPIE 9167 (Spintronics VII 91670U)* arXiv:1506.08193 [cond-mat.mes-hall] (Bellingham: USA)

[2] Ustinov A B, Kolkov P I, Nikitin A A, Kalinikos B A, Fetisov Yu K and Srinivasan G 2011 *Technical Physics* **81** (6) 75

[3] Zvezdin A K and Kotov V A 1997 *Moden Magnetooptics and magnetooptical materials* in: J M D Coey and D R Tilley (London: Institute of Physics Publishing) Studies in Condensed matter physics 381

[4] Smolensky G A, Pisarev R V and Siniy I G 1975 *Soviet Physics Uspekhi* **116** (2) 231

[5] Moskvin A S, Latypov D G and Gudkov V G 1988 *Physics of the Solid State* **30** (2) 413

[6] Telegin A V, Sukhorukov Yu P, Loshkareva N N, Mostovshchikova E V, Bebenin N G, Gan'shina E A and Granovskii A B 2015 *JMMM* **383** (Amsterdam: Elsevier) 104

[7] Bebenin N G 1991 *Semiconductors* **25** 1661

IOP Conf. Series: Journal of Physics: Conf. Series **816** (2017) 012003 doi:10.1088/1742-6596/816/1/012003

[8] Letuk L M, Balbashov A M, Krutogin D G, Gonchar A V, Kudryashkin I G and Salduney A M 1994 *Production technology of magnetoelectronic materials* (Moscow: Metallurgiya)
[9] Danil'kevich M I, Litvinivich G V and Naumenko V I 1976 *Journal of Applied Spectroscopy* **24** 38
[10] Waldron R D 1955 *Physical Review* **99** 1727
[11] Rai R C, Wilser S, Guminiak M, Cai B and Nakarmi M L 2012 *Applied Physics A* **106** 207
[12] Rahman A, Gafur A and Sarker A R 2015 *International Journal of Innovative Research in Advanced Engineering* **2** (1) 99
[13] Bozorth R M, Tilden E F and Wiliams A J 1955 *Physical Review* **99** 1788

IOP Conf. Series: Journal of Physics: Conf. Series **816** (2017) 012004 doi:10.1088/1742-6596/816/1/012004

Bulk plasmon-phonon polaritons in n-GaN

E Yu Orlov[1], G A Melentev[1], V A Shalygin[1] and S Suihkonen[2]

[1] Peter the Great St. Petersburg Polytechnic University, 29 Polytechnicheskaya str., St. Petersburg 195251, Russia

[2] School of Electrical Engineering, Aalto University, Otakaari 1, Espoo 02150, Finland

E-mail: shalygin@rphf.spbstu.ru

Abstract. We studied theoretically and experimentally plasmon-phonon polaritons and longitudinal plasmon-phonon oscillations in *n*-GaN epitaxial layers. The studies were carried out on the epitaxial layers with various doping levels. Simulation of the reflectivity spectra and dispersion relations of plasmon-phonon polaritons was performed in a wide frequency range. Reflectivity spectra transformation associated with phonon damping and electron relaxation processes has been revealed. Experimental studies of the reflectivity spectra have been performed in the spectral range of 8–80 meV. The experimental spectra are well fitted by the simulated ones. Results of the study can be used for contactless determination of the electron concentration and mobility in GaN epitaxal layers.

1. Introduction

During recent years, bulk and surface polaritons are extensively researched in order to create mid-infrared and terahertz selective sources [1–3]. Intense incoherent radiation of GaAs plasmon-phonon polaritons was obtained experimentally in the range of 10–15 THz [1]. A plasmon-phonon polariton (PPP) is a "hybrid" quasiparticle emerged in a polar semiconductor under interaction of a photon, a polar phonon and a bulk plasmon. In the framework of the wave theory, the PPP is a transverse electromagnetic wave propagating in a polar semiconductor.

Let us start with a general consideration of electromagnetic waves in an isotropic medium. Propagation of a plane wave is described by the wave equation [4]:

$$\mathbf{q}\cdot(\mathbf{q}\cdot\mathbf{E})-q^2\cdot\mathbf{E}=-\varepsilon(\omega)\frac{\omega^2}{c^2}\mathbf{E}, \quad (1)$$

where $\mathbf{q}$ is the wave vector, $\mathbf{E}$ is the electric field strength, $\varepsilon(\omega)$ is the dielectric permittivity, ω is the angular frequency of the wave.

In accordance with equation (1), two types of electromagnetic waves can propagate in a medium, namely, longitudinal waves and transverse ones. Longitudinal waves ($\mathbf{E} \parallel \mathbf{q}$) can propagate only at frequency values which represent the roots of the following equation:

$$\varepsilon(\omega)=0. \quad (2)$$

In the case of transverse waves ($\mathbf{E}\perp\mathbf{q}$), the dependence of the wave vector on the angular frequency is described by the dispersion relation:

IOP Conf. Series: Journal of Physics: Conf. Series **816** (2017) 012004 doi:10.1088/1742-6596/816/1/C

$$q^2 = \frac{\omega^2}{c^2}\varepsilon(\omega). \tag{3}$$

Let us consider propagation of electromagnetic waves in three particular objects: (i) an *n*-doped nonpolar semiconductor, (ii) a polar semiconductor without free charge carriers (a polar dielectric), and (iii) an *n*-doped polar semiconductor.

(i) The dielectric permittivity of an *n*-doped nonpolar semiconductor can be written as

$$\varepsilon(\omega) = \varepsilon_\infty \left[1 - \frac{\omega_{\rm P}^2}{\omega^2} \right], \tag{4}$$

where $\omega_{\rm P} = \sqrt{\dfrac{4\pi e^2 N_e}{\varepsilon_\infty m_e}}$ is the plasma frequency, N_e is the free electron concentration, e is the elementary charge, m_e is the electron effective mass, ε_∞ is the high frequency permittivity [5]. Equation (4) is related to the simplest model of electron gas (so-called collisionless plasma) when the free electron relaxation time $\tau = \infty$. Because equation (2) has only one root ($\omega = \omega_{\rm P}$), the longitudinal electromagnetic waves can propagate at the frequency $\omega_{\rm P}$ only. These waves represent longitudinal collective oscillations of electron gas (plasmons). Within the framework of the considered approximation, the longitudinal waves have no dispersion, and their behaviour is illustrated by the blue line in figure 1 (a). Besides plasmons, specific transverse electromagnetic waves which are called "plasmon polaritons" can be excited in the considered medium as well. The plasmon polariton dispersion dependence calculated in accordance with equations (3) and (4) is shown in figure 1 (a) by the red line. In the long-wavelength limit (q = 0), plasmons and plasmon polaritons have the same frequency: $\omega = \omega_{\rm P}$. For the region of rather small q when ω is comparable with $\omega_{\rm P}$, it can be shown that the contribution of the electromagnetic field energy to the total plasmon polariton energy is comparable with the contribution of the mechanical energy of transverse electron oscillations. In other words, a plasmon polariton in this region is a proper hybrid of a photon (electromagnetic field) and a plasmon (electron oscillations). Under increase of the wave vector q, the plasmon polariton dispersion $\omega(q)$ asymptotically approaches the straight line $\omega = qc\varepsilon_\infty^{-1/2}$ (at frequencies $\omega >> \omega_{\rm P}$). This behavior indicates that the contribution of the electromagnetic field energy to the plasmon polariton energy is dominant, and the contribution of the mechanical energy of transverse electron oscillations is negligible. As it is known, the linear dispersion dependence $\omega = qc\varepsilon_\infty^{-1/2}$ describes the propagation of a "pure" photon in a medium with the dielectric permittivity ε_∞. It should be emphasized that in the spectral interval $0 \le \omega < \omega_{\rm P}$ a plasmon polariton cannot be excited. In the reflection spectrum one can observe a band of total reflection (reflectivity r = 1) in this frequency range.

(ii) The dielectric permittivity of a polar semiconductor without free electrons can be described by the following expression [4]:

$$\varepsilon(\omega) = \varepsilon_\infty \left[1 + \frac{\omega_{\rm LO}^2 - \omega_{\rm TO}^2}{\omega_{\rm TO}^2 - \omega^2} \right], \tag{5}$$

where $\omega_{\rm LO}$ is the longitudinal optical phonon frequency, $\omega_{\rm TO}$ is the transverse optical phonon frequency. In this medium, equation $\varepsilon(\omega)$ = 0 gives only one root: $\omega = \omega_{\rm LO}$. Subsequently, longitudinal electromagnetic waves can be excited at this frequency only. These waves are associated with longitudinal lattice vibrations (LO phonons). They have no dispersion as it is shown by the blue line in figure 1 (b). Besides, specific transverse electromagnetic waves (phonon polaritons) which represent coupled oscillations of the lattice and electromagnetic field can be excited in a polar semiconductor without free electrons. These transverse oscillations cannot be separated because simultaneously electromagnetic waves excite the TO phonons and oscillating charges radiate electromagnetic waves. The phonon polariton dispersion dependence calculated in accordance with

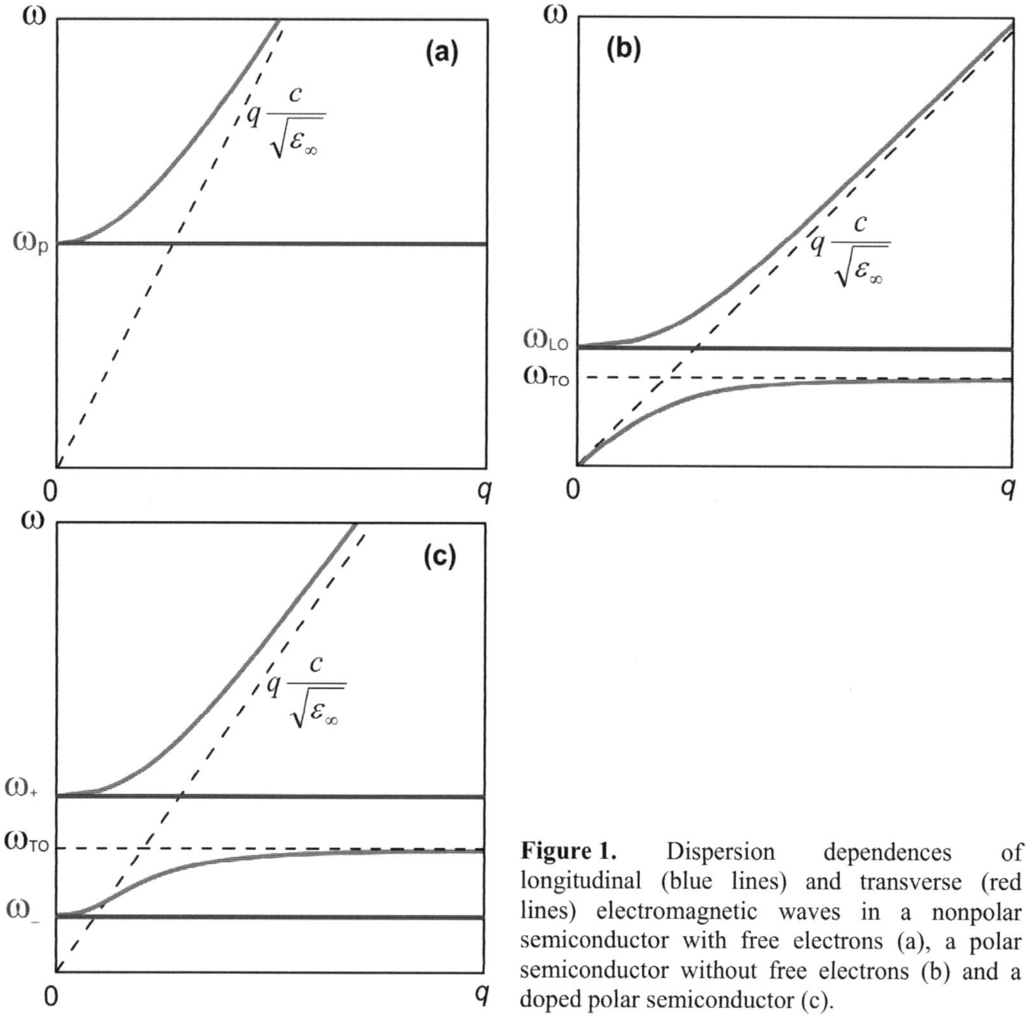

Figure 1. Dispersion dependences of longitudinal (blue lines) and transverse (red lines) electromagnetic waves in a nonpolar semiconductor with free electrons (a), a polar semiconductor without free electrons (b) and a doped polar semiconductor (c).

equations (3) and (5) is shown in figure 1 (b) by the red lines. It consists of two branches. The upper branch starts at the frequency $\omega = \omega_{\rm LO}$. It can be shown that at small q the contribution of the mechanical energy of the lattice oscillations to the total energy of phonon polariton is comparable with the energy of the electromagnetic field. At frequencies $\omega >> \omega_{\rm LO}$, the upper branch of the dispersion dependence of a phonon polariton asymptotically approaches the straight line $\omega = qc\varepsilon_\infty^{-1/2}$. As it was mentioned above, this straight line corresponds to a "pure" photon in a medium with the dielectric permittivity ε_∞. In the high frequency limit, the contribution of the mechanical energy of the lattice oscillations to the total energy of phonon polariton becomes negligible. The lower branch of the phonon polariton dispersion dependence continuously covers the spectral interval from 0 to $\omega_{\rm TO}$. It can be shown that in the limit of large q when $\omega \to \omega_{\mathrm{TO}}$ the contribution of lattice vibrations to the phonon polariton energy dominates the electromagnetic field contribution. In the spectral interval $\omega_{\mathrm{TO}} < \omega < \omega_{\mathrm{LO}}$, transverse electromagnetic waves cannot be excited. Respectively, in the reflection spectrum one can observe total reflection ($r = 1$) in this frequency interval (the so-called Reststrahlen band).

(iii) Finally, let us consider an n-doped polar semiconductor which is the subject of the present work. In this case, the dielectric permittivity includes both the lattice and free electron contributions [4, 5]:

$$\varepsilon(\omega) = \varepsilon_{\infty}\left[1 + \frac{\omega_{\mathrm{LO}}^2 - \omega_{\mathrm{TO}}^2}{\omega_{\mathrm{TO}}^2 - \omega^2} - \frac{\omega_{\mathrm{P}}^2}{\omega^2}\right] \tag{6}$$

and equation (2) gives two different roots (we denote them as ω_+ and ω_-). Subsequently, longitudinal electromagnetic waves can propagate at these two frequencies. These oscillations are known as the coupled plasmon-phonon modes, they have no dispersion (see two blue lines in figure 1 (c)). Besides, specific transverse electromagnetic waves (plasmon-phonon polaritons) can be excited in a doped polar semiconductor. These waves represent coupled transverse oscillations of the electromagnetic field, lattice and electron gas which cannot be separated. The plasmon-phonon polariton dispersion dependence calculated in accordance with equations (3) and (6) is shown in figure 1 (c) by the red lines. It has two branches. The upper branch starts at the frequency $\omega = \omega_+$ and at frequencies $\omega >> \omega_+$ asymptotically approaches the straight line $\omega = qc\varepsilon_{\infty}^{-1/2}$ which corresponds to a "pure" photon in a medium with the dielectric permittivity ε_{∞}. It can be shown that for the upper branch in the limit $q \to \infty$ the contribution of the electromagnetic field energy to the total plasmon-phonon polariton energy is dominant. In contrast to the cases (i) and (ii), here one can see *two* spectral intervals ($0 < \omega < \omega_-$ and $\omega_{\mathrm{TO}} < \omega < \omega_+$) where the transverse electromagnetic waves cannot be excited. Respectively, in the reflection spectrum one can observe total reflection ($r = 1$) in both these frequency intervals.

It should be emphasized that in the above consideration we neglected the optical phonon damping and electron relaxation processes. However, as it will be shown below, the both factors are very important.

The goal of our work is to study theoretically and experimentally plasmon-phonon polaritons and longitudinal plasmon-phonon oscillations in n-GaN epitaxial layers. The studies are carried out on the epitaxial layers with various doping levels. In the paper, we present theoretical consideration taking into account the optical phonon damping and electron relaxation processes. Simulation of the reflectivity spectra and dispersion relations of plasmon-phonon polaritons was fulfilled in a wide frequency range. Experimental studies of the reflectivity spectra were performed in the spectral range of 8–80 meV. It has been shown that the considered theoretical model provides a good match with the experimental results.

2. Longitudinal plasmon-phonon modes in *n*-GaN

To take into account the optical phonon damping and electron relaxation processes, instead of equation (6) we used a more general expression for the dielectric permittivity of a polar semiconductor [4, 5]:

$$\varepsilon(\omega) = \varepsilon_{\infty}\left[1 + \frac{\omega_{\mathrm{LO}}^2 - \omega_{\mathrm{TO}}^2}{\omega_{\mathrm{TO}}^2 - \omega^2 - i\omega\gamma} - \frac{\omega_{\mathrm{P}}^2}{\omega^2 + i\omega\tau^{-1}}\right], \tag{7}$$

where γ is the damping constant for TO phonons, τ is the electron relaxation time determined by the mobility μ: $\tau = \mu m_e / e$. It should be noted that equation (6) can be regarded as a particular case of equation (7) with $\gamma = 0$ and $\tau = \infty$.

We calculated frequencies of the longitudinal plasmon-phonon modes (ω_+ and ω_-) using equations (2), (7) and the Lyddane–Sachs–Teller relation: $\omega_{\rm LO}^2 / \omega_{\rm TO}^2 = \varepsilon_0 / \varepsilon_\infty$, where ε_0 is the static permittivity. The calculations were carried out for bulk GaN at various free electron concentrations N_e in the range from 10^{17} to 10^{20} cm^{-3}. The following parameters of GaN were used [2]: $\varepsilon_\infty = 5.4$, $\varepsilon_0 = 9.5$, $\hbar\omega_{\rm TO} = 69.3$ meV, $\gamma = 7.5 \cdot 10^{11}$ s^{-1}, $m_e = 0.2\, m_0$. The dependence of τ on N_e was derived from the experimental data on the electron concentrations and mobilities in *n*-GaN samples with various doping levels [6].

The resulting dependences of the frequencies ω_+ and ω_- upon electron concentration N_e are shown in figure 2 by the red lines. Besides, we calculated similar dependences ignoring the optical phonon damping and electron relaxation processes ($\gamma = 0$, $\tau = \infty$). The latter dependences are shown in figure 2 by the solid blue lines.

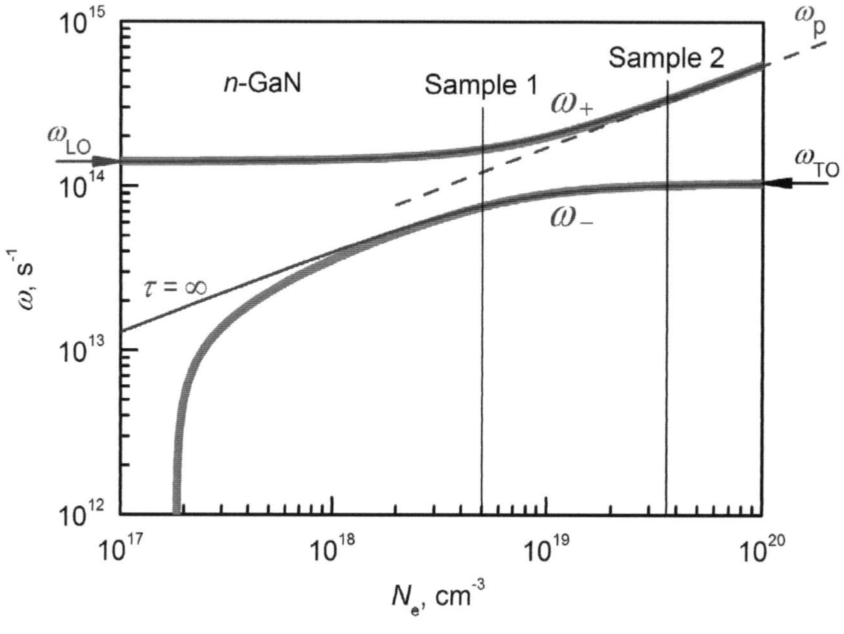

Figure 2. Angular frequencies of the longitudinal plasmon–phonon modes in *n*-GaN as a function of electron concentration. Red lines: calculation taking into account the optical phonon damping and electron relaxation processes. Solid blue lines: calculation ignoring these factors. The dashed blue line shows the concentration dependence of the plasma frequency $\omega_{\rm P}$. Vertical lines denote electron concentrations in the samples under experimental study.

One can see that the dependence $\omega_+(N_e)$ is not affected by the electron relaxation time τ. The dependence approaches the LO phonon frequency $\omega_{\rm LO}$ at low free electron concentrations. At high free electron concentrations (when $\omega_{\rm P} > \omega_{\rm LO}$, i.e. at $N_e > 2 \cdot 10^{19}$ cm^{-3}) $\omega_+(N_e) \approx \omega_{\rm P}(N_e)$. Such a behavior is associated with redistribution of lattice and electron contributions to the dielectric permittivity with doping level increase.

The dependence $\omega_{-}(N_e)$ at high free electron concentrations approaches the TO phonon frequency ω_{TO} (and also is not affected by the electron relaxation time). This is due to electron screening of longitudinal optical oscillations. On the contrary, at low free electron concentrations ($N_e < 10^{18}$ cm^{-3}) the dependence $\omega_{-}(N_e)$ is strongly affected by the electron relaxation time. In the hypothetical case of collisionless plasma ($\tau = \infty$), $\omega_{-}(N_e) \approx \omega_{\text{P}}(N_e)\sqrt{\frac{\varepsilon_\infty}{\varepsilon_0}}$. However, the results of calculation based on the experimental dependence $\tau(N_e)$ drastically differ (see figure 2). In particular, with decreasing electron concentration ω_{-} approaches zero much rapidly than $\omega_{\text{P}}\sqrt{\frac{\varepsilon_\infty}{\varepsilon_0}}$.

3. Dispersion of plasmon-phonon polaritons and simulated reflectivity spectra

Let us consider excitation of the transverse electromagnetic wave (plasmon-phonon polaritons) in a semi-infinite polar semiconductor under normal incidence of the plane electromagnetic wave with a frequency ω at the planar vacuum/semiconductor interface. Under these conditions, the wave excited inside a semiconductor is homogeneous and propagates in the direction perpendicular to the interface. The surfaces of constant phase and constant amplitude are parallel to the interface. If the dielectric permittivity of a semiconductor is characterized by equation (7), propagation of the wave inside a semiconductor can be described by the complex wave vector $q = \frac{\omega}{c}(n + ik)$, where n is the refractive index and k is the extinction coefficient which are determined by the dispersion relation (3). We carried out these calculations for plasmon-phonon polaritons in bulk n-GaN with different electron concentrations and plotted the dependencies of $\text{Re}(q) = \frac{\omega}{c}n$ and $\text{Im}(q) = \frac{\omega}{c}k$ upon photon energy $\hbar\omega$. These two dependencies completely describe the dispersion of plasmon-phonon polaritons. The results of simulation for the n-GaN crystals with electron concentrations of $2\cdot10^{18}$ and $2\cdot10^{19}$ cm^{-3} are presented in figure 3 by the red lines. We assumed the same value of electron mobility (179 cm^2/V·s) for both samples. Other parameters of GaN are specified in section 2. For comparison, the results of dispersion simulation based on a simplified model ignoring dissipation processes ($\gamma = 0$, $\tau = \infty$) are presented in the same figure by the solid blue lines. The reflectivity spectra $r(\hbar\omega)$ calculated for the vacuum/semiconductor interface in the framework of both models are presented in figure 3 as well. The calculation of the reflectivity was performed using Fresnel's formula for the case of normal incidence:

$$r = \frac{(n-1)^2 + k^2}{(n+1)^2 + k^2}. \tag{8}$$

If dissipation processes are ignored, the simulation predicts existence of two spectral ranges where Re(q) is equal to zero: $[0, \hbar\omega_{-}]$ and $[\hbar\omega_{\text{TO}}, \hbar\omega_{+}]$, see the solid blue lines in figure 3 (a, d). Thus, outer illumination of a polar semiconductor in these spectral ranges cannot excite plasmon-phonon polaritons inside the crystal. Obviously, in both ranges the refractive index $n = 0$ and, in accordance with equation (8), the reflectivity is equal to 1 (see the solid blue lines in figure 3 (c, f)).

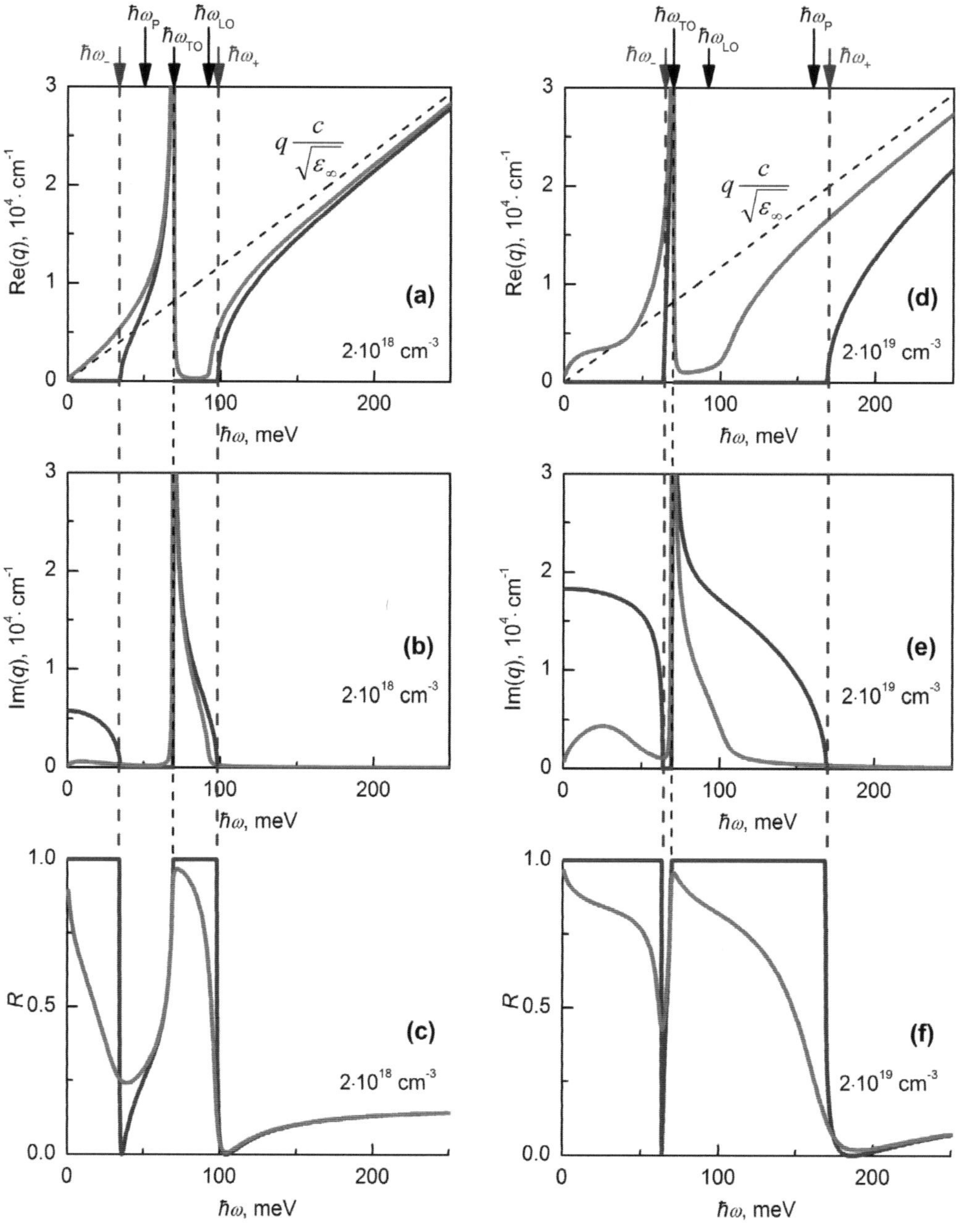

Figure 3. Dispersion dependencies of plasmon-phonon polaritons (a, b, d, e) and reflectivity spectra (c, f) for *n*-GaN. Simulation for two electron concentrations: $N_e = 2{\cdot}10^{18}$ cm^{-3} (a, b, c) and $2{\cdot}10^{19}$ cm^{-3} (d, e, f). Red lines: calculation taking into account the optical phonon damping and electron relaxation processes. Solid blue lines: calculation ignoring dissipation processes.

When accounting for dissipation processes ($\gamma \neq 0$, $\tau \neq \infty$), the quantity Re(q) becomes non-zero at all photon energies (see the red lines in figure 3 (a, d)). Consequently, the reflectivity becomes less than 1 in the whole spectrum (see the red lines in figure 3 (c, f). It means that dissipation allows exciting transverse electromagnetic waves (PPP) at any photon energy, in particular in the ranges $[0, \hbar\omega_-]$ and $[\hbar\omega_{TO}, \hbar\omega_+]$. It should be emphasized that in the spectral range $[\hbar\omega_{TO}, \hbar\omega_+]$ a significant decrease of the reflectivity due to dissipation processes occurs at the high-frequency edge only. However, at the low-frequency edge of this range, the decrease of the reflectivity is rather small and one can see a sharp peak with high reflectivity at $\hbar\omega \approx \hbar\omega_{TO}$. This peak results from the sharp peak in the spectrum of $\mathrm{Im}(q)$ (and, consequently, the spectrum of k), see the red lines in figure 3 (b, e).

Under electron concentration increase from $2\cdot10^{18}$ cm^{-3} (figure 3 (a–c)) to $2\cdot10^{19}$ cm^{-3} (figure 3 (d–f)), the low-frequency dip in the reflectivity spectrum (around $\hbar\omega_-$) becomes significantly narrower but the high-frequency one (around $\hbar\omega_+$) broadens. These peculiarities are associated with redistribution of the phonon and plasmon contributions to the coupled plasmon-phonon modes with doping level increase. Namely, at a lower doping level the mode $\hbar\omega_-$ is mainly formed by the plasmon contribution ($\hbar\omega_- \approx \hbar\omega_P$), and the mode $\hbar\omega_+$ is mainly formed by the phonon contribution ($\hbar\omega_+ \approx \hbar\omega_{LO}$). On the contrary, at a higher doping level the mode $\hbar\omega_-$ is mainly formed by the phonon contribution ($\hbar\omega_- \approx \hbar\omega_{TO}$) and the mode $\hbar\omega_+$ is mainly formed by the plasmon contribution ($\hbar\omega_+ \approx \hbar\omega_P$).

In the considered case of low-mobility electrons, dissipation of the PPP energy due to electron relaxation processes is more intensive in comparison with dissipation caused by phonon damping. This is a key issue for understanding the reflectivity spectrum change under transition from the simplified model (blue curves in figure 3 (c, f)) to general one taking into account dissipation processes (red curves in figure 3 (c, f)). At a higher doping level (figure 3 (f)), corresponding decrease of the reflectivity at photon energy close to $\hbar\omega_+$ is more significant than nearby $\hbar\omega_-$. This is associated with plasmon domination in plasmon-phonon polariton at photon energies close to $\hbar\omega_+$ and phonon domination at photon energies close to $\hbar\omega_-$. At a lower doping level, there is a reciprocal situation (figure 3 (c)).

4. Experimental study of *n*-GaN epilayer reflection

Epitaxial layers of gallium nitride doped with Si were grown by MOVPE on a sapphire substrate. Free electron concentration and mobility in sample 1 were $5\cdot10^{18}$ cm^{-3} and 189 cm^2/V·s, respectively (the thickness of the epilayer was 10 μm). The parameters of sample 2 were $3.6\cdot10^{19}$ cm^{-3}, 122 cm^2/V·s and 6.2 μm, respectively. Sample characterization was carried out by means of the Hall effect and conductivity measurements at room temperature.

The reflectivity spectra were measured by a Fourier spectrometer in the range of 8–80 meV at room temperature. A globar and a pyroelectric detector were used as a source of radiation and a detector, respectively. The incidence angle was about 11°. The experimental spectra are shown in figure 4.

The theoretical calculation of the reflectivity for the samples under experimental study was based on the model considering transmission and reflection of radiation in a three-layer system [7], vacuum/GaN/sapphire in our case. The experimental reflectivity spectra were fitted by the simulated dependences. The free electron concentration and mobility were used as fitting parameters. Good agreement between calculation and experiment (see figure 4) shows that the considered model of plasmon-phonon polaritons adequately describes the *n*-GaN reflection spectra in a wide range of electron concentrations. For sample 1, the best fit is given by the electron concentration $N_e' = 2.8\cdot10^{18}$ cm^{-3} and mobility $\mu' = 180$ cm^2/V·s. For sample 2, the fitting results are the following: $N_e' = 1.9\cdot10^{19}$ cm^{-3} and $\mu' = 179$ cm^2/V·s. One can see that these values slightly differ from the values determined by means of Hall and conductivity measurements. This difference can be explained in the following way.

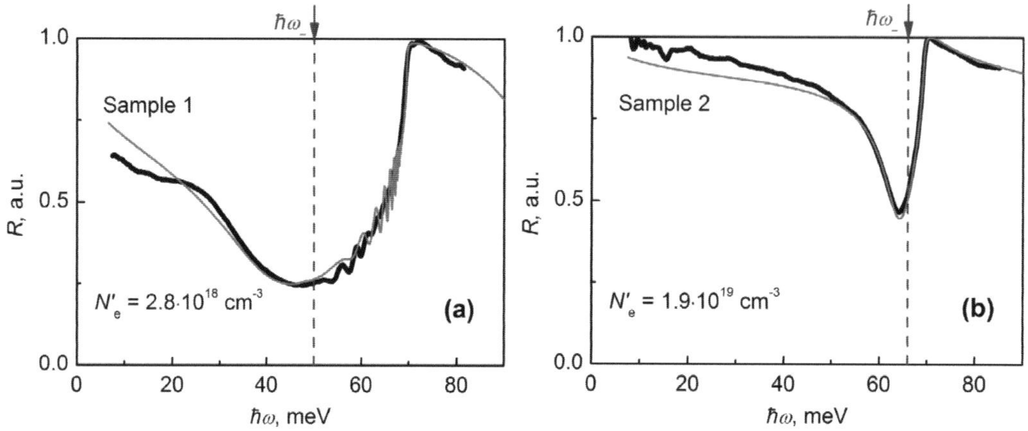

Figure 4. Normalized reflectivity spectra of samples 1 and 2: experiment (black curves) and theoretical fit (red curves). Dashed lines denote the position of the plasmon-phonon mode $\hbar\omega_-$.

The values of concentration and mobility obtained from Hall measurements characterize the epitaxial layer as a whole, and parameters determined from the reflectivity spectra are mainly related to the subsurface region of the epitaxial layer.

The reflectivity spectra demonstrate a series of peculiarities. In every spectrum, one can see a dip close to the position of the longitudinal plasmon-phonon mode $\hbar\omega_-$. The reflectivity spectrum of the low-doped sample 1 demonstrates a broad dip nearby $\hbar\omega_-$ while the heavily-doped sample 2 shows a narrow one at a corresponding spectral position. As mentioned above, such a change in the reflectivity spectrum with doping level increase is associated with redistribution of the phonon and plasmon contributions to the coupled plasmon-phonon mode $\hbar\omega_-$. At a lower doping level, the mode $\hbar\omega_-$ is mainly formed by the plasmon contribution ($\hbar\omega_- \approx \hbar\omega_{\mathrm{P}}$), and at a higher doping level the mode $\hbar\omega_-$ is mainly formed by the phonon contribution ($\hbar\omega_- \approx \hbar\omega_{\mathrm{TO}}$). Besides, at a lower doping level the reflectivity dip at $\hbar\omega_-$ is additionally greatly broadened due to rather fast electron relaxation processes. At higher doping level the corresponding reflectivity dip is only slightly extra-broadened due to rather weak phonon damping.

The experimental spectrum for the low-doped sample 1 (and the simulated spectrum as well) demonstrates the interference pattern in the spectral range of 50–70 meV. In this spectral range, the GaN epilayer is rather transparent that results in multiple reflection of radiation from the GaN/sapphire and GaN/vacuum interfaces. The spectra for sample 2 do not demonstrate any interference pattern because the heavily doped GaN layer in this sample is totally nontransparent in the operating spectral range.

5. Conclusion

To summarize, we studied theoretically and experimentally the bulk plasmon-phonon polaritons in *n*-GaN with various doping levels. The longitudinal plasmon-phonon modes in *n*-GaN were considered as well. The theoretical consideration was carried out taking into account the optical phonon damping and electron relaxation processes. The simulated dispersion dependencies of plasmon-phonon polaritons and reflectivity spectra for semi-infinite *n*-GaN are presented. The reflectivity spectra transformation associated with phonon damping and electron relaxation processes has been revealed. The experimental investigation was performed for the *n*-GaN epitaxial layers grown on a sapphire substrate. The reflectivity spectra were measured in the spectral range of

8–80 meV for *n*-GaN with two electron concentrations: $\sim 3 \cdot 10^{18}$ and $\sim 2 \cdot 10^{19}$ cm^{-3}. The theoretical calculation of the reflectivity for the samples under experimental study was based on the model considering transmission and reflection of radiation in a three-layer system (vacuum/GaN/sapphire). The experimental reflectivity spectra were fitted by the simulated ones. Good agreement between calculation and experiment shows that the considered model of plasmon-phonon polaritons adequately describes the *n*-GaN reflection spectra in a wide range of electron concentrations. It is shown, that reflectivity spectrum analysis can be used for contactless determination of the electron concentration and mobility in GaN epitaxal layers.

Acknowledgements
The authors are sincerely grateful to L.E. Golub and A.V. Poshakinskiy for fruitful discussions. This work was supported by the Ministry of Education and Science of the Russian Federation (state assignment).

References
[1] Širmulis E, Šilėnas A, Poželа K, Poželа J and Jucienė V 2014 *Appl. Phys. A* **115** 199
[2] Melentev G A *et al* 2016 *J. Appl. Phys.* **119** 093104
[3] Melentev G A, Yaichnikov D Yu, Shalygin V A, Vinnichenko M Ya, Vorobjev L E, Firsov D A and Suihkonen S 2016 *J. Phys.: Conf. Series* **690** 012005
[4] Yu P Y and Cardona M 2010 *Fundamentals of Semiconductors* 4th ed (Berlin–Heidelberg: Springer)
[5] Ukhanov Yu I 1977 *Optical Properties of Semiconductors, in Russian* (Moscow: Nauka)
[6] Agekyan V F *et al* 2015 *Phys. Solid State* **57** 787
[7] Born M and Wolf E 2003 *Principles of Optics: Electromagnetic Theory of Propagation, Interference and Diffraction of Light, 7th edition* (Cambridge: Cambridge University Press) pp 735–789

Spontaneous magnetization reversal caused by magnetic fluctuation in GaMnSb thin films

A A Filatov[1], G G Ziborov[1] and A I Dmitriev[2]

[1] Lomonosov Moscow State University, Moscow, 119992, Russia
[2] Institute of Problems of Chemical Physics of Russian Academy of Sciences, Chernogolovka, 142432 Russia

E-mail: aid@icp.ac.ru

Abstract. Time dependences of the magnetic moment $m(t)$ in GaMnSb thin films with MnSb clusters are measured. The $m(t)$ dependences are straightened in semi-logarithmic coordinates $m(\ln t)$. The field dependences of magnetic viscosity $S(H)$ corresponding to the slope of straight lines $m(\ln t)$ are studied. It is demonstrated that the behaviour of dependences $S(H)$ is governed by the magnetic moment fluctuations of MnSb clusters with lognormal distribution of the magnetic anisotropy energy.

1. Introduction

Dilute magnetic semiconductors are characterized by high degrees of spin polarization of charge carriers and may be used in the design of spintronic devices. However, the strength of indirect exchange, which helps maintain the spin-polarized state, is so low that the corresponding Curie temperatures are well below room temperature. The reasons for this are the low solubility of transition-metal impurities in the semiconductor matrix and the resulting formation of various clusters of the secondary phase [1–6]. For example, the theoretically predicted Curie temperature in GaMnSb is 40 K, while the value measured experimentally in thin films grown by molecular beam epitaxy is no higher than 25 K [7–9]. However, the anomalous Hall effect with hysteresis up to room temperatures was observed in GaMnSb thin films grown by pulsed laser deposition [10]. Since long-range ferromagnetism in the GaMnSb films is impossible at room temperature, the authors of [10] interpreted the observed phenomenon as spin polarization of holes tunneling through the MnSb-GaMnSb Schottky barrier. MnSb clusters are the ferromagnetic secondary phase embedded in GaMnSb thin films [10]. The basic magnetic characteristics of thin films (coercive force, blocking temperatures, magnetic anisotropy constants, etc.) were determined in [11, 12]. In the present study, time dependences of the magnetic moment $m(t)$ of GaMnSb thin films with MnSb clusters were measured. The aim of the study was to reveal the effect of magnetic anisotropy distribution of MnSb clusters on spontaneous magnetization reversal of GaMnSb thin films.

2. Experiment

GaMnSb films with a thickness of 90 nm were grown by pulsed laser deposition under high vacuum at a temperature of 100 °C. The GaSb-MnSb eutectic alloy was used as the target. Single-crystal Al_2O_3 served as a substrate. Following deposition, the samples were annealed at a temperature of 350 °C for 30 min. The structure and the composition of thin films were checked with a scanning probe microscope and a scanning electron microscope. The chemical composition of the near-surface layers of films was determined by energy-dispersive spectroscopy. The image of the surface of a GaMnSb

IOP Conf. Series: Journal of Physics: Conf. Series **816** (2017) 012005 doi:10.1088/1742-6596/816/1/0120

thin film (not present) obtained using a magnetic force microscope shows that the films had magnetic nanoinclusions. The technique for sample fabrication and assessment was described in detail earlier in [13, 14]. Time dependences of the magnetic moment $m(t)$ of samples were performed using a SQUID magnetometer MPMS 5XL Quantum Design. The sample was first magnetized in a magnetic field H = 2 kOe (exceeding the strength of the saturation field). The time of exposure of the sample to this magnetic field was 300 s. Following such treatment, the field directed opposite to the magnetic moment vector of the sample was switched on. The time of field reversal was approximately 90 s. After the reverse magnetic field was set, the time dependence of the magnetic moment was measured during 2400 s.

3. Experimental results and discussion

3.1. Time dependences of magnetic moment

Figure 1 shows the magnetic moment increment dependences $\Delta m(t)$ at a temperature T = 8 K in different magnetic fields directed opposite to the magnetic moment vector of the sample on a logarithmic time scale.

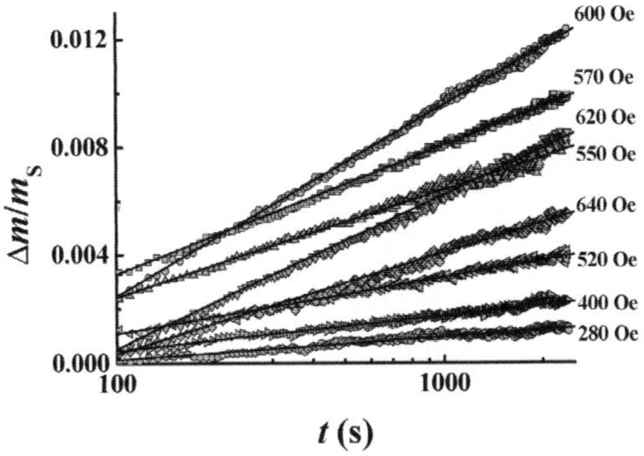

Figure 1. Time dependences of magnetic moment $\Delta m(t)$ normalized on the saturation magnetic moment m_S at T = 8 K in different magnetic fields. Approximations are shown by solid lines.

It can be seen from figure 1 that $\Delta m(t)$ is straightened in semi-logarithmic coordinates. The tangent of the slope of $\Delta m(\ln t)$ dependences corresponds to magnetic viscosity $S = \mathrm{d}m/\mathrm{d}(\ln t)$. The solid lines represent the result of approximation with the following expression:

$$\frac{\Delta m}{m_S} \sim S \ln t, \qquad (1)$$

where m_S = 3.2 μ_B is the saturation magnetic moment. Slopes of $\Delta m(\ln t)$ curves and correspondent magnetic viscosities are non-monotonic functions of magnetic field (figure 1).

3.2. Magnetic viscosity

The shape of the $S(H)$ function depends on volume distribution of ferromagnetic clusters [15, 16]. In case of lognormal distribution, the dependence $S(H)$ is described by the following formula [15, 16]:

$$\frac{S(H)}{m_S} = \frac{T}{25T_B\left(1-\dfrac{H}{H_A}\right)^2}\frac{1}{\sqrt{2\pi}\ln\sigma}\exp\left(-\frac{1}{2}\left(\frac{\ln\left(TT_B^{-1}\left(1-\dfrac{H}{H_A}\right)^{-2}\right)}{\ln\sigma}\right)^2\right), \qquad (2)$$

where T_B is the blocking temperature, H_A is the magnetic anisotropy field, σ is the standard deviation. The blocking temperature T_B = 390 K and the magnetic anisotropy constant $K = H_A m_S/2$ = $7.2\cdot10^4$ erg/cm^3 at T = 8 K were extracted from the approximation of $S(H)$ dependence by Equation

(2). Field dependence $S(H)$ at $T = 8$ K shows a maximum at the coercive field $H_C = 600$ Oe (figure 2) corresponding to the extremum of Equation (2): $H_C = H_A\left(1 - \sqrt{TT_B^{-1}\sigma^{-\ln\sigma}}\right)$.

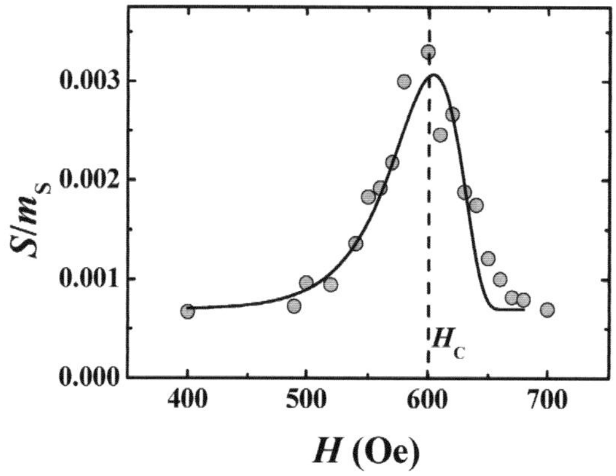

Figure 2. Field dependence of the normalized magnetic viscosity S/m_S at $T = 8$ K. The approximation described in the text is shown by a solid line. The coercive field is marked by a vertical dashed line.

As follows from Equation (2), the slope of the $\Delta m(\ln t)$ curve (magnetic viscosity S) is a non-monotonic function of temperature. Function $S(T)$ has a maximum at $T = T_B\left(1 - HH_A^{-1}\right)^2$. The temperature dependence $S(T)$ was discussed in [17].

Figure 3 shows the size (diameter D) distribution of MnSb clusters derived from the magnetic force microscopy data and magnetic anisotropy energy distribution $\rho(E_A)$ obtained as a result of the approximation of $S(H)$ dependence according to Equation (2). Function $\rho(E_A)$ corresponds to a lognormal distribution in Equation (2), where $H = 0$ Oe, $E \sim k_BT$ and $E_A \sim 25k_BT_B$ (here k_B is Boltzmann constant). It follows from figure 2 and figure 3 that the field dependence of the magnetic viscosity $S(H)$ is governed by the lognormal distribution of the magnetic anisotropy energy of MnSb clusters.

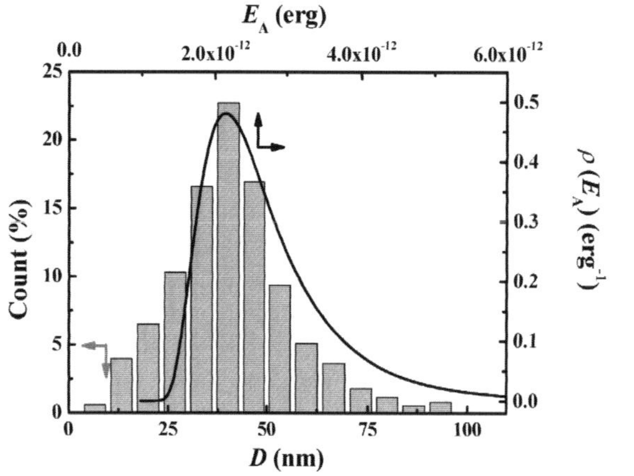

Figure 3. Size distribution of MnSb clusters derived from the magnetic force microscopy data and $\rho(E_A)$ dependence obtained as a result of analysis of time dependences of the magnetic moment of GaMnSb thin films.

4. Summary

The correlation between the size distribution of MnSb clusters, which was derived from the magnetic force microscopy data, and the distribution of MnSb clusters over the magnetic anisotropy energy, which was obtained as a result of analysis of time dependences of the magnetic moment $m(t)$ of GaMnSb thin films, was found. This correlation suggests that $m(t)$ dependences and magnetic viscosity are governed by the lognormal size (magnetic anisotropy energy) distribution of MnSb clusters.

Acknowledgments
Authors are grateful to R B Morgunov for discussion. The work was supported by the Grant of President of Russian Federation, Project No. MK-5754.2016.3.

References
[1] Rovezzi M, D'Acapito F, Navarro-Quezada A, Faina B, Li T, Bonanni A, Filippone F, Bonapasta A A and Dietl T 2009 *Phys. Rev. B* **79** 195209
[2] Bonanni A *et al* 2008 *Phys. Rev. Lett.* **101** 135502
[3] Morgunov R B and Dmitriev A I 2009 *Phys. Solid State* **51(10)** 1985
[4] Dmitriev A I, Morgunov R B, Kazakova O L and Tanimoto Y. 2009 *J. Exp. Theor. Phys.* **108 (6)** 985
[5] Dmitriev A I, Talantsev A D, Zaitsev S V, Danilov Yu A, Dorokhin M V, Zvonkov B N, Koplak O V and Morgunov R B 2011 *J. Exp. Theor. Phys.* **113(1)** 138
[6] Dmitriev A I, Morgunov R B and Zaitsev S V 2011 *J. Exp. Theor. Phys.* **112(2)** 317
[7] Matsukura F, Abe E and Ohno H 2000 *J. Appl. Phys.* **87** 6442
[8] Matsukura F, Abe E, Ohno Y and Ohno H 2000 *Appl. Surf. Sci.* **159-160** 265
[9] Abe E, Matsukura F, Yasuda H, Ohno Y and Ohno H 2000 *Physica E* **7** 981
[10] Rylkov V V, Aronzon B A, Danilov Yu A, Drozdov Yu N, Lesnikov V P, Maslakov K I and Podol'skii V V 2005 *J. Exp. Theor. Phys.* **100(4)** 742
[11] Talantsev A D, Koplak O V and Morgunov R B 2015 *Phys. Solid State* **57 (2)** 322
[12] Koplak O V, Polyakov A A, Davydov A B, Morgunov R B, Talantsev A D, Kochura A V Fedorchenko I V, Novodvorskii O A, Shorokhova A V and Aronzon B A 2015 *J. Exp. Theor. Phys.* **120(6)** 1012
[13] Marenkin S F, Novodvorsky O A, Shorokhova A V, Davydov A B, Aronzon B A, Kochura A V Fedorchenko I V, Khramova O D and Timofeev A V 2014 *Inorg. Mater.* **50(9)** 897
[14] Lotin A A, Novodvorsky O A, Parshina L S, Khaydukov E V, Zuev D A, Khramova O D and Panchenko V Ya, 2011 *Appl. Phys. B* **104**, 565
[15] Tejada J, Zhang X X and Hernandez J M 1996 *Proc. of NATO ASI on Magnetic Hysteresis in Novel Materials (Mykonos)* vol 338 (Amsterdam: Kluwer Academic Publishers) p 221
[16] Lyberatos A, Chantrell R W and O'Grady K 1993 *Proc. of NATO ASI on Nanophase Materials: Synthesis-Properties-Applications (Corfu)* vol 260 (Amsterdam: Kluwer Academic Publishers) p 653
[17] Dmitriev A I and Filatov A A 2016 *Phys. Solid State* **58(10)** 2005

Electrophysical properties of PbTe doped with CdSe

A M Karmokov, R M Kalmykov and Z V Shomakhov

Kabardino–Balkarian State University, 175 Chernyshevsky str., Nalchik 360004, Russia

E-mail: kalmykov.rustam@yandex.ru

Abstract. The temperature dependences of electrical conductivity, thermoelectric power and thermoelectric power factor of solid solutions of the PbTe system doped with up to 3 mol% CdSe have been studied in the present work. The dependences show that these parameters reach their maximum values at a CdSe concentration of 3 % in the temperature range of 420-450 K with $\alpha = 310$ μV/K, $\sigma = 210$ (Ohm·cm)$^{-1}$ and $P = 15.1$ μW/(cm·K^2).

1. Introduction

To form thermoelectric converters with high figure of merit in low and medium temperature ranges and infrared receivers of wide range, semiconductor compounds based on lead chalcogenides (IV-VI), in particular PbTe–based ones, are used [1–5]. Using solid solutions based on PbTe as thermoelectric converters requires dopants which can increase its thermoelectric efficiency. It is of particular interest to study the behavior of impurities whose physical and chemical nature is similar to that of the initial component of a certain chemical compound. Doping of PbTe with Tl up to 2 at% [6] increases the thermoelectric figure of merit ZT of the material and is mainly used at low temperatures. Doping with nanoparticles of Na up to 1 at% [7] also increases ZT at high temperatures, and doping with In [8] improves sensitivity of photodetectors of thin films in the terahertz spectral range. Within the temperature range of 600 – 800 K, it is most technologically and economically advantageous to use impurities of semiconductor compounds, in particular of cadmium selenide, in lead telluride.

Peculiarities of the phase diagram of lead telluride make it significantly difficult to prepare homogeneous crystals that are similar in composition to the stoichiometric ones [1]. The maximum melting temperature of lead telluride is shifted towards that of chalcogen, and during solidification of lead telluride its composition changes as follows: first p–type crystals precipitate (with excess of tellurium), then the crystal composition changes towards a higher lead content and a transition from a p-type to an n-type conductivity can be observed. Only at a certain composition of the solid solution, which corresponds to the maximum melting temperature, the liquid phase retains its composition during solidification («invariant point»). Only for this composition, homogeneous crystals can be prepared by any method of crystallization from the melt. At all other compositions, including the stoichiometric one, a gradual change in crystal composition occurs during solidification.

The present work studies effects of doping lead telluride with cadmium selenide in the range of solid solutions (up to 3 mol% CdSe [9]) on temperature dependence of electrical conductivity and thermoelectric power in the temperature range of 300 – 480 K.

2. Experiment

For our investigations, we prepared samples of lead telluride with different molar concentrations of CdSe impurity (0.5, 1, 2, 3 mol%). As the starting components, we used lead of high purity (99.999 %), tellurium of high purity (99.999 %) and cadmium selenide of high purity as a disperse powder with a particle size of 28 – 35 μm. The synthesis process was performed under the high temperature flux in atmospheric environment at 1183 K. The resulting melt was subjected to continuous mixing by acoustic waves of 22 kHz and 400 W. Then, the melt was kept for 15 minutes at

T = 1203 K. Next, the melt was cooled at a rate of 130 K/min to T = 823 K, after which it was subjected to homogenizing annealing for 8 hours.

The studied samples had the shape of a disk of 14–17 mm in diameter and 3–4 mm thick, which we then carefully polished using diamond paste and etched in the solution of CrO_3 (3g) + H_2SO_4 (25ml) + H_2O (75ml) for 3 minutes.

We studied electrical conductivity and thermoelectric power of the resulting semiconductor compounds in the temperature range of 300 – 480 K.

3. Results and discussion

We used the four–probe method to measure the specific electrical conductivity, σ, of the semiconductor materials. The resulting temperature dependences of specific electrical conductivity for the PbTe–CdSe system show that in the temperature range of 300 – 350 K the undoped lead telluride compound has the highest value of σ, equal to ~ 250 $(Ohm\cdot cm)^{-1}$ at T = 340 K (figure 1, line 1). Adding small amounts of cadmium selenide (up to 2 mol%) results in a decrease of the specific electrical conductivity in the same temperature range, showing the values σ = 110 $(Ohm\cdot cm)^{-1}$ for the sample with 1 mol% CdSe (line 3) at T = 340 K, and σ = 72 $(Ohm\cdot cm)^{-1}$ for the sample with 2 mol% CdSe (line 4) at T = 340 K. For the solid solution containing 0.5 mol% CdSe (line 2), it is characteristic to have a slight change of the specific electrical conductivity over the entire temperature range. We got the maximum value of the specific electrical conductivity at a CdSe concentration of 3 mol% at T = 470 K, that is σ = 210 $(Ohm\cdot cm)^{-1}$ (line 5).

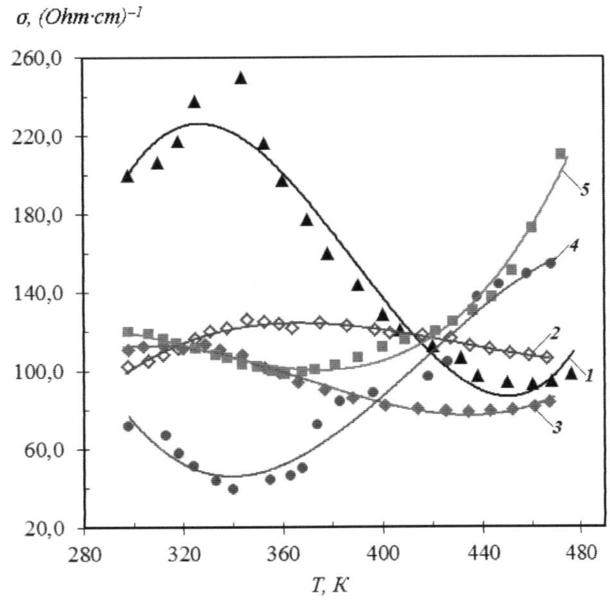

Figure 1. Temperature dependences of electrical conductivity of PbTe doped with CdSe at different concentrations: ***1*** – 0 mol%, ***2*** – 0.5 mol%, ***3*** – 1 mol%, ***4*** – 2 mol%, ***5*** – 3 mol%.

We used the standard differential method to measure the temperature dependence of thermoelectric power, α, in vacuum with a pressure of 1×10^{-4} mmHg in the temperature range of 300 – 380 K. Figure 2 presents the obtained results.

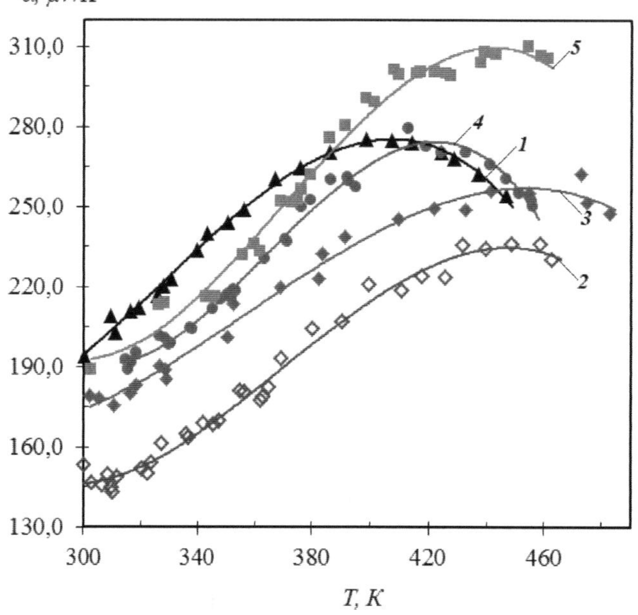

Figure 2. Temperature dependences of thermoelectric power of PbTe doped with CdSe at different concentrations: ***1*** – 0 mol%, ***2*** – 0.5 mol%, ***3*** – 1 mol%, ***4*** – 2 mol%, ***5*** – 3 mol%.

The data obtained show that for all the compounds under study an increase in temperature leads to an increase in the thermoelectric power coefficient. The maximum value for the starting undoped lead telluride compound (line 1) is $\alpha = 275$ μV/K at $T = 410$ K. When we added about 0.5 mol% CdSe (line 2), we observed a decrease in the thermoelectric power, and its maximum value shifts to higher temperatures ($\alpha = 236$ μV/K at $T = 450$ K).

The thermoelectric power increases with temperature at CdSe concentrations of 1 mol% (line 3) and 2 mol% (line 4). The maximum values of the thermoelectric power are the following: $\alpha = 262$ μV/K at $T = 470$ K and $\alpha = 270$ μV/K at $T = 430$ K. With increasing cadmium selenide concentration to 3 mol% (line 5), the thermoelectric power reaches the maximum value $\alpha = 310$ μV/K at $T = 450$ K, which is highly important for devices operating in the high temperature range.

Using the obtained experimental values of the specific electrical conductivity and thermoelectric power for all the compounds under study, we calculated the specific thermoelectric energy (thermoelectric power factor), P, in the temperature range of 300 – 450 K by the formula

$$P = \alpha^2\sigma.$$

Figure 3 presents the dependences of the thermoelectric power factor of the studied solid solution on the content of cadmium selenide. Analysis of the concentration dependences of the thermoelectric power factor of the investigated semiconductor compounds shows that at $T = 450$ K (line 5) the maximum corresponds to the solid solution containing 3 mol% of cadmium selenide, $P = 15.1$ μW/(cm·K^2), while for the starting undoped compound of lead telluride $P = 5.6$ μW/(cm·K^2), which is almost 3 times lower. In our opinion, this result is highly important in terms of practical use.

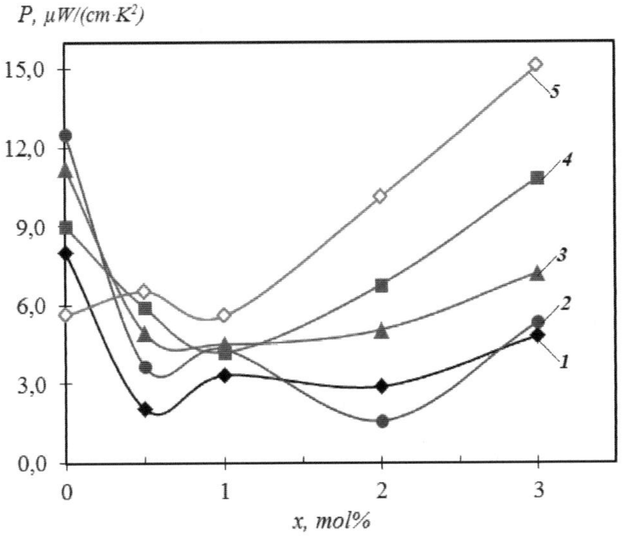

Figure 3. Dependences of the thermoelectric power factor of PbTe on CdSe content : *1* – 300 K, *2* – 340 K, *3* – 380 K, *4* – 420 K, *5* – 450 K.

We believe that to explain this character of temperature dependences of the investigated kinetic coefficients, one can use the so–called model of two valence bands [10]. According to this model, at low temperatures the band of heavy holes is lower than the edge of the band of light holes with a splitting $\Delta\varepsilon \approx 0.17$ eV. Therefore, at low temperatures only light holes play a dominant role. The bandgap ε_g increases with increasing temperature, and the splitting $\Delta\varepsilon$ decreases. When the temperature reaches 400–500 K, the splitting completely disappears, and the tops of the bands of light and heavy holes coincide with each other. Yet further increase in temperature lowers the band of light holes even more, while the distance between the conduction band and the band of heavy holes does not change. Thereby, at high temperatures the valence band edge is the edge of the band of heavy holes, and the one of light holes can be disregarded in the electrical conductivity. As a result, the average hole effective mass increases, which leads to increasing of thermoelectric power coefficient. Apparently, adding CdSe changes the character of temperature and concentration dependences of the thermoelectric power factor. In this case, light holes stop dominating in the temperature range of 300 – 340 K, and the valence band edge is formed by the heavy holes. Thus, figure 3 shows decrease in the power factor at 300 – 340 K, while at higher temperatures the power factor increases.

4. Conclusions

Thus, the study of temperature dependences of electrical conductivity, thermoelectric power and thermoelectric power factor of solid solutions of PbTe–CdSe system has showed that they have maximum values at a CdSe concentration of 3 mol% in the temperature range of 420 – 450 K with α = 310 μV/K, σ = 210 $(\text{Ohm}\cdot\text{cm})^{-1}$ and P = 15.1 μW/(cm·K^2).

Acknowledgments

The work is being carried out with financial support of the Fund for Assistance to Small Innovative Enterprises in Scientific and Technical spheres of Russian Federation. Grant №0017817.

The authors are grateful to Centre of Collective Usage «X–ray diagnostics of materials» for doing out X–ray analyses of the investigated materials.

References

[1] Ravich Yu I, Efimova B A and Smirnov I A 1968 *Research methods of semiconductors applied to lead chalcogenides PbTe, PbSe and PbS* (Moscow: Science) 384

[2] Abrikosov N Kh and Shelimova L E 1975 *Semiconductor materials on the base $A^{IV}B^{VI}$ compou-nds* (Moscow: Science) 195

[3] Zimin S P and Gorlachev E S 2011 *Nanostructured lead chalcogenides* (Yaroslavl: Yaroslavl State University) 232

[4] Aleksandrova O A, Maksimov A I, Moshnikov V A and Chesnova D B 2008 *Chalcogenides and oxides of elements of group IV. Obtaining, research and application* (Saint–Petersburg: Technolit) 240

[5] Karmokov A M, Karamurzov B S, Molokanov O A, Usaev A A and Kalmykov R M 2012 *Materials of V International scientific and technical conference «Micro–and nanotechnology in electronics»* (Russia: Nalchik*)* 45

[6] Prokofeva L V, Shabaldin A A, Korchagin V A, Nemov S A and Ravich Yu I 2008 *Semiconductors* **42** 1161

[7] Dmitriev A V and Zvyagin I P 2010 *Physics - Uspekhi* **53** 8 789

[8] Ryabova L I and Khokhlov D R 2014 *Physics - Uspekhi* **57** 959

[9] Rajewski S D, Zbigli K R, Kazak G F and Prunich M D 1983 *Inorganic materials* **6** (19) 889

[10] Aliev S A and Zulfugarov E I 2009 Thermomagnetic and thermoelectric phenomenon in science and technique (Baku: ELM) 325

Luminescent property modification of SiO_x/Al_2O_3 multilayers by annealing and hydrogenation

D A Grachev[1], A M Legkov[1], I I Chunin[2] and A V Ershov[1]

[1] Department of Semiconductor Physics and Optoelectronics, Lobachevsky University, Nizhny Novgorod 603950, Russia
[2] Research Institute of Physics and Technology, Lobachevsky University, Nizhny Novgorod 603950, Russia

E-mail: grachov@phys.unn.ru

Abstract. We have studied the composite nanoperiodic (3-8 nm) SiO_x/Al_2O_3 structures fabricated by physical deposition from separate sources. The structures annealed in nitrogen atmosphere at 900–1100 °C included Si-NCs that showed intense room-temperature photoluminescence at 1.6 eV. The quantum-confinement peak shift was observed under varying the SiO_x layer thickness. Up to fourfold intensity enhancement was observed after hydrogenation of the structures in the range of 400–450 °C for 2 hours.

1. Introduction

In recent years there has been a significant growth of interest in the study of silicon nanocrystals (Si-NCs) in various dielectric layers, matrices alternative to silicon oxide [1–4]. We present the results of an optical properties study of the composite dielectric films obtained by high-temperature annealing of nanoperiodic SiO_x/Al_2O_3 sandwich structures and containing ordered arrays of Si-NCs. According to the transparency range, alumina is similar to silica but it has a higher refractive index (1.7 vs. 1.5 [5]). Thus, the variation of the dielectric matrix material may be an additional tool to control the effective refractive index of the composites as the active medium for silicon photonics.

Apart from preparing samples, a problem of enhancement of the Si-NCs photoluminescence (PL) intensity by hydrogenation has drawn our attention. The principal possibility of controlling the intensity by passivation of the defect centers [6,7] is fundamentally important in terms of understanding the PL mechanism as well as exciton localization in Si-NCs in particular [8,9].

2. Preparation of samples and measurements procedure

The SiO_x/Al_2O_3 multilayer structures with layer thicknesses of 3-8 nm and a number of layers up to 60 were fabricated by physical deposition on silicon substrates (table 1). To obtain ordered arrays of Si-NCs, as described in refs. [10–12], the samples were annealed at 1000 – 1100 °C in a dry nitrogen atmosphere for 2 hours.

Hydrogenation was carried out in an environment of molecular hydrogen (1 atm.) at 400 – 550 °C for 2 hours. Subsequent dehydrogenation was performed in nitrogen atmosphere at temperatures of 600 – 700 °C but for just one hour. Figure 1 as well as table 2 clarifies the details.

Table 1. Sample description

Sample	Thicknesses of a bilayer (nm\nm)	Number of layers
α1	8\3	19
α2	4\3	56
α3	2\3	60

The stationary PL measurements were carried out at room temperature using a Nanometrics RPM PL Wafer Mapping System in the range of 580 – 1100 nm, the samples were pumped with the radiation of a Nd:YAG laser at λ = 532 nm with a power of 26 mW (a power density of 325 mW/cm2). We used a CCD 512-TE detector in tandem with a 570LP filter.

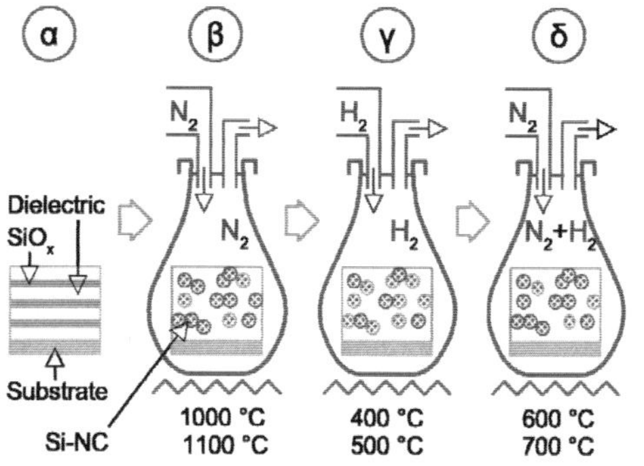

Figure 1. Schematic representation of the SiO_x/Al_2O_3 nanostructure modification. Circled letters are keys that symbolize modification steps: α is forming, β is annealing, γ is hydrogenation, δ is dehydrogenation.

The IR-transmission spectra in the range of 400–1400 cm^{-1} were recorded with a spectrometer (Varian 4100 Excalibur, a spectral resolution of 2 cm^{-1}) by the method of multiple signal accumulation.

Table 2. Treatment parameters

	Annealing – β, 2h		Hydrogenation – γ, 2h				Dehydrogenation – δ, 1h		
Temperature (°C)	1000	1100	400	450	500	550	600	650	700
Sample designation	β0	β1	γ40	γ45	γ50	γ55	δ60	δ65	δ70

3. Experimental results and discussion

3.1. The optical properties of the structures under annealing

Figure 2 shows the typical IR transmittance spectra for sample α1 annealed at different temperatures. There are a few absorption peaks [13] in the range from 400 to 1400 cm^{-1}. The 800 cm^{-1} band associated with ring-like configurations of silicon atoms separated from each other by oxygen atoms is observed when the temperature is not higher than 500 °C. Upon annealing at 600 °C, a broad absorption peak emerges at a frequency of 810 cm^{-1} corresponding to the bending vibrations of Si-O-Si groups [3,14]. As to the 1050 cm^{-1} peak associated with the asymmetric vibrations of Si–O–Si groups, it becomes stronger and shifts to higher frequencies. The 435 cm^{-1} band corresponding to the transverse optical rocking vibrations of Si–O–Si groups in SiO_2 does likewise. Thus, the absorption bands specific to the Si–O–Si vibration modes in silicon monoxide are transformed into the absorption bands typical of vibration modes of silicon dioxide (e.g. $2SiO_x \rightarrow xSiO_2 + (2–x)$ Si reaction takes place).

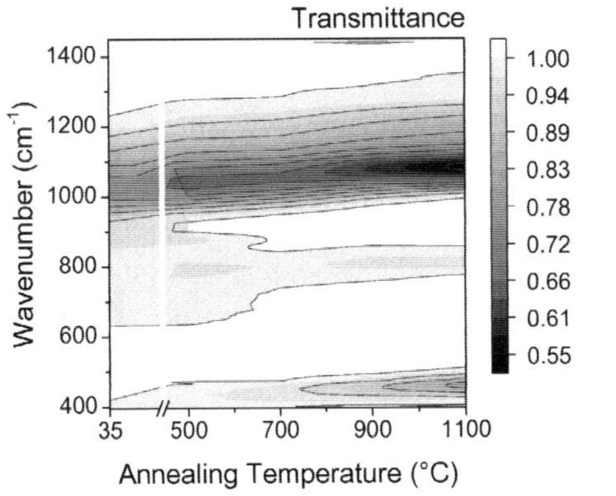

Figure 2. FTIR counterplot spectra of the annealed nanostructure α1 (8nm\3nm).

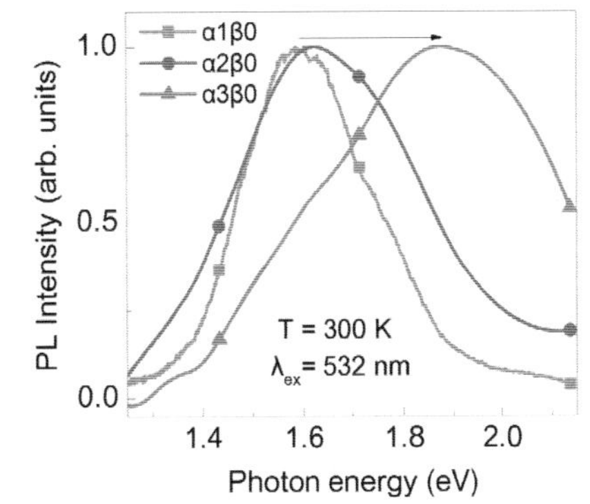

Figure 3. Normalized RT PL spectra of the structures annealed at 1000 °C that demonstrate a shift of the Si-NC peak.

The PL spectra are shown in figures 3 and 4. Depending on the temperature, we observe two intensive PL bands. The first one at 1.8 eV appears when the temperature is about 600 °C. We can assume that the band is caused by defects, since its position on the energy scale is constant for all the studied systems. Referring to the discussion of the IR-transmittance results, it can be concluded that a source is the dangling bonds of silicon-oxygen (810 cm^{-1}). Aluminium is expected to have a little effect on the oxidation process, since, as well-known, silicon is more electronegative than aluminium.

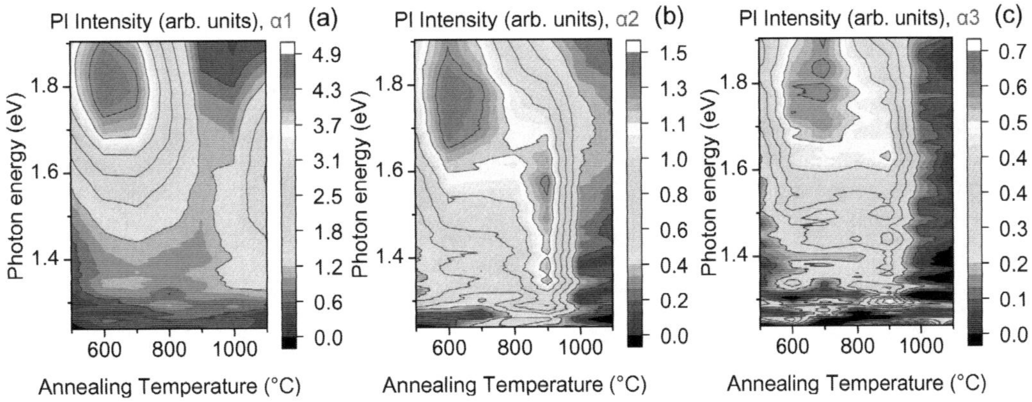

Figure 4. RT PL counterplot of the annealed nanostructures: (a) α1, (b) α2, (c) α3.

The second band around 1.6 eV seems to be associated with Si-NCs [6,10,15]. According to the spectra, the preferable temperature for forming is about 900–1100 °C (lower than for SiO_2 [13]), and it is determined by the thicknesses of the initial SiO_x layers. Thus, the thicker the base layer, the higher the annealing temperature is needed. This fact indicates that the 1.6 eV band is related to the diffusion process, and this is a proof that the source is Si-NCs. In order to compare the systems, we considered them at the same annealing temperature of 1000 °C (β0). In this case, based on the slight PL shift (figure 3, X-section of the items in figure 4 at 1000 °C is not clear enough) due to the confinement phenomena, there are the samples with different size Si-NCs, wherein the Si-NC size is proportional [10,13] to the thickness of the SiO_x layers (table 1).

3.2. *The optical properties of the structures under hydrogenation and dehydrogenation*

Figure 5 shows the typical PL spectra before and after hydrogenation at different temperatures. Annealing the samples at 400–550 °C in the hydrogen atmosphere leads to several-fold enhancement of the PL intensity [16], and here we observe a weak shift of the PL band due to the passivation of the dangling bonds, which is more effective for comparatively large Si-NCs [6]. As the temperature increases, the process comes to naught because of the growth of the hydrogen effusion [17]. Subsequent dehydrogenation of the samples in hydrogen-free environment at temperatures of 600-700 °C leads to a significant decrease of the PL intensity (figure 6) as a result of the hydrogen effusion. This proves the fact that the increase in the PL intensity should be associated with the passivation of the dandling bonds by hydrogen.

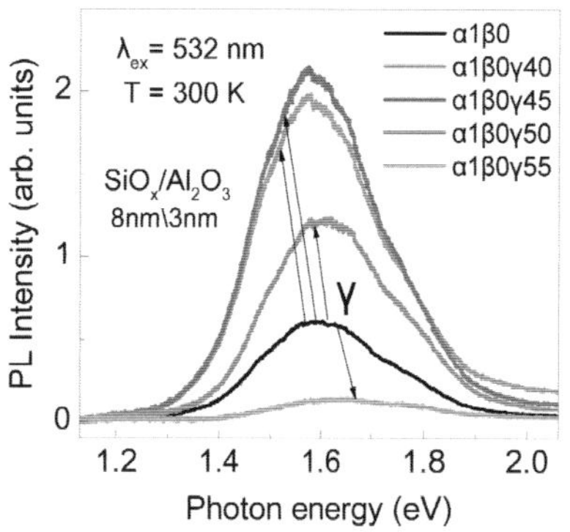

Figure 5. RT PL spectra of the α1 nanostructure (8nm\3nm) under various temperatures of hydrogenation.

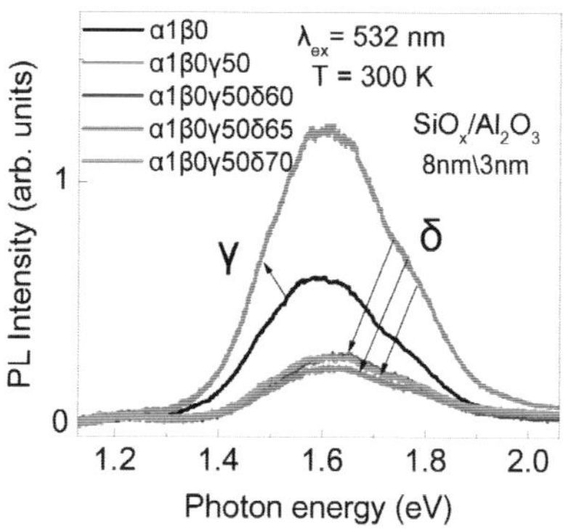

Figure 6. RT PL spectra of the α1 nanostructure (8nm\3nm) under various temperatures of dehydrogenation.

Figure 7 shows the curves of the normalized PL intensity versus the hydrogenation temperature. It is easy to see that the optimal temperature is in the range of 400 – 450 °C. Although sample α1 (8nm\3nm) shows the highest integrated PL intensity, sample α2 (4nm\3nm) demonstrates a better enhancement under hydrogenation.

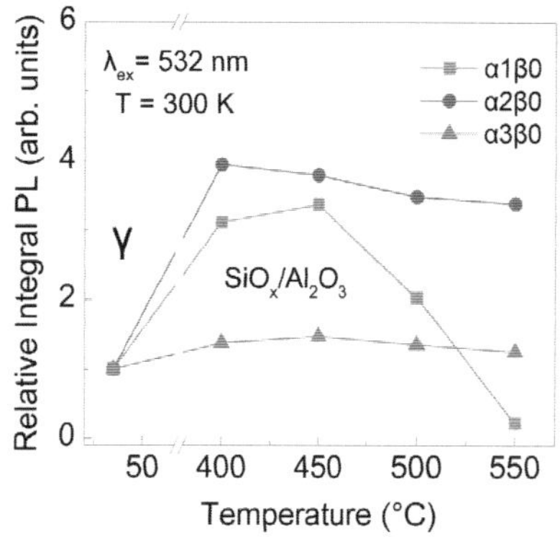

Figure 7. Normalized RT PL intensity of the nanostructures under various temperature of hydrogenation.

4. Summary

We have studied the composite nanoperiodic SiO_x/Al_2O_3 structures fabricated by physical deposition from separate sources. Depending on the SiO_x layer thicknesses, the structures included Si-NCs after annealing in the temperature range of 900 – 1100 °C that is lower than for SiO_x/SiO_2 counterparts. The structures showed intense room-temperature photoluminescence, the band position of which demonstrated the quantum-confinement phenomena. Under hydrogenation in the range of 400 – 450 °C, the photoluminescence intensity of the samples was enhanced several fold.

The study was partially supported by RFBR (grants #14-02-00119 and #15-02-05086) and by the Ministry of Education and Science of the Russian Federation (state assignment #3.285.2014/K).

References

[1] Turishchev S Y, Terekhov V A, Koyuda D A, Pankov K N, Ershov A V, Grachev D A, Mashin A I and Domashevskaya E P 2013 *Semiconductors* **47** 1316–23
[2] Ershov A V., Pavlov D A, Grachev D A, Bobrov A I, Karabanova I A, Chugrov I A and Tetelbaum D I 2014 *Semiconductors* **48** 42–5
[3] Mikhaylov A N, Belov A I, Kostyuk A B, Zhavoronkov I Y, Korolev D S, Nezhdanov A V, Ershov A V, Guseinov D V, Gracheva T A, Malygin N D, Demidov E S and Tetelbaum D I 2012 *Phys. Solid State* **54** 368–82
[4] Grachev D A, Garakhin S A, Belolipetsky A V, Nezhdanov A V and Ershov A V 2016 *J. Phys. Conf. Ser.* **741** 12129
[5] Palik E D 1998 *Handbook of Optical Constants of Solids II* (Academic Press)
[6] Garcia C, Garrido B, Pellegrino P, Ferre R, Moreno J A, Morante J R, Pavesi L and Cazzanelli M 2003 *Appl. Phys. Lett.* **82** 1595–7
[7] Bi L and Feng J Y 2006 *J. Lumin.* **121** 95–101
[8] Gusev O B, Ershov A V, Grachev D A, Andreev B A and Yablonskiy A N 2014 *J. Exp. Theor. Phys.* **118** 830–7
[9] Gert A V and Yassievich I N 2015 *Semiconductors* **49** 492–7
[10] Zacharias M, Heitmann J, Scholz R, Kahler U, Schmidt M and Bläsing J 2002 *Appl. Phys. Lett.* **80** 661
[11] Turishchev S Y, Terekhov V A, Koyuda D A, Spirin D E, Parinova E V, Nesterov D N, Grachev D A, Karabanova I A, Ershov A V, Mashin A I and Domashevskaya E P 2015 *Semiconductors* **49** 409–13
[12] Ershov A V, Pavlov D A, Grachev D A, Bobrov A I, Karabanova I A, Chugrov I A and Tetelbaum D I 2014 *Semiconductors* **48** 42–5
[13] Ershov A V, Chugrov I A, Tetelbaum D I, Mashin A I, Pavlov D A, Nezhdanov A V, Bobrov A I and Grachev D A 2013 *Semiconductors* **47** 481–6
[14] Tsu D V, Lucovsky G and Davidson B N 1989 *Phys. Rev. B* **40** 1795–805
[15] Vaccaro L, Spallino L, Zatsepin A F, Buntov E A, Ershov A V, Grachev D A and Cannas M 2015 *Phys. Status Solidi* **252** 600–6
[16] Li Y, Liang P, Hu Z, Guo S, You Q, Sun J, Xu N and Wu J 2014 *Appl. Surf. Sci.* **300** 178–83
[17] Kim K-H, Johnson E V. and Roca i Cabarrocas P 2016 *Jpn. J. Appl. Phys.* **55** 72302

IOP Conf. Series: Journal of Physics: Conf. Series 816 (2016) 012008 doi:10.1088/1742-6596/816/1

Thermodynamic and kinetic roughening: Monte Carlo simulation and experiment on GaAs

D M Kazantsev[1,2], I O Akhundov[1,2], A S Kozhukhov[1] and V L Alperovich[1,2]

[1] Rzhanov Institute of Semiconductor Physics, 630090 Novosibirsk, Russia
[2] Novosibirsk State University, 630090 Novosibirsk, Russia

E- mail: dmkazantsev@isp.nsc.ru

Abstract. GaAs thermal smoothing at temperatures $T \leq 650°C$ in the conditions close to equilibrium yields surfaces with atomically smooth terraces separated by steps of monatomic height. At higher temperatures $T \geq 700°C$, surface smoothing is changed to roughening. In the present paper, possible reasons of surface roughening at elevated temperatures are studied by means of Monte Carlo simulation and compared with the experimental results on GaAs. It is proved that the thermodynamic roughening transition, which consists in spontaneous generation of atomic steps due to decrease in the step line tension down to zero, cannot explain the experiment because it should occur at temperatures $T \sim 1800$ - $2000°C$, i.e. much higher than in the experiment. Kinetic instabilities caused by deviations from equilibrium towards growth or sublimation are shown to cause GaAs roughening at elevated temperatures. The microscopic mechanisms of kinetic-driven roughening are discussed.

1. Introduction

Fundamental surface science, nanostructures fabrication and device applications require high-quality, atomically flat crystal surfaces [1-5]. A technique for GaAs surface smoothing by allowing mass transport at elevated temperatures was developed in [6,7]. To avoid surface depletion with arsenic, the anneals were performed in conditions close to equilibrium between the surface and Ga and As vapors, when neither growth nor sublimation takes place. This technique yields GaAs(001) surfaces with atomically smooth terraces separated by steps of monatomic height (figure 1(a-c)).

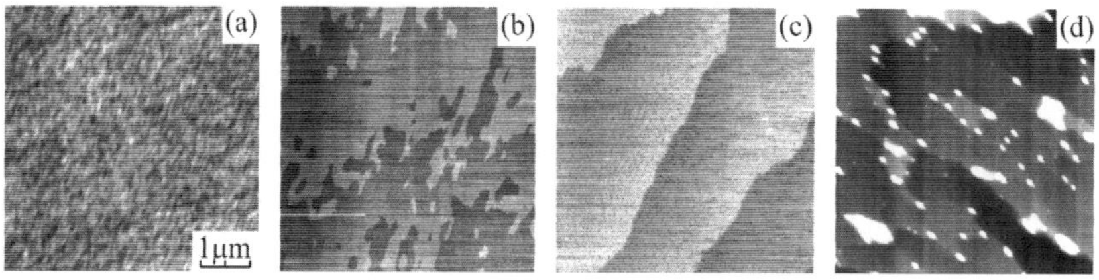

Figure 1. AFM images of an epi-ready GaAs substrate before (a) and after anneals for 2 h at 575°C (b), 650°C (c) and 775°C (d) [6].

Increasing the annealing temperature speeds up surface mass transport and, thus, facilitates the smoothing process. However, at temperatures $T \geq 700°C$, GaAs surface smoothing is replaced by roughening which consists in the formation of step bunches, multilayer islands and pits, and destruction of step-terraced morphology (figure 1(d)) [6]. This roughening restricts the annealing temperature and, thus, the maximal initial surface roughness that can be smoothed within a reasonable time. In principle, surface roughening can be caused by the thermodynamic roughening transition, when the step line tension decreases down to zero due to the entropy term in the surface free energy and spontaneous step generation becomes energetically favorable [8,9]. The roughening temperature, however, is typically close to or above the crystal melting point [8-10]. Alternatively, surface roughening can be caused by kinetic instabilities which, in their turn, may arise due to deviations of annealing conditions from equilibrium towards growth or sublimation. This paper is aimed at elucidating the reason of GaAs surface roughening at elevated temperatures by means of Monte Carlo simulation of thermodynamic and kinetic roughening.

2. Method

Initial step-terraced GaAs(001) surfaces used in this experiments were obtained by means of annealing "epi-ready" substrates at $T = 600 - 650$°C in the conditions close to equilibrium between the surface and Ga and As vapors, when atomic fluxes to the surface and from the surface are equilibrated so neither surface sublimation, nor growth occurs [6,7]. The equilibrium conditions were provided in a quasi-closed container by the presence of the saturated Ga-As melt. The anneals at higher temperatures $T = 700 - 775$°C were performed in the same container. The morphology of the initial and annealed GaAs surfaces was studied *ex situ* by atomic force microscopy (AFM).

Monte Carlo simulations of surface roughening were performed in a one-component model on the Kossel crystal. GaAs(001) surface smoothing and step-terraced morphology formation were previously described by this method [11,12]. The model parameters, which include the diffusion activation energy $E_d = 1.3 \pm 0.05$ eV, the lateral bond energy $E_b = 0.32 \pm 0.02$ eV and the adatom desorption energy $E_{des} = 2.1 \pm 0.05$ eV, were obtained earlier from the description of the experimental GaAs smoothing kinetics by Monte Carlo simulation. The same parameters were used in the present study for the surface roughening simulation.

3. Results and discussion

3.1. Thermodynamic roughening transition

At the first stage of the present study, we performed Monte Carlo simulation of the thermodynamic roughening transition on the initially flat Kossel crystal surface. Simulation of anneals at $T \leq 800$°C did not lead to substantial surface roughness which could explain the experiment. Notable surface roughening occurred at much higher temperatures $T \geq 1700$°C. Shown in figure 2(a-c) are the simulated surface images corresponding to various temperatures and an annealing time of 0.01 s.

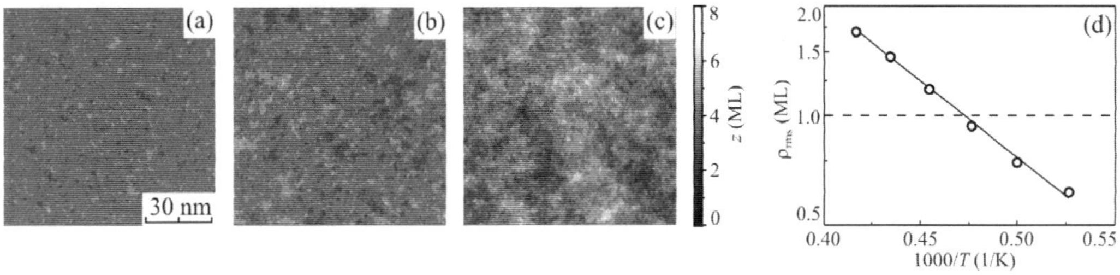

Figure 2. Simulated images of the initially flat surface after annealing for 0.01 s at 1725°C (a), 1925°C (b), 2125°C (c) and the rms roughness of the annealed surface as a function of inverse temperature (d). Dashed line in (d) shows monatomic step height. Solid line is the Arrhenius law approximation.

The roughening simulations are much more time-consuming than smoothing simulations performed in [11,12] due to increased temperature. Therefore, the annealing time in the simulation is limited by restricted computational resources. With increasing the temperature, the transition from a "smooth" surface (figure 2(a)) to an entirely disordered "rough" surface with undulations of about ± 8 ML (figure 2(c)) takes place. An estimated roughening transition temperature of the GaAs(001) surface is about $T_r \approx 1900°C$. The question remains: how to characterize quantitatively the transition from the smooth to the rough surface?

The simplest and straightforward characteristic of a rough surface is root mean square (rms) roughness. The Arrhenius plot of the rms roughness ρ_{rms} of the simulated surfaces is shown in figure 2(d). It is seen that ρ_{rms} is described by the Arrhenius dependence on the temperature with an activation energy of 2.6 E_b. The obtained activation energy value shows that a surface atom has to break 2 or 3 lateral bonds (2.6 on the average) to become an adatom. The activation dependence has no visible bend point or other features which could be considered as a roughening transition temperature. Assuming that the roughening transition corresponds to $\rho_{rms} = 1$ ML, we obtain the roughening temperature $T_r \approx 1825°C$.

A more strict criterion for surface roughening is asymptotic behavior of the surface height correlation function $G(\rho) = \langle(h_i - h_{i+\rho})^2\rangle_i$, where h_i is the surface height at point i, ρ is an arbitrary vector in the surface plane [8,9]. If the function diverges with increasing ρ, the surface is rough. If $G(\rho)$ remains finite at large ρ, the surface is considered as smooth, although there are islands and pits. The height correlation functions calculated for the simulated surfaces are plotted in figure 3(a) in a semi-logarithmic scale. It is seen that the asymptotic behavior of the height correlation function is not clear enough due to limited annealing time and, possibly, due to the finite size of the model surface area. Still, the roughening transition temperature estimated from figure 3(d) is about $T_r \approx 2000°C$.

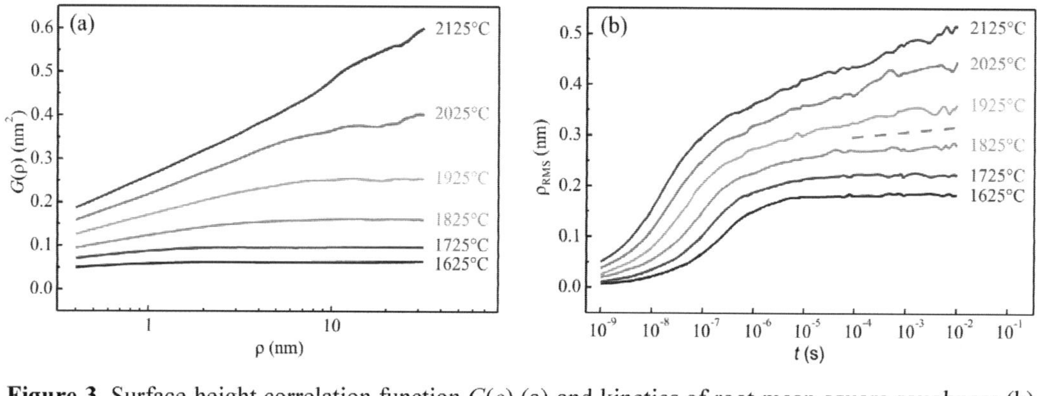

Figure 3. Surface height correlation function $G(\rho)$ (a) and kinetics of root mean square roughness (b) calculated at various annealing temperatures. The dashed line in (b) separates the "smooth" and "rough" surfaces.

To discriminate between smooth and rough surfaces more clearly, the rms roughness kinetics can be used. Figure 3(b) shows this kinetics for the simulated surfaces annealed at various temperatures. It is seen that, within the simulation accuracy, $\rho_{rms}(t)$ saturates for $T \leq 1825°C$ (the surfaces remained smooth) and keeps increasing for $T \geq 1925°C$ (the surfaces became rough). One can speculate that smooth surfaces reached thermodynamic equilibrium during the simulation, while above the roughening transition the equilibrium morphology is not reached, and the surface roughness continues to build up with increasing annealing time. The roughening transition temperature estimated from the kinetics is $T_r = 1925 \pm 50$ °C.

The estimated value of the thermodynamic roughening transition temperature T_r is much higher than the temperatures at which the experimental GaAs surface roughening is observed; T_r is even higher than the GaAs melting point T_m = 1240°. As the equilibrium surface roughness is determined by the competition between the step line tension and step thermal fluctuations [8,9], one needs to decrease the lateral bond energy by the factor of 2, down to E_b = 0.16 eV in order to fit the experimental roughening by the simulation, which seems unreasonable. Therefore, the experimentally observed GaAs surface roughening is likely due to kinetic instabilities which are considered in the next section.

3.2. *Kinetic roughening*

To prove that GaAs surface disordering at elevated temperatures is due to kinetic factors, an additional experiment was performed. Two GaAs samples with step-terrace morphology were annealed simultaneously with one of them being covered by another GaAs slab in order to form a narrow slit ("capillary") between two adjacent GaAs surfaces [6]. Shown in figure 4 are the initial step-terraced surface (preliminary obtained by annealing at T = 650°C) and the surfaces of the two samples after annealing at T = 775°C. It is seen that, as a result of the high-temperature anneal, multilayer high islands (white dots in figure 4(b)) and multilayer deep pits (black dots in figure 4(c)) have been formed on the open and covered surfaces, respectively. The qualitative difference in morphology between the open and covered samples proves that the disordering is due to kinetic instabilities caused by deviations of annealing conditions from equilibrium towards sublimation (for the open sample) or growth (for the covered one), because one would expect a universal morphology of the disordered surface for the thermodynamic roughening.

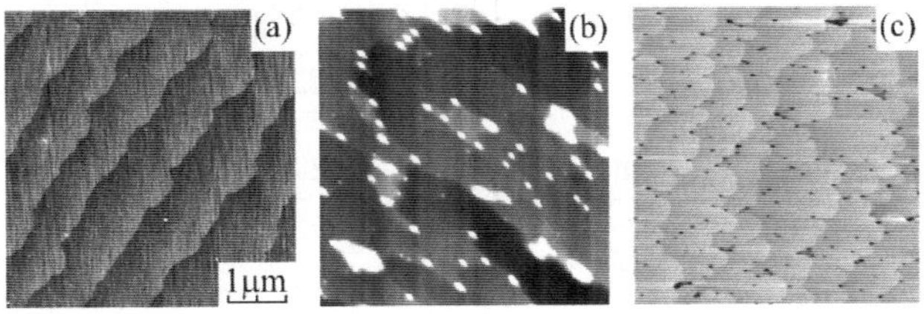

Figure 4. Surface roughening under different annealing conditions. AFM images of the initial step-terraced surface (a), and of the open (b) and covered (c) samples after annealing at T = 775°C.

The formation of multilayer islands and pits can be explained by monoatomic step flow around defect- or contamination-induced surface spots at which the sublimation and growth are inhibited. This explanation is confirmed by Monte Carlo simulation of steps interaction with a single defect spot. Shown in figure 5 is the simulated sublimation of a step-terraced surface with one spot in the center at which the sublimation is forbidden. After approaching the spot (figure 5(a)), a monatomic step is stopped by the spot. The step flows around the spot and, thus, forms an inlet (figure 5(b)). The inlet becomes longer and thinner (figure 5(c)), and, eventually, the step detaches from the spot (figure 5(d)). It is seen from figure 5(d) that the next step approaches the spot. Each time a monatomic step passes through the spot, the spot height is increased by one monolayer (ML) with respect to the terrace level. Thus, the islands of multilayer height are formed on the surface under sublimation.

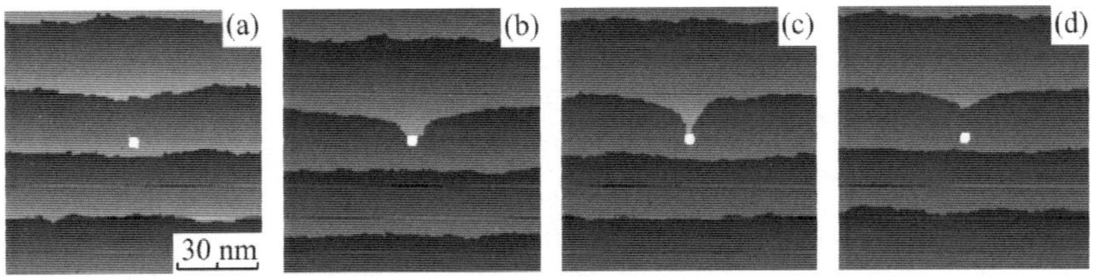

Figure 5. Consequent stages of simulated step flow through a defect-induced spot during sublimation.

4. Conclusions

In summary, possible reasons of GaAs surface roughening, which was observed experimentally at elevated annealing temperatures $T \geq 700°C$, are studied by Monte Carlo simulations. According to the simulations with model parameters obtained from the description of surface smoothing at lower temperatures $T \leq 650°C$, the thermodynamic roughening transition can not explain the experiment because it should occur at too high temperatures $T > 1800°C$. It is proved both experimentally and by means of Monte Carlo simulations that GaAs surface roughening is due to kinetic instabilities caused by deviations from equilibrium towards surface growth or sublimation. In particular, the formation of multilayer islands and pits on terraces are explained by steps flowing round some defect-induced surface spots at which the sublimation and growth are inhibited.

Acknowledgements

The annealing experiments and Monte Carlo simulations were supported by RFBR (Grant No. 14-02-00352). The surface relief AFM measurements were supported by RSF (Grant No. 14-22-00143).

References

[1] Crommie M F, Lutz C P and Eigler D M 1993 *Science* **262** 218–20
[2] Teichert C 2002 Phys. Rep. **365** 335–432
[3] Ibach H 2006 *Physics of Surfaces and Interfaces* (Heidelberg: Springer-Verlag)
[4] Karkare S and Bazarov I 2011 *Appl. Phys. Lett.* **98** 094104
[5] Cao R X, Miao B F, Zhong Z F, Sun L, You B, Zhang W, Wu D, Hu A, Bader S D and Ding H F 2013 *Phys. Rev.* B **87** 085415
[6] Alperovich V L, Akhundov I O, Rudaya N S, Sheglov D V, Rodyakina E E, Latyshev A V and Terekhov A S 2009 *Appl. Phys. Lett.* **94** 101908
[7] Akhundov I O, Alperovich V L, Latyshev A V and Terekhov A S 2013 *App. Surf. Sci.* **269** 2–6
[8] Lapujoulade J 1994 *Surf. Sci. Rep.* **20** 191–249.
[9] Pimpinelli A and Villain J 1998 *Physics of crystal growth* (Cambridge: Cambridge University Press)
[10] Zhdanov V P 1997 *Surf. Sci.* **388** L1126–9
[11] Kazantsev D M, Akhundov I O, Karpov A N, Shwartz N L, Alperovich V L, Terekhov A S and Latyshev A V 2015 *App. Surf. Sci.* **333** 141–6
[12] Kazantsev D M, Akhundov I O, Shwartz N L, Alperovich V L and Latyshev A V 2015 *App. Surf. Sci.* **359** 372–9

Ga focused ion beam etching of a Si_3N_4/GaN substrate for submicron selective epitaxy

M I Mitrofanov[1,2,3], S N Rodin[1], I V Levitskii[1,2,3], S I Troshkov[1], A V Sakharov[1], W V Lundin[1] and V P Evtikhiev[1]

[1] Ioffe Institute, 26 Politekhnicheskaya, St Petersburg, 194021, Russian Federation
[2] Submicron Heterostructures for Microelectronics, Research & Engineering Center, RAS 26 Politekhnicheskaya, St Petersburg, 194021, Russian Federation Submicron
[3] ITMO University 49 Kronverksky pr., St. Petersburg, 197101, Russian Federation

E-mail: maxi.mitrofanov@gmail.com

Abstract. Authors represent utilizing UHV Ga FIB for preparing a Si_3N_4/GaN substrate for submicron selective-area epitaxy. In result, GaN submicron stripes were grown in the 100, 200 and 500 nm windows of a Si_3N_4 mask layer. SEM investigations show good crystalline perfection of the grown stripes.

1. Introduction

There is a need for submicron patterned structures formation in the technology of III-N photonic and electronic devices. A topical issue is the formation of nonalloyed submicron contacts to the source and drain regions of a HEMT transistor [1]. A promising method is a selective epitaxial growth process [2]. It requires such complex, difficult and expensive methods as electron beam and extreme-ultraviolet lithography. An alternative method at the device development stage is nanolithography by focused ion beam (FIB). Among the works devoted to the selective epitaxy, the majority of them refer to the growth of quantum dots on substrates without masking coating [3, 4]. In this case, preliminary defined places are usually formed by focused ion beam on the substrate surface. Using substrates without masks, one can grow multilayer structures with a predetermined position of quantum dots [3]. However, this approach does not prevent the occurrence of epitaxial film islands between areas of deliberate crystallization, which are, as a result, the defects of structure. Utilization of masking coatings allows one to practically eliminate unintended growth [5-8]. The comparison of masks quality formed by various methods in a Ti film and the results of selective epitaxial growth of GaN layers [8] showed that the use of ion-beam lithography allows one to achieve substantially the same results as in the case of electron beam lithography but excludes a complicated step of removing the photoresist.

In our work, the possibility of submicron windows formation in a Si_3N_4 masking layer using direct ion lithography with a Ga focused ion beam is demonstrated. The results of studying the selectively grown epitaxial structures show good crystalline perfection.

2. Experiment

A 3 μm thick lightly Si-doped GaN layer was epitaxially grown on the (0001) sapphire wafer by metalorganic vapour phase epitaxy (MOVPE). The concentration of Si was $1\text{-}3\cdot10^{17}$ cm^{-3}. A 5 nm thick amorphous Si_3N_4 layer was deposited in the same technological process. SiH_4 and NH_3 were used as precursors, and the deposition temperature was close to 1000°C. A selective-area epitaxial mask was formed on Si_3N_4/GaN by our own designed ultra-high vacuum (UHV) Ga FIB equipped by an Orsay Physics "Cobra" ion gun. The windows of the mask were 200 μm long stripes of different widths: 100, 200 and 500 nm. The FIB etching parameters were 30 keV beam energy, 450 pA probe current and 10 nm beam step. The total exposure dose was close to $8 \cdot 10^{16}$ cm^{-2}. For the etching rate enhancement, preventing the etched material redeposition as well as appearing of gallium drops, XeF_2 was used as a precursor gas. The atomic force microscopy (AFM) measurements (see figure 1) of etched stripes showed an etching depth of 6-7 nm and an average roughness (Ra) of 0.53 nm. AFM and SEM images show good quality of the Si_3N_4 mask surface without any impurities after FIB etching.

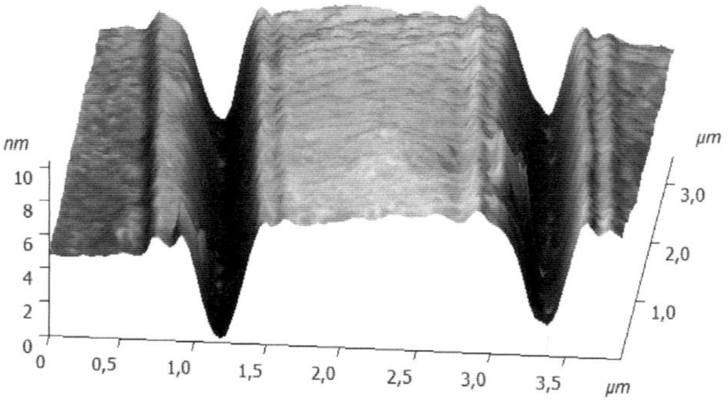

Figure 1. AFM image of the 500 nm wide FIB etched stripes on a Si_3N_4/GaN substrate.

Si+ doped GaN stripes were selectively grown on the FIB prepared substrates. The secondary electron microscopy (SEM) images of the stripes are shown in figure 2. Inset in figure 2 (a) shows the formation of the facets of the stripe.

(a) 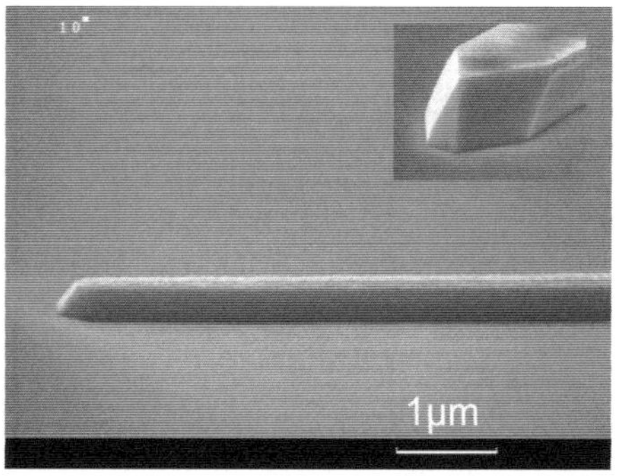

(b) 

Figure 2. SEM images of: (a) a selective-area epitaxilly grown submicron GaN stripe, inset shows the formation of the stripe facets; (b) a cleaved facet of the stripe.

3. Micro photoluminescence study

The micro photoluminescence (micro PL) technique was used for the estimation of doping and crystalline quality of the grown stripes. Figure 3 shows room temperature (RT) micro PL spectra of the Si-doped GaN epitaxial layer in the area between the stripes (a solid curve) and a highly Si-doped

GaN stripe grown in a 200 nm width window (a dashed curve). A HeCd laser was used for optical pumping.

The luminescence spectrum of the GaN epitaxial layer contains a typical for GaN near-band-edge luminescence (363 nm) peak and also blue band (430 nm) and yellow band (550 nm) peaks. The long-wave edge of the spectrum is modulated due to interference of light reflected from both the top and bottom boundaries of the GaN layer.

The luminescence spectrum of a stripe contains a highly broadened and long-wave shifted GaN interband luminescence peak and a bright yellow band (550 nm) luminescence. Broadening, long-wave shifting of band edge luminescence and a yellow band are typical for GaN epitaxial layers heavily doped by Si.

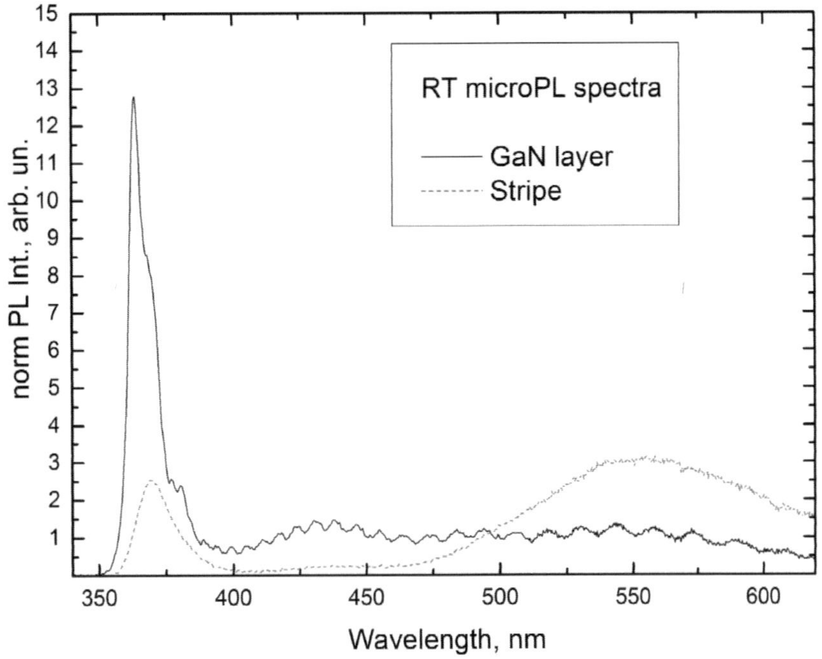

Figure 3. Micro PL spectra of the undoped GaN epitaxial layer and the highly Si-doped GaN stripe grown in a 200 nm wide window.

4. Conclusion

We have shown that FIB nanolithography can be used as a rapid technique for preparing masked substrates intended for selective-area epitaxy. Substrate surface investigations show good quality of the Si_3N_4 mask without any impurities after FIB etching. Formation of the stripe facets observed in the SEM images confirms good crystalline perfection of selectively grown material. The obtained results could be utilized for prototyping of submicron photonic and electronic structures, especially for nonalloyed ohmic contacts to HEMT transistors.

Acknowledgments

Investigation was supported by DNIT RAS fundamental research program # 5 "Fundamental problems of physics and technology of heterostructures and heterostructure-based devices";
RAS fundamental research program # 1 "Nanostructures: Physics, Chemistry, Biology and Fundamentals of Technology";

Russian Science Foundation grant # 16-12-10503 "Purcell effect in photonics nanostructures".

References

[1] Denninghoff D J, Dasgupta S, Lu J, Keller S and Mishra U K 2012 *IEEE Elect. Dev. Lett.* **33** 6 785-7
[2] Li SF and Waag A 2012 *J. Appl.Phys.* **11** 7 071101
[3] McKay H, Rudzinski P, Dehne A and Millunchick J M 2007 *Nanotechnology* **18** 45
[4] Martin A J, Saucer T W, Rodriguez G V, Sih V and Millunchick J M 2012 *Nanotechnology* **23** 13
[5] Gierak J 2009 *Semicond. Sci. and Tech.* **24** 4
[6] Kitslaara P *et al* 2006 *Microelectronic Eneneering* **83** 811–14
[7] Wang Y L, Temkin H, Hamm R A, Yadvish R D, Ritter D, Harriott L H and Panish M B 1991 *Electronics Letters* **27** 15 1324-26
[8] Barbagini F, Bengoechea-Encabo A, Albert S, Martinez J, Garcia M, Trampert A and Calleja E 2011 *Nanoscale Res. Lett.*, **6** 1-9

IOP Conf. Series: Journal of Physics: Conf. Series 816 (2017) 012010 doi:10.1088/1742-6596/816/1/0120

Plasmon resonance induced photoconductivity of ZrO_2(Y) films with embedded Au nanoparticles

D A Liskin[1], D O Filatov[2], O N Gorshkov[2], A P Gorshkov[1,2], I N Antonov[2], M E Shenina[2], S Y Zubkov[2] and D S Sinutkin[1]

[1] Department of Physics, Lobachevsky State University of Nizhny Novgorod, Nizhny Novgorod 603950, Russia

[2] Research and Educational Center for Physics of Solid State Nanostructures, Lobachevsky State University of Nizhny Novgorod, Nizhny Novgorod 603950, Russia

Email: dmitry_liskin@mail.ru

Abstract. We report on the experimental observation of photoconductivity (PC) in ultrathin ($\approx$ 40 nm thick) ZrO_2(Y) films with single layered arrays of Au nanoparticles (NPs) of 1 to 3 nm in diameter. The samples were prepared by the deposition of islanded Au films of ~1 nm in thickness sandwiched between two ZrO_2(Y) layers by Magnetron Sputtering followed by annealing. The effect of PC was attributed to the photoexcitation of the collective plasmon oscillations in dense Au NP arrays. The temperature dependencies and kinetics of PC have been studied. At 300 K, the PC has been found to originate mainly from heating of ZrO_2(Y) due to plasmon optical absorption in Au NPs (the bolometric effect). At 77 K, the photoresponse has been attributed to plasmon-assisted electron transport between NPs via the vacancy α-band in ZrO_2(Y) barriers.

1. Introduction

The investigations of photoconductivity (PC) of the nanocomposite materials with the metallic nanoparticle (MNP) arrays embedded into the dielectric matrices due to the optically excited collective plasmonic oscillations in the MNPs have attracted an increasing interest in recent years [1]. The effect of PC in such systems is considered to be promising for various applications in novel integrated optics and photonic devices, such as photodetectors, optical switches, etc. [2]. There were many publications devoted to the investigations of collective plasmon excitations in MNP arrays [3]. However, the details of the PC mechanisms in the dielectric films with embedded MNPs, namely how the energy of the plasmon oscillations in the MNPs is converted into the change of the electrical conductivity of the nanocomposite films is not clear yet. Several models of the PC mechanism have been proposed in the literature. Thus, in [4] the PC in two-dimensional (2D) arrays of closely-packed Au nanoparticles (NPs) of 2 to 3 nm in diameter in the alkane-thiol films prepared on the SiO_2/Si substrates by the Langmuir-Blodgett technique has been studied. A pronounced peak at the plasmon resonance (PR) wavelength $\lambda_p \approx 600$ nm in the MNP array has been observed in the PC spectrum. Therefore, the authors concluded the PC to originate from the excitation of the surface plasmon-polariton oscillations in MNP arrays [5]. The effect was attributed to the increasing of the conductivity of the alkane matrix due to heating because of the plasmonic optical absorption in MNPs, i. e. the PC had a bolometric nature. In [6], the PC in 2D Ag NP arrays in a porous Al_2O_3 film was investigated. The PC was attributed to the electron tunneling between MNPs across thin (~ 1 to 10 nm) Al_2O_3 barriers stimulated by surface plasmon-polaritons in Ag NPs. In [7], the collective plasmon excitation induced PC in Au NP arrays dispersed in the porphyrin (dithiol-PZn_3) has been studied. In that case, the dielectric matrix was able to manifest the PC itself, and the role of Au NPs was limited to the enhancement of the light energy concentration between MNPs due to collective plasmon oscillations. This, in turn, enhanced the PC in the porphyrin-filled gaps between Au NPs.

Summing up the above, one can conclude the PC mechanism in the nanocomposite films to depend strongly on the properties of the dielectric matrix. In the present study, the PC in ultrathin yttria stabilized zirconia ZrO_2(Y) films with embedded Au NP arrays (ZrO_2(Y):NP-Au films) has been investigated. It is worth noting that the PC in transition metal oxides with embedded MNPs has not been studied to date. An important distinctive feature of ZrO_2 is that the oxygen vacancies generate the deep donor states in the energy gap of this material with the energy $E_c - 0.3$ eV where E_c is the conduction band edge [8]. In ZrO_2(Y) with the cubic elementary cell, the oxygen vacancy is a crystal structure element, and the vacancy concentration is defined by the one of Y atoms [8]. When the Y molar fraction is ≥ 0.1, the oxygen vacancy related donor levels create a defect band (α-band), and the electron transport in this band plays an important role in the electron conductivity mechanism in ZrO_2(Y).

The goal of the present work was to study the mechanisms of the collective plasmon excitation inducted PC in the interacting Au NP arrays embedded into thin ZrO_2(Y) films.

2. Experimental details

The ZrO_2(Y) films were deposited by high frequency (HF) Magnetron Sputtering in Ar-O_2 gas mixture environment (50:50 % mol.) at the pressure of ~10^{-2} Torr from the powder oxide targets using a Torr International® MSS-3GS vacuum system for thin film deposition. The molar fraction of the stabilizing oxide Y_2O_3 in the target material was ≈ 0.12. The substrate temperature was $T_g \approx 300$ °C.

The nanocomposite ZrO_2(Y)-based films were deposited onto the fused silica substrates polished to the optical quality. For the PC measurements, the Au(20 nm)/Cr(20 nm) strip-wise electrodes with the width of ~ 1 mm with the butterfly-wise gaps were formed on the substrate surface by standard optical photolithography prior to the deposition of the nanocomposit films. A schematic representation of the electrode configuration is presented in figure 1(a). The initial Au films with Cr sublayers were deposited by direct current (DC) Magnetron Sputtering in Ar ambient at $T_g = 200$ °C. The width of the gaps between the electrode tips was ~1 μm. Some areas on the substrate surface were left free of electrodes for the optical transmission spectra measurements. First, the underlying ZrO_2(Y) layers of ≈20 nm in thickness were deposited onto the silica glass substrates with the Au/Cr electrodes. Then, the islanded Au films with the nominal thickness $d_{Au} = 0.5$ to 2.0 nm were deposited by dc Magnetron Spattering at $T_g = 200$ °C.

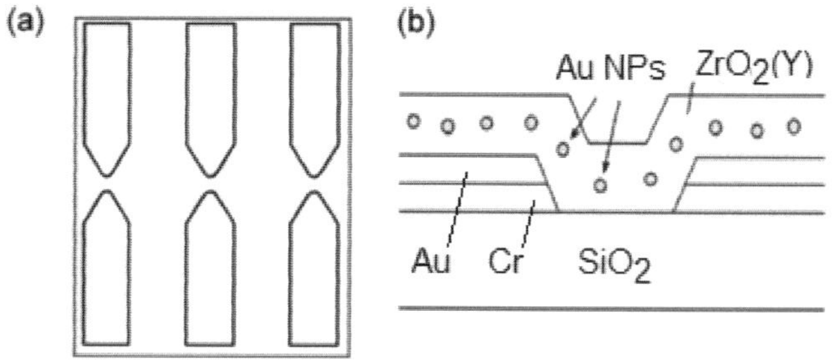

Figure 1. Schematic representation of a ZrO_2(Y):NP-Au/SiO_2 sample for PC measurements. (a) electrode configuration (top view); (b) **a** cross-section along a strip electrode axis.

IOP Conf. Series: Journal of Physics: Conf. Series **816** (2017) 012010 doi:10.1088/1742-6596/816/1/012010

In turn, the islanded Au films were capped by the ≈ 20 nm thick cladding ZrO_2(Y) layers deposited in the same conditions as the underlying one. Finally, the samples were annealed in Ar ambient at 450°C during 1 hour. The resulting structure of the nanocomposite film in the gap between the tips of the Au/Cr electrodes is shown schematically in figure 1(b). Earlier, High Resolution Cross-sectional Transmission Electron Microscopy (HR X-TEM) studies have shown the islanded Au films in the ZrO_2(Y)/Au/ZrO_2(Y) stacks prepared as described above to coagulate into Au NPs during annealing under certain conditions [9]. The nearly spherical Au NPs with a diameter, *D*, of 1 to 3 nm were arranged almost in a single sheet inside ZrO_2(Y) films. The average spacing between Au NPs in the 2D arrays, ℓ, was found to be from 3 to 4 nm, and the density parameter of the arrays $a = D/\ell$ was from 0.3 to 0.6.

Along with the ZrO_2(Y)-based samples, the samples based on ZrO_2 and GeO_2 films with Au NPs formed under the same conditions have been fabricated. GeO_2 is featured by dominating covalent-type chemical bonding, different crystal structure, and the electronic properties of the structural defects as compared to ZrO_2(Y). Particularly, there is no α-band in ZrO_2 as well as in GeO_2. Besides, the oxygen ion mobility in GeO_2 is much lower than the one in ZrO_2(Y). A comparative study of the PC in the samples based on different materials was expected to help clarifying the details of the PC mechanism in ZrO_2(Y), particularly, the role of the α-band in this one. Also, the pure films from the above materials of ≈ 40 nm in thickness without MNPs as were fabricated to serve as the reference.

The optical transmission spectra of the nanocomposite films were measured at 300 K using a Varian® Cary™ 6000i spectrophotometer. More details on the technique and the results of the optical transmission spectroscopy of the nanocomposite films similar to the ones investigated in the present work can be found in [9].

The PC measurements were carried out by the standard lock-on technique. A number of continuous wave semiconductor laser diodes (LDs) with the emission wavelengths λ ranging from 405 nm to 660 nm were applied as monochromatic photoexcitation sources. The LD emission was focused onto the gaps in the strip-wise electrodes on the sample surfaces in the spots of ~1 mm in size. The LD beam was chopped by a Stanford Research® SR-540 chopper with a frequency, *f*, of 6 to 350 Hz. The LD emission power was up to 1.5 W and was attenuated by a set of the neutral (grey) light filters. The bias voltage V_b = 10 V was applied to the electrodes, and the samples were loaded onto the current input of a Stanford Research® SR-810 digital lock-on detector. The dark current at V_b = 10 V measured with an Agilent® B1500A semiconductor device analyzer was ≈ 0.5 pA. To measure the temperature dependencies of the PC and of the dark current, the samples were immersed into a glass Dewar flask filled with liquid nitrogen.

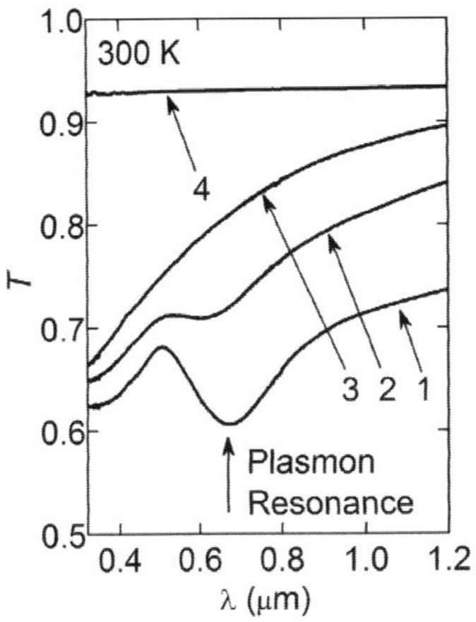

Figure 2. Optical transmission spectra of ZrO_2(Y):NP-Au/SiO_2 films. Initial Au nominal thickness d_{Au}, nm: *1* — 2.0; *2* — 1.0; *3* — 0.5. *4* — reference (SiO_2 substrate).

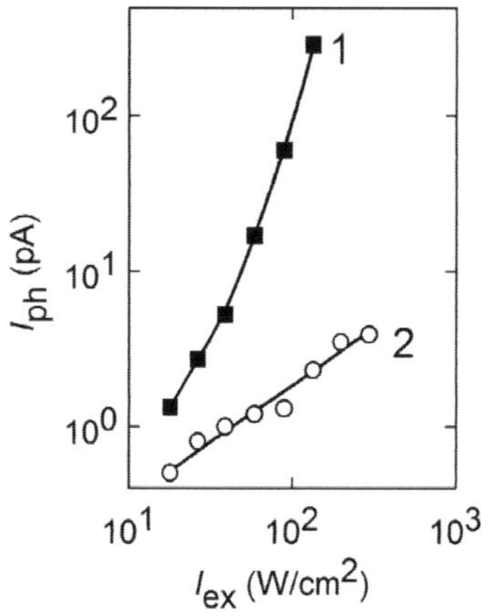

Figure 3. Dependence of PC in $ZrO_2(Y)$:NP-Au film (d_{Au} = 2.0 nm) on the photoexcitation intensity I_{ex} at λ = 660 nm. *T*, K: *1* – 300; *2* – 77.

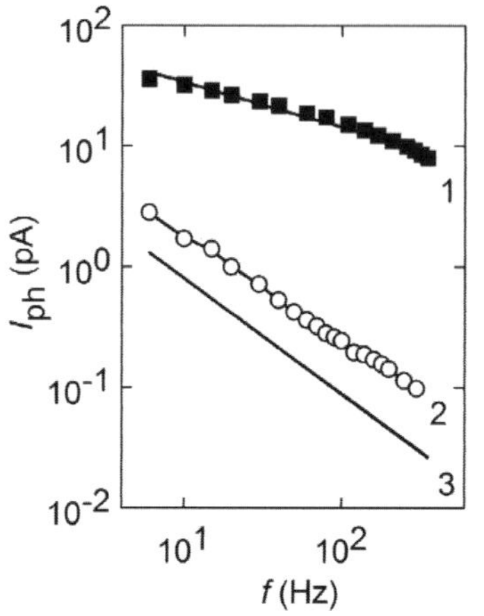

Figure 4. Dependence of PC in $ZrO_2(Y)$:NP-Au films (d_{Au} = 2.0 nm) on the chopper frequency *f* (300 K): *1, 2* – experiment; *3* – calculation. I_{ex} = 65 W/cm^2, λ = 660 nm. Y molar fraction: *1* – 0.12, *2* – 0.

3. Results and discussion

Figure 2 shows the optical transmission spectra of the $ZrO_2(Y)$:NP-Au films with different values of d_{Au}. A well expressed plasmon absorption peak was observed at λ ≈ 680 nm for d_{Au} = 2.0 nm (figure 2, curve 1). The intensity of the plasmon absorption peak decreased with decreasing d_{Au} (figure 2, curves 2 and 3). It is worth noting that the PR wavelength calculated for the individual Au NPs with the sizes specified above in the $ZrO_2(Y)$ matrix according to the Mie scattering theory [10], λ_p = 590 to 600 nm. The red shift of the PR observed in the experiment relative to the calculated values of λ_p for the non-interacting NPs was attributed to the collectivization of plasmon excitations in dense Au NP arrays.

The values of photocurrent I_{ph} measured in $ZrO_2(Y)$: NP-Au films were within the range of 1 to 100 pA that was well above the dark current as well as the noise level (~ 0.1 pA). The PC depended on the photoexcitation intensity I_{ex} superlinearly at 300 K (figure 3). On the other hand, we haven't observed any PC above the noise level in the reference $ZrO_2(Y)$ films without Au NPs. All the above allowed us to attribute the PC in $ZrO_2(Y)$: NP-Au films to the plasmon excitations in Au NPs.

The PC at λ = 660 nm (300 K) decreased with increasing chopping frequency *f* (figure 4, curve 1). Such a behavior is typical for the case when PC is measured by the lock-on technique, and $f^{-1} \ll \tau$ where τ is the PC decay time. Assuming the exponential PC decay law, the relation between the steady state PC I_0 (at $f \to 0$) and the amplitude of the 1st harmonic of the PC I_{ph} registered by the lock on detector is given by the following formula [11]:

$$I_{ph} = I_0 \mathrm{th}\left(\frac{1}{4f\tau}\right) \quad (1)$$

The decay time τ > 1 s could originate from the stray capacitance of the measurement circuit, *C*, which was ~10 pF. Indeed, for I_{ph} ~ 10 pA at V_b = 10 V one has the effective sample resistance $R = V_b/I_{ph}$ ~ 10^{12} Ω that gives *RC* ~ 10 s. However, at 77 K the PC didn't depend on *f*. To explain the frequency dependence of PC, we suggested the PC to originate from heating of the film material due to the plasmonic light absorption in NPs [12], i. e., the PC of $ZrO_2(Y)$:NP-Au films had a bolometric nature as in [5]. In order to confirm this hypothesis, we have solved the time-dependent one-dimensional heat transfer equation. The NP layer was considered as a δ-type heat source. The relation between the time dependent temperature and resistance of the ZrO_2(Y):NP-Au film was extracted from the results of the dark current temperature dependence measurements. The model $I_{ph}(f)$ dependence is presented in figure 4 (curve 3). One can see a qualitative agreement between the model $I_{ph}(f)$ dependence and the measured one. Within the framework of the bolometric mechanism, the independence of PC on f at 77 K was explained by a negligible dark current because it was found to obey the Mott law that, in turn, is typical for the hopping conductivity via the α-band in ZrO_2(Y) [8]. However, one can see in figure 4 that heating is not enough to explain the PC magnitude observed in the experiment. Moreover, the slope of the model $I_{ph}(f)$ dependence (figure 4, curve 3) is greater than the one of the dependence measured experimentally (figure 4, curve 1). In addition, the $I_{ph}(I_{ex})$ dependence at 77 K in figure 3 is sublinear (the power index was ≈ 0.73) while the one measured at 300 K is essentially superlinear. The latter agrees with the proposed bolometric mechanism as the heat transfer rate, in general, increases with increasing heat generation intensity in Au NPs. Nevertheless, all the observations presented above point to different PC mechanisms at various temperatures. So far, other PC mechanisms should be considered along with the bolometric one.

Potentially, the PC in ZrO_2(Y):NP-Au films could be attributed to the internal photoemission of the electrons from the Fermi level in Au NPs into the conduction band of ZrO_2(Y). However, the potential barrier height at the Au/ZrO_2(Y) interface is ≈ 2.4 ± 0.1 eV at 300 K [13] (see the calculated band diagram of Au NPs in the ZrO_2(Y) matrix in figure 5). At the same time, the photon energy $h\nu$ corresponding to $\lambda = 660$ nm is ≈ 1.9 eV that is not enough for the internal photoemission.

On the other hand, $h\nu \approx 1.9$ eV is almost enough to excite the electrons from the Fermi level in Au NPs into the α-band in ZrO_2(Y) barriers. The excitation of electron into the α-band can also be associated with the electron tunneling through the triangle barrier at the Au/ZrO_2(Y) interface, as shown schematically in figure 5. Taking into account the PR in Au NPs to occur during the illumination with $\lambda = 660$ nm, one can expect the mechanism of the optical excitation of the electrons from Au NPs into the α-band in ZrO_2(Y) barriers in the conditions of the plasmon resonance in Au NPs to be an efficient one [6].

It should be stressed here that we haven't observed any PC signal in ZrO_2:NP-Au films above the noise level at 77 K that can be explained by the absence of the α-band in ZrO_2, in contrary to ZrO_2(Y). At 300 K, the observed values of PC in a ZrO_2:NP-Au film with $d_{Au} \approx 2.0$ nm (figure 4, curve 2) were much closer to the expected values for the bolometric mechanism (figure 4, curve 3) than the ones for the ZrO_2(Y):NP-Au film (figure 4, curve 1). In addition, the slope of the experimental dependence $I_{ph}(f)$ was close to the slope of the model one (cf. curves 2 and 3 in figure 4). All the above allowed us to conclude the PC observed in ZrO_2:NP-Au films to have a pure bolometric nature.

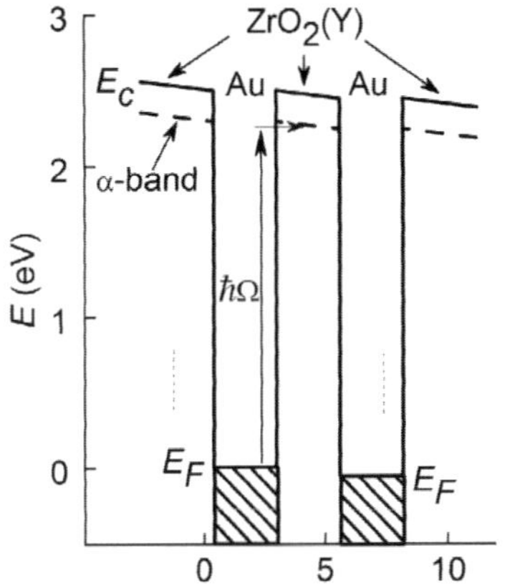

Figure 5. Calculated band diagram (300 K) of Au NPs in ZrO_2(Y) film in the external electric field applied between the Au/Cr electrodes (figure 1). $D = 3$ nm; $V_b = 10$ V; $\hbar\Omega$ is the collective plasmon-polariton energy in the Au NP array.

This fact, in turn, supports the proposed mechanism of PC in ZrO_2(Y):NP-Au films related to the plasmon-assisted excitation of the electrons from the Fermi level in Au NPs into the α-band in ZrO_2(Y) barriers and the electron transport to the adjacent Au NPs via the α-band.

We haven't observed any PC in GeO_2-based films (neither with nor without Au NPs). This fact can be explained by the following. The potential barrier height at the Au/GeO_2 interface can be estimated as ≈ 3.5 eV according to [14] whereas the PR in the optical transmission spectra of the GeO_2:NP-Au film has been observed at $h\nu \approx 2.3$ eV (for $d_{Au} \approx 2.0$ nm) that is surely not enough for the plasmon-assisted tunneling between NPs.

4. Conclusion

The results of the present study have demonstrated the nanocomposite ZrO_2(Y) films with embedded single sheet arrays of Au NPs to exhibit the PC under excitation at the wavelength corresponding to the plasmon optical absorption resonance. The effect was attributed to the optical excitation of the collective plasmonic oscillations in the dense Au NP arrays. The investigations of the PC kinetics have revealed two components of the photoresponse dominating at different temperatures. The slow component related to the bolometric effect due to heating of the ZrO_2(Y) matrix between NPs because of the plasmon optical absorption in Au NPs predominates at 300 K. The bolometric effect is absent at 77 K, when the fast component related to the plasmon-assisted excitation of electrons into the oxygen vacancy related α-band in ZrO_2(Y) barriers and the electron transport to the adjacent NPs via the α-band dominates.

Thus it was established that the nanocomposite ZrO_2(Y) films with embedded single sheet arrays of Au NPs demonstrate the PC under the excitation wavelength corresponds to the plasmon optical absorption resonance. The effect is attributed to the optical excitation of the collective plasmonic oscillations in the dense Au NP arrays.

Acknowledgments

The authors gratefully acknowledge the financial support from Ministry of Education and Science of the Russian Federation (Project # 3.2441.2014/K).

References

[1] Hwang J D, Wang F H, Kung C Y, Lai M J and Chan M C 2014 *J. Appl. Phys.* **115** 173110
[2] Atwater H A and Polman A 2010 *Nature Mater.* **9** 205
[3] Pelton M, Aizpurua J and Bryant G 2008 *Laser & Photon.* **2** 136
[4] Hashimoto T, Fukunishi Y, Zheng B, Uraoka Y, Hosoi T, Shimura T and Watanabe H. *Appl. Phys. Lett.* **102** 083702
[5] Mangold M A, Weiss C, Calame M and Holleitner A. W 2009 *Appl. Phys. Lett.* **94** 161104
[6] Huang C-H, Lin H-Y, Lau B-C, Liu C-Y, Chui H-C and Tzeng Y 2010 *Opt. Express.* **18** 27891
[7] Banerjee P, Conklin D, Nanayakkara S, Park T-H, Therien M J, Bonnell D A 2010 *ACS Nano.* **4** 1019
[8] Abbas H A 2012 *Stabilized Zirconia for Solid Oxide Fuel Cells or Oxygen Sensors. Characterization of Structural and Electrical Properties of Zirconia Doped with Some Oxides* (New York: LAP Lambert Academic)
[9] Gorshkov O N, Antonov I N, Filatov D O, Shenina M E, Kasatkin A P, Pavlov D A, Bobrov A I 2016 *Tech. Phys. Lett.* **42** 36
[10] Mie G 1908 *Ann. Phys.* **3**, 377
[11] Ryvkin S M 1950 *Sov. Phys. JETP* **20** 139
[12] Baffou G, Quidant R and Garcia de Abajo F J 2010 *Nano Lett.* **4** 709
[13] Filatov D O, Guseinov D V, Antonov I N, Kasatkin A P and Gorshkov O N 2014 *RSC Advances* **4** 57337
[14] Shaposhnikov A V, Perevalov T V, Gritsenko V A, Cheng C H and Chin A 2012 *Appl. Phys. Lett.* **100** 243506

Room-temperature formation of Pt_3Si/Pt_2Si films on poly-Si substrates

S A Mironov, V P Dubkov, K V Chizh, and V A Yuryev

A. M. Prokhorov General Physics Institute of the Russian Academy of Sciences,
38 Vavilov Street, Moscow, 119991, Russia

E-mail: sa.mironov@kapella.gpi.ru

Abstract. We have observed a phenomenon of formation of a thin bilayer Pt_3Si/Pt_2Si film at room temperature on poly-Si substrates in the process of Pt magnetron sputtering and wet etching, obtained such a film and investigated its structure and phase composition. By direct X-ray photoelectron-spectroscopic measurements, we have verified our previous observation of the Pt_2Si layer formation between Pt and poly-Si films as a result of Pt magnetron sputtering at room temperature. This layer likely appears due to a high enough temperature of Pt ions in the magnetron plasma, sufficient for the chemical reaction of the silicide film formation on the Si surface. The Pt_3Si layer likely forms from the Pt–Pt_3Si layer ($Pt_{95}Si_5$), which arises under a Pt film during magnetron sputtering, as a result of Pt removal by wet etching.

1. Introduction

Platinum silicides have attracted attention of researchers for a number of decades due to their exceptional prospects in microelectronics and silicon-based microphotonics [1–3]. PtSi infrared detector arrays represented a qualitative breakthrough and opened a new era in the infrared imaging technology [4, 5]. Now Pt silicides are considered as metals for Schottky-barrier formation to poly-Si:P in thin-film diode bolometers [6, 7]. There is no doubt about the prospects of their application as ohmic contacts and submicron lines in microelectronics especially taking into account their low formation temperatures that is very important for CMOS and especially for nanoelectronic devices [8–10]. In addition, low-temperature silicides, e.g. Pt_2Si, are expected to be suitable for formation of uniform Schottky contacts in devices of power electronics in which the sizes of contacts reach millimeters and nonuniformity of the barrier height causes a significant increase in reverse current [11]. Pt silicide ohmic contacts as well as Schottky barrier formation on polycrystalline [12] or amorphous silicon is of special interest for photovoltaic and sensor technology [7]. So, this class of materials should be considered as one of the friendliest ones to silicon technology. However, as the used silicide films become thinner and reach tens or even units of nanometers, their resistivity becomes of primary importance. In this connection, Pt_3Si, which has the lowest sheet resistance among the Pt silicides (18.9 $\Omega/\square$ in comparison with 2.6 $\Omega/\square$ of Pt, 31.8 $\Omega/\square$ of Pt_2Si and 57.6 $\Omega/\square$ of PtSi) [13], starts to play the main role. So, the development of simple processes of formation of thin Pt_3Si films becomes more and more important.

This letter describes a phenomenon of formation of a thin bilayer Pt_3Si/Pt_2Si film at room temperature on a poly-Si substrate as a result of Pt magnetron sputtering and wet etching, which itself is of high scientific interest, and introduces a simple process of obtaining of such films.

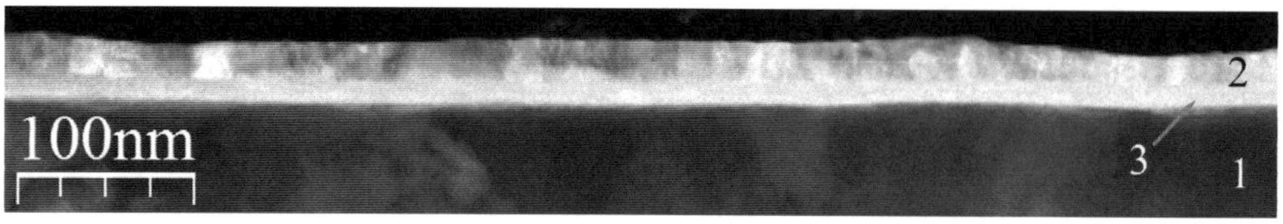

Figure 1. STEM image of the as-deposited Pt/poly-Si structure shows a poly-Si layer (1), a layer of Pt (2) and an interfacial layer (3) consisting of a $Pt_{95}Si_5$ layer ($Pt_{19}Si$) on a layer of Pt_2Si [6].

2. Experimental Details

2.1. Samples

A 35-nm thick film of platinum was deposited by magnetron sputtering at room temperature on a 125-nm thick poly-Si:P layer formed on a $Si_3N_4/SiO_2/Si(001)$ artificial substrate (figure 1). After deposition, platinum was removed by chemical etching in a warm aqueous solution of *aqua regia* ($H_2O : HCl : HNO_3$ [4 : 3 : 1]; the reaction is $3Pt + 18HCl + 4HNO_3 \rightarrow 3H_2[PtCl_6] + 4NO\uparrow + 8H_2O$). Details of the sample preparation process can be found in Ref. [6].

2.2. Equipment and techniques

The samples were studied by means of the X-ray photoelectron spectroscopy (XPS). The measurements were carried out using a Riber EA 150 cylindrical mirror electron energy analyzer [14] installed in an ultrahigh-vacuum chamber; the residual gas pressure in the chamber did not exceed 7×10^{-7} Pa. Non-monochromatic Al K_α X-rays ($\hbar\omega$ = 1486.7 eV) were used for photoexcitation of electrons. Survey spectra were scanned at a resolution (FWHM) better than 1.8 eV; high resolution spectra of specific elements were obtained at a resolution not worse than 0.96 eV. XPSPEAK 4.1 peak fitting program was utilized for treatment of spectra. It was taken into the account during peak deconvolution that the Al K_α band consists of two lines, K_{α_1} and K_{α_2}, with an energy difference of $\hbar\Delta\omega$ = 0.5 eV. Shifts of peaks related to elements in chemical compounds were compared with the NIST X-ray Photoelectron Spectroscopy (XPS) Database [15, 16]. Relative concentrations of atoms were estimated from ratios of normalized areas under corresponding peaks. The inelastic mean free path (IMFP) was estimated using the TPP-2M equation and the NIST Database [17, 18]. The spectrometer was calibrated against the Si^{4+} signal (E_b = 103.6 eV) since the C 1*s* XPS peak recorded at high resolution had a low signal-to-noise ratio (figure 2) that did not allow us to obtain reliable data on its energy position.

The scanning transmission electron microscopy (STEM) image was obtained using a Carl Zeiss Libra-200 FE HR transmission electron microscope; the WSxM software was used for image processing [19].

3. Results and Discussion

Figure 1 demonstrates a STEM image of the as-deposited Pt/poly-Si structure. This structures was previously analyzed in details using STEM, X-ray diffraction and reflection, and XPS. It is composed of a poly-Si layer, a layer of Pt and an interfacial layer consisting of a $Pt_{95}Si_5$ layer formed on a layer of Pt_2Si [6]. This structure was subjected to wet etching to remove Pt.

Photoelectron peaks of Pt, Si, O and C are observed in the survey photoelectron spectrum of the film obtained after Pt removal (figure 2). Monotonic growth of signal is observed in the spectrum on the right (at the higher E_b side) of the Pt 4*f* doublet that is explained by inelastic scattering of the photoexcited 4*f* electrons of Pt. This allows us to make a conclusion that the film containing Pt is overlaid by a layer of a different composition in which the inelastic scattering happens [20].

A high-resolution photoelectron spectrum of Pt 4*f* consists of three components (figure 3 (a)). The first one is described by a doublet with E_b = 71.5 eV and a distance between peaks ΔE = 3.38 eV which may be attributed to Pt_3Si [16, 21–23]; the next one, with E_b = 72.1 eV and ΔE = 3.38 eV, obviously corresponds to Pt_2Si [16, 21–23]. The third component lays at higher energies (E_b = 73.62 eV,

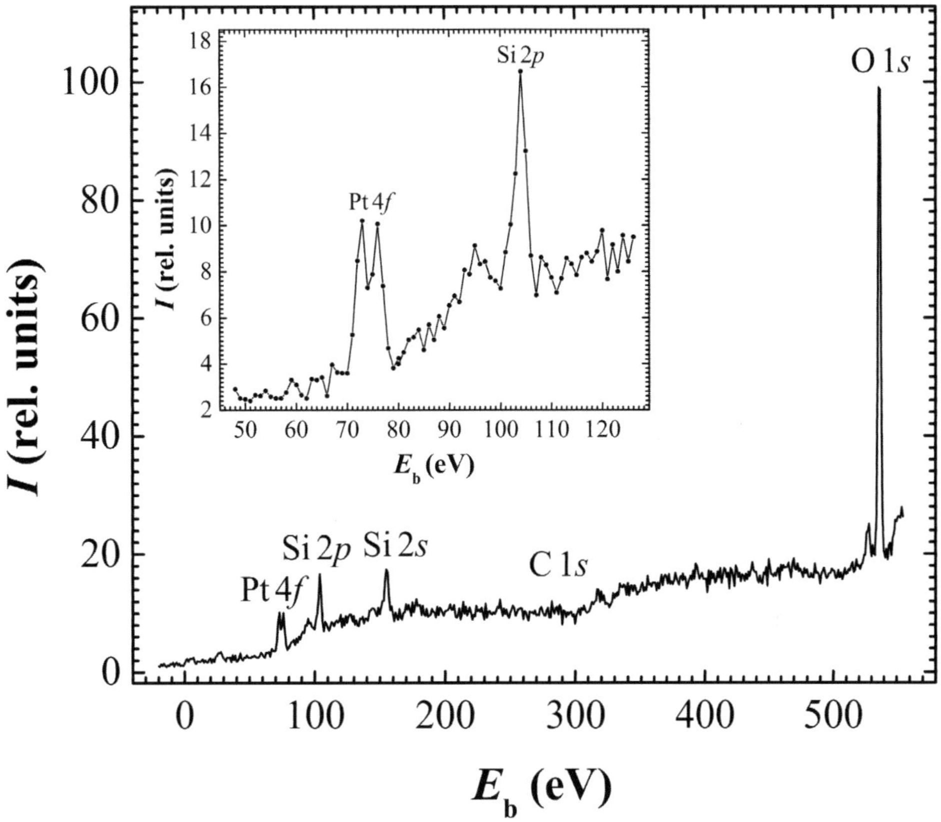

Figure 2. Survey XPS spectrum of the Pt-silicide/poly-Si film. The insert demonstrates a magnified region around the Pt 4f and Si 2p peaks.

ΔE = 3.38 eV) and may be attributed to both platinum oxide PtO_x (E_b = 72.8 to 74.6 eV), likely PtO (E_b = 73.8 eV [16]), and $PtCl_2$ (E_b = 73.6 eV) [16] or $[Pt(NH_3)_2(NO_2)_2]$ (CAS Registry No. 14286023, E_b = 73.7 eV [16]) which may arise as a result of Pt etching in the *aqua regia* solution. The ratio of peak areas for Pt_3Si, Pt_2Si and PtO_x is 64 : 28 : 8.

The main contribution to the signal of silicon (figure 3 (b)) is made by Si^{4+} (E_b = 103.6 eV) obviously related to silicon dioxide. In addition, a Si^{3+} peak can be detected by deconvolution which can be explained by superposition of photoelectron peaks related to Pt_3Si, Pt_2Si (E_b = 102.6 eV) and SiO_x compounds having a broad range of binding energy values (E_b = 100.4 to 103.6 eV) [16]. An estimate of contribution of Si atoms contained in Pt_3Si and Pt_2Si to this peak, made in assumption of uniformity of the upper film and taking into account empirical sensitivity factors of platinum and silicon, gives the value of ~ 16 %.

Notice that an additional cycle of etching in the *aqua regia* solution changed neither a ratio of the photoelectron peak intensities nor their energy positions.

It should be noted also that a small photoelectron peak at E_b = 99.5 eV is also observed in the Si 2p spectrum; it corresponds to pure Si of the polycrystalline substrate that allows us to roughly estimate the thickness of the formed film by 3 to 5 values of the inelastic mean free path of electrons, i.e. by the value from 5 to 10 nm.

A more accurate estimation of the film thickness can be made from the ratio of areas under the Pt 4f

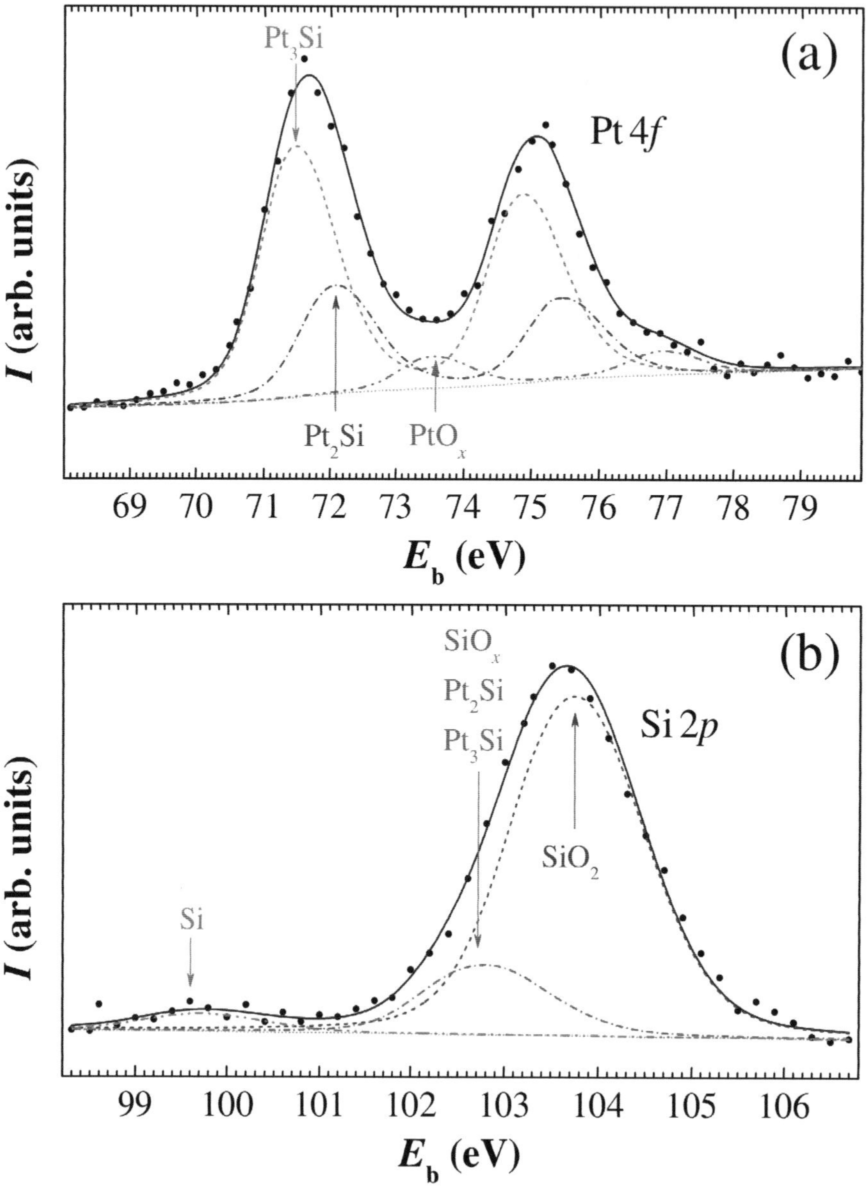

Figure 3. (Color online) High-resolution XPS spectra of Pt 4*f* (a) and Si 2*p* (b) obtained from the resultant Pt silicide film on the poly-Si substrate; measured bands are shown by solid lines, peaks obtained as a result of deconvolution are drawn by dotted and dashed lines.

peaks of Pt_3Si and Pt_2Si $\sigma \approx 2.3$. Let us suppose that the Pt_3Si layer overlays the Pt_2Si one. This assumption is based on the X-ray phase analysis according to which the Pt_2Si and $Pt_{95}Si_5$ layers, 5.6 and 10 nm thick, respectively, arise between poly-Si and Pt as a result of Pt sputtering [6]. Considering the IMFP value [18] of Pt_3Si, we obtain the thickness of the Pt_3Si layer h_1 from the values of densities and molar weights of Pt_3Si and Pt_2Si, σ and the thickness of the Pt_2Si layer h_2: $h_1 \approx 2.6$ nm. (The value of h_1 is calculated from a recursive equation: $h_1 = h_{IMFP} \times \ln[2\sigma\mu_{Pt_3Si}\varrho_{Pt_2Si}h_2/3\mu_{Pt_2Si}\varrho_{Pt_3Si}h_1]$, where h_{IMFP} is the inelastic mean free path of electrons in Pt_3Si [18], μ_{Pt_2Si}, ϱ_{Pt_2Si}, μ_{Pt_3Si} and ϱ_{Pt_3Si} are molar weights and densities of Pt_2Si and Pt_3Si, respectively.) So, the total thickness of the bilayer Pt-silicide film $h_1 + h_2 \approx 8.2$ nm that is in good agreement with the above rough estimate.

Now, if we assume that the $Pt_{95}Si_5$ layer previously detected under Pt [6] is a mixture of Pt and Pt_3Si ($16Pt + Pt_3Si$) we may also suppose that Pt is etched away from this layer during processing in the *aqua regia* solution and the Pt_3Si layer is deposited atop Pt_2Si. We can estimate h_1 from the thickness of $Pt_{95}Si_5$ and the densities of Pt, $Pt_{95}Si_5$ (20.5 g/cm^3) [6] and Pt_3Si: $h_1 \approx 2.4$ nm. Thus, the silicide film thickness estimate of 8.0 nm obtained from the X-ray phase analysis [6] corresponds with that obtained from XPS.

4. Conclusion

In summary, we can make the following conclusions.

(i) We have verified our previous observation of Pt_2Si formation in between Pt and poly-Si films as a result of Pt magnetron sputtering at room temperature [6]. This phenomenon probably occurs due to high enough kinetic energy of Pt ions reaching the Si surface, sufficient to give rise to chemical reaction of the silicide film formation. It opens a pathway to room-temperature process of formation of thin Pt silicide films on Si.

(ii) We have discovered a phenomenon and proposed a way of formation of thin bilayer Pt_3Si/Pt_2Si films on poly-Si substrates at room temperature by magnetron sputtering of Pt followed by etching in a warm aqueous solution of *aqua regia* and obtained such films using this completely CMOS compatible process.

(iii) Using the X-ray photoelectron spectroscopy we have verified the structure and composition of the formed Pt_3Si/Pt_2Si films; we have estimated the thickness of the Pt_3Si layer of the films and found it to be equal to ~ 2.6 nm; the thickness of the Pt_3Si/Pt_2Si films has been found to be ~ 8.2 nm. The obtained estimates practically coincide with those made on the basis of our previous data of the X-ray phase analysis [6].

(iv) We assume that the Pt_3Si layer is formed from the $Pt_{95}Si_5$ layer, which appears under a Pt film during magnetron sputtering [6], as a result of Pt removal by wet etching.

Acknowledgments

This research was funded by RFBR (grant No. 16-32-00854). Equipment of the Center for Collective Use of Scientific Equipment of GPI RAS was used for the study. We thank Ms N V Kiryanova for her contribution to management of this research, Mr O V Uvarov for obtaining the STEM images and Ms L A Krylova for chemical treatments of the samples.

References

[1] Samsonov G V, Dvorina L A and Rud' B M 1979 *Silicides* (Moscow: Metallurgia) in Russian
[2] Murarka S P 1983 *Silicides for VLSI Applications* (New York: Academic Press)
[3] Murarka S P 1995 *Intermetallics* **3** 173–186
[4] Rogalski A 2012 *Opto-Electron. Rev.* **20** 279–308
[5] Kimata M 2001 *Infrared Detectors and Emitters: Materials and Devices* ed Capper P and Elliott C T (Dordrecht: Kluwer Adamic Publishers) chap 4, pp 77–98
[6] Yuryev V A, Chizh K V, Chapnin V A, Mironov S A, Dubkov V P, Uvarov O V, Kalinushkin V P, Senkov V M, Nalivaiko O Y, Novikau A G and Gaiduk P I 2015 *J. Appl. Phys.* **117** 204502
[7] Yuryev V A, Chizh K V, Chapnin V A and Kalinushkin V P 2015 *Proc. SPIE* **9519** 95190K
[8] Xu D X, McCaffrey J P, Das S R, Aers G C and Erickson L E 1996 *Appl. Phys. Lett.* **68** 3588–3590
[9] Gambino J P and Colgan E G 1998 *Mater. Chem. Phys.* **52** 99–146
[10] Yuryev V A and Arapkina L V 2011 *Nanoscale Res. Lett.* **6** 522

[11] Komarov F, Milchanin O, Kovalyova T, Solovjov J, Turtsevich A and Karwat C 2011 *9th International Conference "Interaction of Radiation with Solids", September 20–22, 2011* (Minsk, Belarus) pp 365–367
[12] Colgan E G, Gambino J P and Hong Q Z 1996 *Mater. Sci. Eng., R* **16** 43–96
[13] Streller F, Qi Y, Yang J, Mangolini F, Rappe A M and Carpick R W 2016 *J. Phys. Chem. Lett.* **7** 2573–2578
[14] Hafner H, Arol Simpson J and Kuyatt C E 1968 *Rev. Sci. Instrum.* **39** 33–35
[15] Lee A Y, Blakeslee D M, Powell C J and Rumble, Jr J R 2002 *Data Science Journal* **1** 1–12
[16] Naumkin A V, Kraut-Vass A, Gaarenstroom S W and Powell C J 2012 *NIST X-ray Photoelectron Spectroscopy Database. NIST Standard Reference Database 20, Version 4.1* (Gaithersburg, MD, USA: National Institute of Standards)
[17] Powell C J and Jablonsky A 2010 *J. Electron Spectrosc. Relat. Phenom.* **331** 178–179
[18] Powell C J and Jablonsky A 2010 *NIST Electron Inelastic-Mean-Free-Path Database. Version 1.2, SRD 71* (Gaithersburg, MD, USA: National Institute of Standards)
[19] Horcas I, Fernandez R, Gomez-Rodriguez J M, Colchero J, Gomez-Herrero J and Baro A M 2007 *Rev. Sci. Instrum.* **78** 013705
[20] Tougaard S 1996 *J. Vacuum Sci. Technol. A* **14** 1415–1423
[21] Streller F, Wabiszewski G E, Mangolini F, Feng G and Carpick R W 2014 *Adv. Mater. Interfaces* **1** 1300120
[22] Streller F, Agarwal R, Mangolini F and Carpick R W 2015 *Chem. Mater.* **27** 4247–4253
[23] Fryer R T and Lad R J 2016 *J. Alloys Compd.* **682** 216–224

Variation of the resistivity and chemical composition of CVD graphene under annealing in a reductive atmosphere

O V Tolochko, T V Larionova, T S Koltsova, M V Kozlova, E V Bobrynina, O A Martynova and V E Gasumyants

Peter the Great St. Petersburg Polytechnic University, 29 Polytechnicheskaya str., St. Petersburg, 195251 Russia

E-mail: ol_tol@hotmail.com

Abstract. Graphene samples synthesized by chemical vapor deposition, transferred onto Si/SiO$_2$ substrates by the standard technology using PMMA, and then annealed in a reductive atmosphere have been characterized by XPS observations and measurements of the resistance temperature dependence. It was shown, that most of as-transferred samples exhibit a metallic-like resistivity dependence but some of them can demonstrate a non-metallic one. A comparative analysis of the results obtained by the XPS study and resistance measurements allowed us to conclude that the conduction process in CVD graphene is directly affected by functional groups adsorbed on the sample surface. Annealing in a mixture of H$_2$ and Ar at T = 250–750°C is shown to result in a cleaning of the graphene surface from adsorbed contaminations and in a modification of the resistance temperature dependence which demonstrates a higher slope in case of a metallic-like behavior for as-transferred samples and a suppression of the activation-type dependence in case of a non-metallic one.

1. Introduction

Modern two-dimensional (2D) single layer structures, such as graphene, phosphine, dichalcogenides of transition metals attract growing interest due to their electronic, optical, mechanical, and other properties promising various applications. Graphene has recently become one of the most popular 2D materials for nano- and optoelectronic applications [1, 2]. Graphene is characterized by very high mobility of charge carriers at room temperature and a manifestation of a number of quantum effects (the Hall effect, the ballistic transport effect, etc.) at low temperatures, as well as by high transparency, electrical and thermal conductivity [3]. This makes it an ideal candidate for using in future electronics and energy harvesting devices, such as conducting electrodes, switches, sensors, etc. However, impurities and topological defects arising during the graphene synthesis and/or transfer process can substantially affect its electronic properties and the performance in these devices, so the quality of graphene layers is crucial for these applications [2].

Chemical vapor deposition (CVD) of graphene utilizing copper as a catalyst is one of the most promising methods for producing continuous layers of a high quality graphene over a large area [4]. CVD synthesis is determined by a large number of technological parameters, and since some of them are difficult to control, the samples synthesized at the same specific parameters may differ essentially from each other. The transfer of the graphene layer onto a nonconductive substrate is a multistep procedure, which is fraught with introducing chemical and morphological defects. For this reason, controlling the graphene quality during the synthesis and transfer procedure is a difficult task, especially when working in ambient conditions. X-ray photoelectron spectroscopy (XPS) is one of the basic analytical methods used for the characterization of carbon-based materials, especially graphene-related ones. The XPS study allows to identify the sp^2-hybridized carbon which characterizes C-C bonds in graphene layers itself, the sp^3-hybridized diamond-like carbon in defect sites of graphene, carbides, vacancy-type defects, and a carbon bound to oxygen in various functional groups [5–7]. Except for the above, the integral intensity of the core level C1s peak gives an information on an average thickness of a continuous carbon layer [8].

Another way to control the graphene sample quality is the measurements of the resistivity, ρ, temperature dependence. Usually, CVD graphene is characterized by a granular structure so that the grain boundaries give a contribution to the measured resistivity value. As a result, in most cases the resistivity demonstrates a non-metallic behavior (i.e., its value increases with decreasing temperature), especially in case of a few-layer graphene [9]. A metallic-like $\rho(T)$ dependence for both few-layer [10] and monolayer [11] graphene samples can be obtained when measuring with a distance between electrodes of several microns thus excluding the grain boundary influence.

It is reasonable to assume that an annealing in a reductive atmosphere can result in a removal of a polymer residue and functional groups adsorbed from the ambient atmosphere on the graphene surface, leading to the subsequent changes in the electronic and structural properties of samples. For this reason, the aim of this work is to trace the correlation of the chemical composition evolution of transferred CVD graphene under annealing and its electrical properties on a macroscopic scale.

2. Methods

Graphene samples were synthesized using the CVD technique from a gas mixture of methane, hydrogen, and argon to copper foil substrates (Alfa Aesar; thickness is 25 microns). Before synthesis, the polished copper substrates were annealed in a mixture of argon and hydrogen for 1 h. Synthesis was carried out for 15 min at a temperature of 1000°C. The process was conducted at a reduced pressure; the flow rates of hydrogen, methane, and argon were 5 ml/min, 50 ml/min, and 250 ml/min, respectively. The synthesized graphene layers were transferred onto a nonconductive substrate by the standard technology, i.e., the graphene film on one side of the Cu foil was coated with poly(methyl methacrylate) (PMMA) and placed in an aqueous solution of iron chloride to dissolve Cu on the unprotected side. The PMMA film that carries the graphene layer was rinsed several times with distilled water to minimize the chemical contamination, and then transferred onto a silicon substrate covered with thermally grown SiO_2 of 285 nm thickness. After the transfer, the sample was allowed to dry, and the PMMA was dissolved in acetone at a temperature of 200°C. After characterization, the samples were additionally annealed at temperatures of 250–750°C in a mixture of H_2 and Ar at a ratio of 1:9 for 30 min.

XPS analysis was performed on a Thermoscientific K-alpha using Al Ka radiation (1486.6 eV). Measurements of the samples on nonconductive substrates were made using an electric flood gun for charge compensation. The spectra were collected in vacuum (10^{-9} mbar) with a spot diameter of 400 μm. The peaks were fitted with Voigt functions having the 70 % Gaussian and 30 % Lorentzian character, after a Shirley background subtraction. The peak of sp^2 of C–C bond was fitted using an asymmetric shape. The peak assignments were performed basing on the data reported earlier (see, e.g., [12, 13]).

The sheet resistance was measured in the temperature range from 77 K to 300 K by the standard ac (20 Hz) four probe method in the van der Pauw geometry with using silver paste contacts. The distance between contacts in our measurements was approximately equal to 3–4 mm.

3. Results and Discussion

A typical microphotograph of a copper substrate surface with a graphene layer on the top is shown in figure 1 (a). As it is known, an optical visualization of graphene synthesized on a copper substrate can be easily achieved by post-synthesis annealing in air at a temperature of about 200°C. This leads to an oxidation of copper that is not covered with graphene and, as a result, to a distinct optical contrast [14]. In figure 1 (a), the graphene layer is continuous, fully repeating the copper surface relief. The graphene layer transferred to Si/SiO_2 is presented in figure 1 (b). The inset shows an isolated graphene island allowing to evaluate the graphene grain size, which is equal to 30–40 μm.

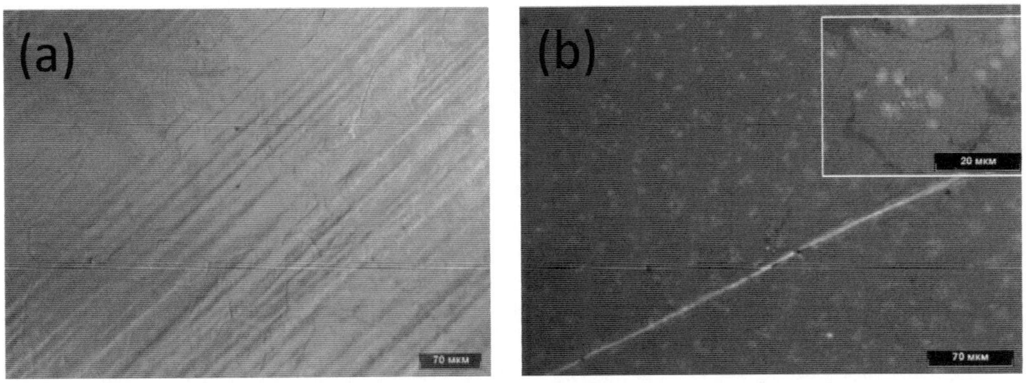

Figure. 1. The optical images of as-synthesized graphene on a copper substrate (a) and graphene transferred onto a Si/SiO_2 substrate (b). Inset: an island of graphene.

High resolution XPS spectra of graphene transferred to Si/SiO_2 before and after annealing at $T = 750°C$ are shown in figure 2. There are at least 4 components distinguished in the C1s core level spectrum, among them are sp^2-hybridized graphitic carbon (at about 284.2 eV), sp^3-hybridized carbon (at about 284.9 eV), and peaks corresponding to the bonding energies of species in functional groups adsorbed on graphene from the ambient atmosphere or during the transfer procedure. As seen in figure 2 (d), the peaks attributed to carbon bounds to oxygen disappear completely after sample annealing and the intensity of the sp^2 peak becomes higher. This indicates that annealing leads to a cleaning of the graphene surface from adsorbed contaminations resulted from their interactions with hydrogen and the formation of volatile products. The simultaneous increase in the intensities of the Si2p and O1s peaks (see figure 2 (e), (f)) indicates a considerable thinning of the coating on the Si/SiO_2 surface.

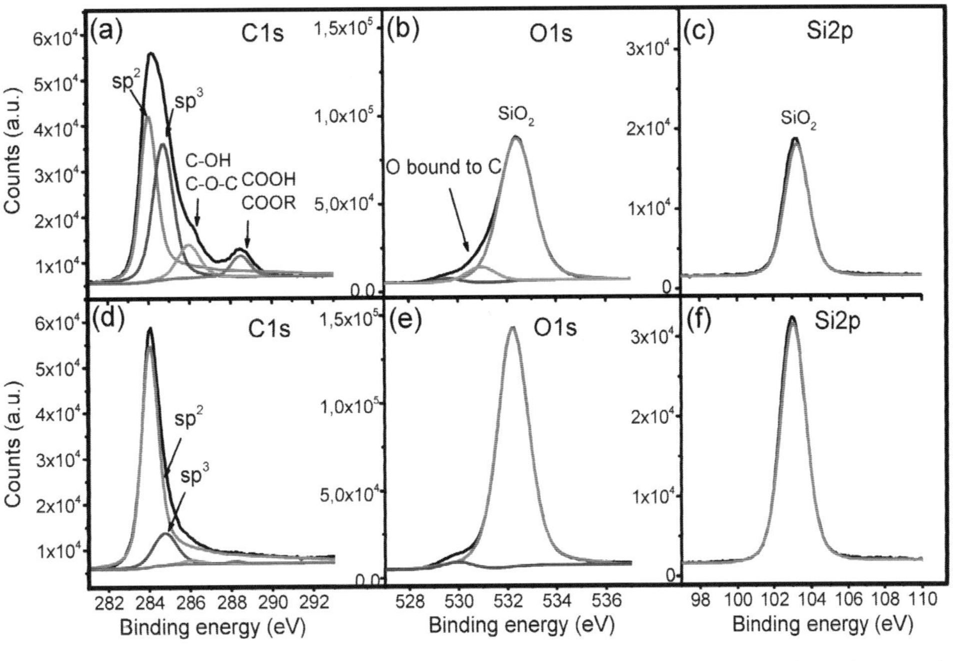

Figure 2. High resolution C1s (a, d), O1s (b, e), and Si2p (c, f) XPS spectra of graphene on the Si/SiO_2 substrate before (a – c) and after (d – f) annealing at 750°C in H_2+Ar atmosphere for 30 min.

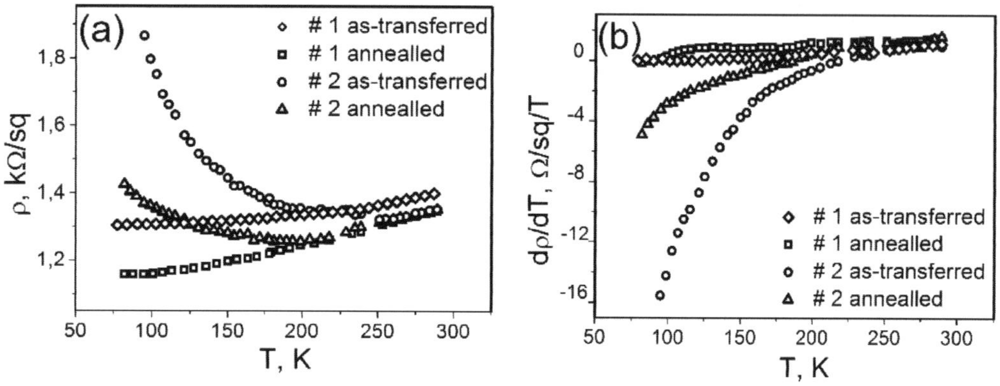

Figure 3. Temperature dependences of the sheet resistance (a) and its temperature derivative (b) for some of the studied samples.

Let us now consider the results obtained by the resistance measurements. Figure 3 shows typical temperature dependences of the sheet resistance of our CVD graphene samples, as well as the variation of the $^{d\rho}/_{dT}$ value with temperature. The room-temperature value of the sheet resistance for all the samples is in the interval of 1–1.5 kΩ/sq. For most of our as-transferred samples, the ρ(T) dependence measured at the macroscopic scale is metallic-like (see results for sample # 1 in figure 3 (a)). With decreasing temperature, the resistance decreases weakly over the whole investigated temperature range with an almost unchanged temperature derivative (excluding temperatures lower than 100 K, see figure 3 (b)). Such a resistance behavior coincides well with that obtained for CVD graphene when measuring at the micron scale [10, 11]. Thus, the granular sample structure (see insert in figure 2) influences insignificantly the conduction process, if any, that confirms a good quality of our CVD graphene samples. The resistance of the annealed sample also demonstrates a metallic-like behavior but with a higher slope of the ρ(T) dependence. Obviously, like the case of the observed modification of the XPS spectrum, this is a result of an improved graphene layer quality due to the annealing influence on the previously adsorbed contaminations on its surface. However, the ρ(T) change under annealing indicates additionally that these contaminations affect directly the conduction process in graphene samples.

For some samples, the resistance decreases with decreasing temperature at high temperatures but rises at low temperatures with a continuously increasing slope (see results for sample # 2 in figure 3). In such cases, in the low temperature range the ρ(T) curves can be fitted well by simple exponential functions. However, the calculated activation energy E_a is found to be extremely small (for example, for sample # 2 $E_a = 2.7$ meV). In our opinion, such a resistance behavior can be attributed not to the grain boundary influence but rather to the effect of chemical and structural defects formed during the synthesis and transfer process. In other words, in case of sample # 2 we can observe the same direct effect of contaminations on the conduction process as discussed above for as-transferred and annealed sample # 1 but the extent of their influence is higher that can be related to slight changes in the sintering and/or transfer processes.

In order to check this assumption, we have measured the high resolution C1s XPS spectra for the samples demonstrating a non-metallic ρ(T) dependence. The results obtained for sample # 2 are shown in figure 4 (a). One can see that the component of spectrum attributed to the C-OH or C-O-C bounds has a higher intensity as compared to the samples demonstrating a metallic-like resistance behavior (see figure 2 (a)). This gives strong grounds to state that it is these groups that result in the observed resistance behavior. Moreover, as seen in figure 4 (b), analogously to the case of sample # 1 the additional annealing of sample # 2 at 450°C in H_2+Ar atmosphere for 30 min leads to a reduction of the corresponding peak thus pointing to a decrease in the amount of carbon-based functional groups on the sample surface but they are still existing in the annealed sample. This finding correlates well with the observed change of the ρ(*T*) dependence under annealing. Indeed, as seen in figure 3, the annealed sample demonstrates a transition to an activation-type temperature dependence of the resistance at a lower temperature as compared to the as-transferred sample and the slope of the ρ(*T*) curve increases with decreasing temperature much weaker (the corresponding activation energy is equal to $E_a = 0.7$ meV, i.e. four times less).

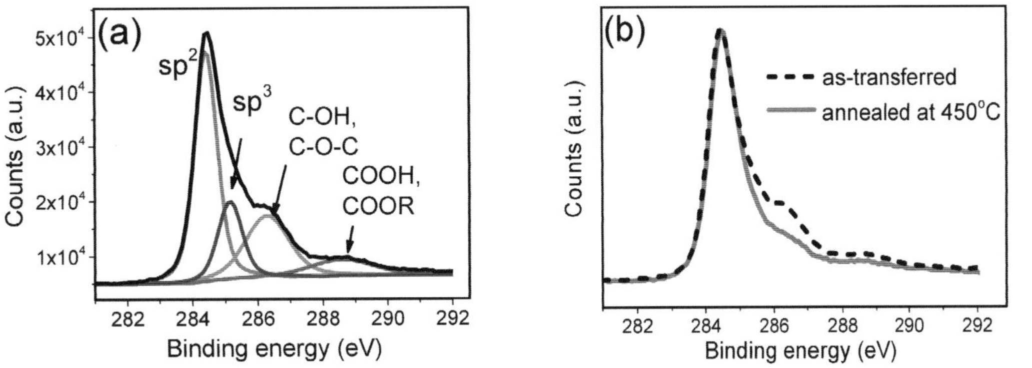

Figure 4. Deconvolution of the C1s XPS spectra of the sample with a non-metallic resistance behaviour (a) and the evolution of the spectrum under annealing at 450°C in H_2+Ar atmosphere for 30 min (b).

Thus, for both types of the ρ(*T*) dependences obtained on our CVD graphene samples, we can see the analogous positive effect of an additional annealing on their chemical composition and electrical properties.

4. Conclusion

In summary, in this paper we present the results of the comparative study of the high resolution XPS spectra and the temperature dependences of the resistance for the graphene samples synthesized using the CVD technique and then transferred onto the nonconductive Si/SiO_2 substrates. According to the XPS data, our as-transferred graphene samples can be characterized by the existence of functional groups adsorbed on the surface. These adsorbed contaminations affect directly the conduction process in graphene samples. Nevertheless, we were able to produce graphene layers of a large area and a good quality that is confirmed by a high value of the conductivity and a metallic-like behavior of the resistance measured at the macroscopic scale. It was also shown that the sample annealing in a mixture of H_2 and Ar at $T = 250$–750°C results in a cleaning of the graphene surface from adsorbed contaminations and in a modification of the ρ(*T*) dependence which demonstrates a higher slope in case of a metallic-like behavior for the as-transferred sample and a suppression of the activation-type dependence in case of a non-metallic one. Thus, changes in the chemical composition and electrical properties of transferred CVD graphene correlate with each other and an additional annealing in the H_2+Ar atmosphere can be proposed as a way to improve the graphene sample quality.

Acknowledgments
We acknowledge funding from The Ministry of Education and Science of the Russian Federation (Project RFMEFI58114X0006) for providing financial support to this project.

References

[1] Bonaccorso F, Sun Z, Hasan T and Ferrari A C 2010 *Nat. Phot.* **4** 611
[2] Bin Mohd Yusoff A R 2015 *Graphene-based energy devices* (Weinheim: Wiley-VCH)
[3] Castro Neto A, Guinea F, Peres N M R, Novoselov K and Geim A 2009 *Rev. Mod. Phys.* **81** 109

[4] Mattevi C, Kim H and Chhowall M *J. Mater. Chem* **21** 3324
[5] Ganguly A, Sharma S, Papakonstantinou P and Hamilton J 2011 *J. Phys. Chem.* **115** 17009
[6] Lin Y C, Lu C C, Yeh C H, Jin C and Chiu P W *Nano Lett.* **12** 414
[7] Yulaev A, Cheng G, Hight Walker A N, Vlassiouk I V, Myers A, Leite M S and Kolmakov A 2016 *RSC Adv.* **6** 83954
[8] Rudskoy A I, Koltsova T S, Larionova T V, Smirnov A N, Vasil'eva E S and Nasibulin A G 2016 *Met. Sci. Heat Treat.* **58** 40
[9] Babichev A V, Gasumyants V E and Butko V Y 2013 *J. Appl. Phys.* **113** 076101
[10] Gonnelli R S, Paolucci F, Piatti E, Sharda K, Sola A, Tortello M, Nair J, Gerbaldi C, Bruna M and Borini S 2015 *Sci. Rep.* **5** 9554
[11] Butko A V and Butko V Yu 2015 *Phys. Solid State* **57** 1048
[12] Wang X, Ruslinda R A, Ishiyama Y, Ishii Y and Kawarada H 2011 *Diam. Relat. Mater.* **20** 1319
[13] Desimoni E, Casella G I, Morone A and Salvi A M 1990 *Surf. Interface Anal.* **15** 627
[14] Duong D L *et al* 2012 *Nature* **490** 235

Electrical conductivity of Cu/ZnO/Si heterostructures

A O Zotov[1], A A Perevalov[1], A V Egorov[2], A N Belov[1] and A V Solnyshkin[1,2]

[1] National Research University of Electronic Technology, Shokin square, 1, Zelenograd, Moscow, 124498, Russia

[2] Tver State University, Zhelyabova 33, Tver, 170100, Russia

e-mail: nanointech@mail.ru

Abstract. The electrical conductivity studies of the Cu/ZnO/Si thin film heterostructures were carried out by the current-voltage (*I-V*) characteristics. It was found that the *I-V* characteristics were asymmetric and showed weak rectifying properties. The most probable mechanism of electrical conductivity was determined. The concentration of trapping levels and the carrier mobility were calculated.

1. Introduction

Zinc oxide possesses a unique combination of physical and chemical properties: a high melting point and thermal conductivity, photosensitivity [1], piezoelectric [2] and pyroelectric [3] effects, the adsorption of gases on the surface [4]. This material is widely used in microelectronics, chemistry and medicine [5]. Single crystals, thin films and diode structures based on the ZnO material are the subjects for scientific studies and applications.

An anisotropic crystal structure, a non-stoichiometric composition, a wide band gap, semiconductor parameters varied by doping and other properties make this material attractive for comprehensive research. Material characteristics largely depend on preparation conditions and various external factors. Such dependences are especially significant in the case of ZnO thin films, which arouse now great interest, for example, in the semiconductor sensor technology. Recently, the preliminary measurements of electrical conductivity were carried out on ZnO films deposited on silicon wafers by the RF-magnetron sputtering [6]. In this paper, we present further research of the electrical properties of ZnO films and perform a more detailed analysis of their parameters. Thus, the purpose of this work is to produce ZnO thin film structures and to study the electrical conductivity by determining their characteristics.

2. Experiment

The samples were Cu/ZnO/Si heterostructures. A ZnO film was deposited on a *p*-type silicon substrate (100) by atomic layer deposition. Argon (99.998 %) was used as a purging and carrier gas. The process temperature was 150°C. Diethylzinc and deionized water were used as precursor. The pulse length for diethylzinc and water was 200 ms, the purge time was 1 s. Figure 1 presents the results of structure and composition investigations of ZnO films.

The thickness of the ZnO film on a Si substrate is equal to 200 nm. Copper circular electrodes with a diameter of 3 mm were sputtered on the film surface through a shadow mask by thermal evaporation. According to the data obtained by scanning electron microscopy, the films have a crystalline structure.

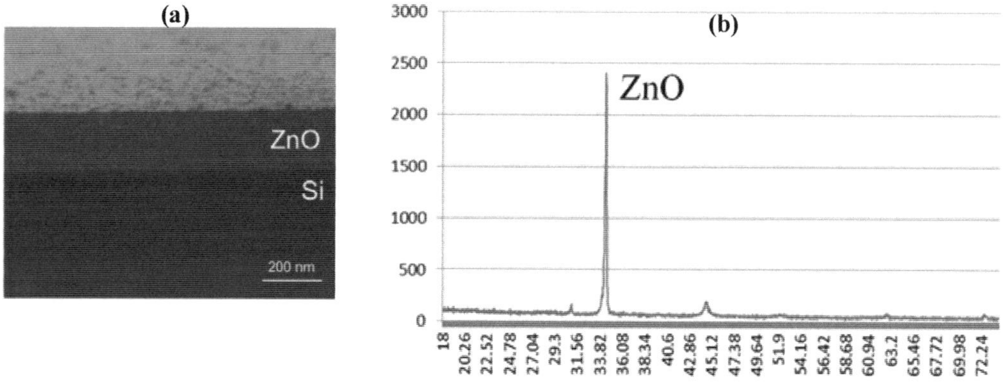

Figure 1. Scanning electron microscopy image (a) and x-ray diffraction spectra (b) of a ZnO film formed on a silicon substrate.

The conductivity measurements were carried out by the current-voltage (*I-V*) characteristics method using an immittance meter E7-20. This device allows applying the bias to the sample from −40 to +40 V in steps of 0.02 V in the range of −4 to +4 V and 0.2 V at higher voltages. The voltage applied to the film is accepted positive if a positive potential is applied to the top electrode, and vice versa.

3. Results and discussion

I-V characteristics were obtained at room temperature to determine the conductivity mechanisms of the ZnO film deposited on Si. Figure 2 shows a typical dependence of electric current on a voltage applied to a Cu/ZnO/Si heterosructure.

In case of a positive potential applied to the top electrode, the electric current increases nonlinearly with the voltage except for the initial voltage range from 0 to 4 V. Moreover, the *I-V* characteristics are asymmetric and show weak rectifying properties with a rectification coefficient $K = 5$ (at $U = 2.5$ V) and the non-ideality coefficient of diode structure $n = 5.55$. On the basis of the linear section of the current-voltage characteristics, the electrical conductivity of a ZnO layer is calculated according to the relation:

$$\sigma = \frac{I \cdot d}{U \cdot S}, \tag{1}$$

where I is the current, d is the layer thickness, U is the applied voltage, S is the top electrode area. The value of the ZnO layer conductivity is equal to $4.5 \cdot 10^{-6}\ \Omega^{-1}\mathrm{m}^{-1}$.

Rectifying properties of the structure are associated with the presence of the potential barrier at the interface ZnO/Si. The asymmetry of the current-voltage characteristics of the barrier is typical for such structures. A dependence of current on voltage in these structures is caused by a change in the number of majority carriers taking part in charge transfer processes. A role of external voltage is to change the number of free carriers transferred from one to another part of the barrier structure. However, in this case, the barrier height is not uniform over the boundary of the zinc oxide due to the presence of surface states. Thus, current may leak through the lowered barrier areas. This explains the nonideal rectifying properties and the presence of the high reverse current.

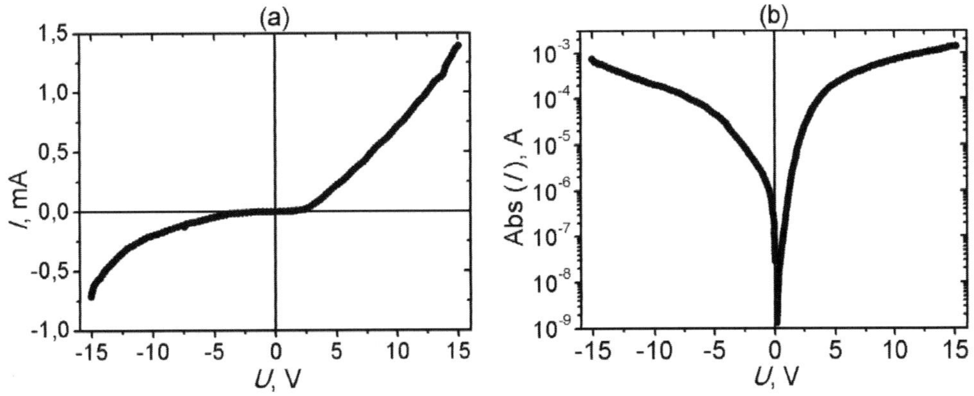

Figure 2. Current-voltage characteristics of the ZnO film deposited on silicon on: (a) a linear scale, (b) a semi-logarithmic scale.

It should be noted that the forward and reverse trajectories of current-voltage characteristics are practically the same that indicates the absence of the resistive switching effect in the present ZnO films.

To explain the observed behavior of the current-voltage characteristics, several conduction mechanisms of film structures can be considered. It is well known [7, 8] that the leakage current in dielectric or semiconductor films may be attributed to several conduction mechanisms including the Poole-Frenkel emission, the Schottky emission, the Fowler-Nordheim tunnelling and the space charge limited current (SCLC).

In the voltage region from 5 to 14 V, the current observed in the ZnO films increases in accordance with a quadratic law, i.e. $I \sim U^2$ (figure 3). Such a behavior can be ascribed to the SCLC mechanism. For the space charge limited current in the case when a single discrete trap level exists in the band gap, the current density is given by equation (2) [9]:

$$j = \frac{9}{8}\varepsilon\varepsilon_0\mu\frac{U^2}{d^3}, \tag{2}$$

where j is the current density, ε is the material dielectric permittivity, ε_0 is the vacuum dielectric permittivity, μ is the carrier mobility.

As previously mentioned, the linear segment of the I-V characteristics preceded the quadratic region that corresponded to the SCLC model. At low voltages, the injection level is low and a concentration of injected carriers does not exceed the concentration of free equilibrium carriers. In this case, not all traps are filled, so the current obeys Ohm's law

$$j = en_0\mu\frac{U}{d}, \tag{3}$$

where n_0 is the concentration of free equilibrium carriers, e is the elementary charge.

The transition from linear to quadratic parts of I-V characteristics occurs when the concentration of injected carriers becomes comparable to the concentration of free equilibrium carriers. In this case, all traps are filled and do not affect the flow of space-charge-limited current. On the one hand, the voltage of the transition from linear to quadratic parts of I-V characteristics is given by

$$U_{tr} = \frac{4}{3}\frac{en_0d^2}{\varepsilon}. \tag{4}$$

On the other hand, the voltage is

$$U = \frac{eN_t d^2}{2\varepsilon}, \tag{5}$$

where N_t is the concentration of trapping levels.

Taking into account that the dielectric permittivity of ZnO was equal to 8.5 and the transition occurred at a voltage of ≈ 4.5 V, we used equation (5) and found the concentration of trapping levels $N_t \approx 5.3 \cdot 10^{18}$ cm^{-3}. We determined $n_0 \approx 2 \cdot 10^7$ cm^{-3} from equation (4). We substituted the n_0 concentration in equation (3) and found $\mu \approx 190$ cm^2/(V·s).

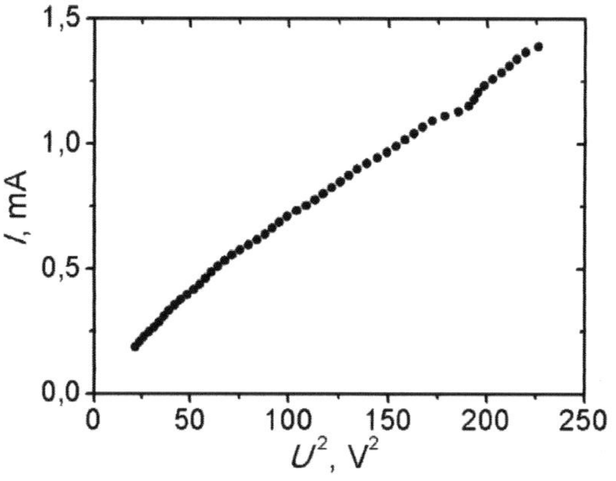

Figure 3. Dependence of the current on square voltage applied to Cu/ZnO/Si.

For the negative voltage part of *I-V* characteristics, the most probable mechanism of electrical conductivity is the Poole-Frenkel emission. The current dependence on the applied voltage modulus is shown in figure 4 in the Poole-Frenkel representation.

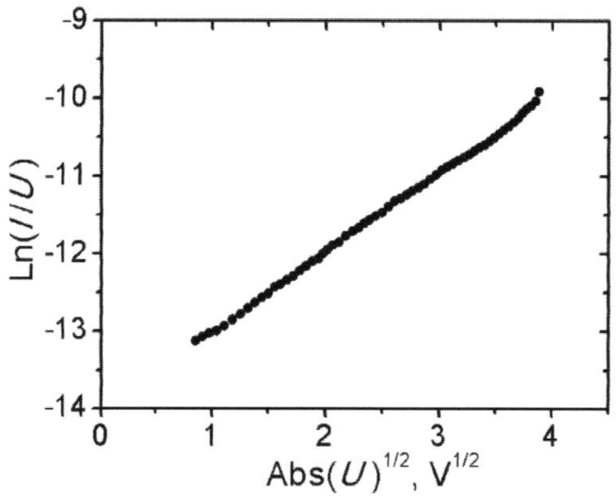

Figure 4 Current–voltage characteristic of the structure on the base of ZnO according to the Poole-Frenkel mechanism.

This dependence is nearly linear that gives evidence of the Poole-Frenkel emission contribution to the current. According to this mechanism, a field applied to the sample changes the shape of potential barriers for charge carriers between atoms in the lattice. It leads to an increase in the number of free carriers in the sample due to overcoming the potential barrier.

4. Conclusion

The ZnO thin films were fabricated on the *p*-type silicon substrate (100) by atomic layer deposition to make the ZnO/Si heterogeneous structure. For the electrical response measurements, Cu-electrodes were deposited on a free film surface by means of thermal sputtering. To determine conductivity mechanisms of the ZnO films, the current-voltage characteristics were measured at room temperature. It was found that the *I-V* characteristics are asymmetric and nonlinear with the exception of the initial voltage range from 0 to 4 V. The conductivity of the ZnO layer calculated for a weak field is equal to $4.5\cdot10^{-6}$ $\Omega^{-1}m^{-1}$. The main mechanisms of the electrical conductivity in the Cu/ZnO/Si heterostructure were revealed. For the positive voltage region (~ 4 – 15 V), the electric current obeyed the quadratic law that corresponded to the space-charge-limited current model. At the negative potential applied to the top Cu-electrode, the *I-V* characteristic is described by the Poole-Frenkel emission. On the basis of the *I-V* characteristic measurements, the concentration of free equilibrium carriers, the concentration of trapping levels and the free carrier mobility were calculated. The value of n_0 is nearly equal to $2\cdot10^7$ cm^{-3}, N_t is about $5.3\cdot10^{18}$ cm^{-3} and $\mu \approx 190$ $cm^2/(V\cdot s)$. To study the properties associated with the presence of the barrier in the heterostructure, additional measurements are needed such as capacitance-voltage characteristics, dielectric dispersion etc. This is the subject of further research.

Acknowledgments

This work was supported by the Russian Science Foundation, grant no. 15-19-00138.

References

[1] Lee J, Jung J, Kang M and Koo S 2013 *J. Nanosci. Nanotech.* **13** 7033
[2] Xua T, Wub G, Zhang B and Haob Y 2003 *Sensors and Actuators A: Physical* **104** 61
[3] Heiland G and Ibach H 1966 *Solid State Comm.* **4** 353
[4] Shao C, Chang Y and LongY 2014 *Sensors and Actuators B: Chemical* **204** 667
[5] Coleman V and Jagadish C 2006 *Zinc Oxide Bulk, Thin Films and Nanostructures* ed Jagadish C and Pearton S (Amsterdam: Elsiever)
[6] Egorov A V, Solnyshkin A V, Bogomolov A A, Silibin M V, Gavrilov S A 2012 *Proceedings of the International Conference* "INTERMATIC-2102" Moscow: MIREA, *in Russian* **2** 164
[7] Solnyshkin A V, Troshkin A S, Bogomolov A A, Raevski I P, Sandjiev D N and Shonov V Yu 2009 *Int. Ferroelectrics* **106** 61
[8] Kamenshchikov M V, Solnyshkin A V, Bogomolov A A and Pronin I P 2011 *Physics of the Solid State* **53** 2080
[9] Kao K S and Hwang W 1981 *Electrical Transport in Solids* (Oxford, New York, Toronto: Pergamon Press)

Research and development of the method of graphene oxide thin films local reduction by modulated laser irradiation

N S Struchkov, V A Kondrashov, R Yu Rozanov and V K Nevolin

National Research University of Electronic Technology, 1 Shokin square, Zelenograd, Moscow, 124498, Russian Federation

E-mail: vkn@miee.ru

Abstract. This article describes the reduction process of graphene oxide films deposited on various substrates by 445 nm laser irradiation with exposure time from $4{\cdot}10^{-3}$ to $3{\cdot}10^{-1}$ s. AFM images of the deposited films are presented. To obtain a better quality of the reduced graphene oxide films on substrates, we found the optimal ratio of the partial pressures of air and argon in the background environment. Variation of laser irradiation exposure time allowed us to determine its influence on sheet resistance and the ratio of Raman peaks intensities I_D/I_G, I_{2D}/I_G of the reduced material. As a result, we demonstrated the possibility of graphene conductive traces formation with a sheet resistance of 188 Ohm/□ at 0.1 um film thickness.

1. Introduction

Graphene is a promising conductive carbon material for the integration into micro- and nanoelectronics processes. Carbon atoms form a two-dimensional hexagonal lattice with sp^2-hybridized bonds [1]. It is known that graphene has a high thermal conductivity of ~ $5{\cdot}10^3$ W/m·K, a transparency of ~ 97.7%, an elasticity of 1 TPa, a mobility of charge carriers of ~ $2{\cdot}10^5$ cm^2/V·s. This material can be applied in the fields of optoelectronics, sensor technology, spintronics.

There are several methods of graphene production, however, they have a number of technological problems. Micromechanical exfoliation allows to obtain graphene flakes of small dimensions. CVD synthesis uses various temperatures and catalysts for graphene production on a metal surface, but the structure transferring from the metal produces a large number of defects in a graphene film. Epitaxial growth on the silicon carbide substrate with 6H-SiC orientation allows obtaining high quality graphene, but the complexity and the high cost of silicon carbide plates preparing make the process uneconomical and difficult to integrate into standard electronics processes.

Nowadays, the reduction from graphene oxide (GO) is one of the cheapest and fastest methods of graphene production. Graphene oxide is a promising material for electronics, optoelectronics and sensorics due to a possibility of its mass production.

There are several methods of graphene oxide production, they involve oxidation and intercalation of graphite with strong oxidants ($KMnO_4$, H_2SO_4, etc.). The main groups of methods are named after the authors: Staudenmaier, Hammers and Brodie. Graphene oxide is stable in water and in different organic solvents. When its dispersed carboxyl and hydroxyl groups are ionized, as a result, electrostatic repulsive forces prevent adhesion of the particles providing stability of the suspension. After coating a surface with the suspension and removing the solvent, several monolayers of the material remain on the substrate forming a mosaic film with a thickness varying from a few atomic layers to a thick film. Graphene oxide contains sp^3 hybridized carbon areas with functional groups that violate the electrical conductivity making graphene oxide dielectric with a resistance of ~ $2 \cdot 10^3$ Ohm·m [2]. Reduction of graphene oxide is a process of the functional groups removal and C-C bonds in sp^2 lattice restoration. It allows to receive the reduced graphene oxide layers with lateral dimensions equal to the initial graphene oxide sheets. Despite the fact that the lattice of reduced graphene oxide has many defects and unremoved groups, multilayer films of reduced graphene oxide have a low resistivity, which depends on the method of reduction and may vary from 10^{-1} to 10^{-5} Ohm·m [2-4]. Graphene oxide reduction methods should be separated into chemical, thermal and photo ones [4]. Treating in strong reducing agents such as hydrazine and its derivatives, sulfur-containing compounds can be classified as chemical methods. Thermal reduction method involves heating a graphene oxide suspension or deposited material by means of annealing [5]. The last group of methods uses femtosecond lasers as a light irradiation source [6]. A femtosecond pulse causes multiphoton scattering without phonon generation. There is an *ab initio* prediction of energy required for the graphene oxide reduction [7]. Lasers with different wavelengths are widely used for photothermal reduction by local heating (UV: 248 nm [8], visible spectrum: 532 nm [9], IR: 1064 nm [10]).

The reduced graphene is investigated by X-ray photoelectron spectroscopy, Raman spectroscopy. Raman spectroscopy is widely used for studying carbon materials, such as nanotubes, graphene and graphene oxide. The position, width and intensity of the I_D (~ 1350 cm^{-1}), I_{2D}(~ 2700 cm^{-1}) and I_G (~ 1580 cm^{-1}) peaks are determined by composition, defects and numbers of graphene layers. High-quality graphene is characterized by the absence of the I_D peak and a small width at half-height of the I_{2D} peak on Raman spectra. The number of layers is estimated by the ratio of the I_{2D}/I_G peaks intensities [11].

Development of methods for local reduction of graphene oxide thin films by laser irradiation will solve the existing problems.

2. Experimental

In this paper we used the graphene oxide suspension produced by «SIC Graphene» Ltd. This material was prepared by the modified Hammers method [12]. We used monocrystalline silicon plates KDB-7.5 with (100) orientation with natural oxide (< 5 nm) and polyethylene naphthalate films (PEN) as substrates. Deposition was performed by drying 80 µl of 0.8 mg/ml graphene oxide suspension on a square of 1 cm^2 at 22°C for 24 hours. Vertical withdrawal of the substrate from the suspension with a speed of 0.1–10 mm/min was performed to obtain monolayer GO films. Modification of GO was carried out in a climatic chamber filled with a mixture of air and argon at atmosphere pressure and room temperature. The sample was treated with a 2 W solid-state laser working at the 445 nm wavelength with the 5 nm spectral width. Output laser power was regulated by PWM of DPSS laser current with a 30 kHz sampling rate. The laser power was calibrated with a silicon photodiode FDS-155 Rus, the calibration curve was plotted. The laser was focused on the samples by microscope lens with a numerical aperture of 0.11 and 4.7x optical multiplicity in a 30 µm circle. For the automated processing, the samples were fixed on a motorized XY table. To study the samples by Raman spectroscopy, a Centaur U HR spectrometer «Nano Scan Technologies» Ltd. was used. The quality of graphene oxide films reduction was estimated by the relative intensity of the I_D, I_G, I_{2D} peaks that characterize graphene. To achieve representativeness of the data, we obtained Raman spectra from different areas of samples and averaged out the results. Measurements of the reduced graphene oxide sheet resistance were performed by the four-probe station and a digital multimeter Agilent 34401A. The deposited graphene oxide films were investigated by an AFM (Solver Pro company «NT-MDT»).

3. Results

Withdrawal of the substrate from the suspension at a speed of 0.1–10 mm/min allowed to receive GO films deposited on the substrate. Atomic Force Microscopy of the deposited material confirmed the presence of a significant number of single-layer flakes of graphene oxide with lateral dimensions greater than 1 μm and a thickness of about 1.2 nm (figure 1 (a)) that is consistent with experimental data from other works [13]. The optimal speed providing a high density of oxide graphene monolayers on surface (~ 90% covering) was determined to be ~ 5 mm/min (figure 1 (b)).

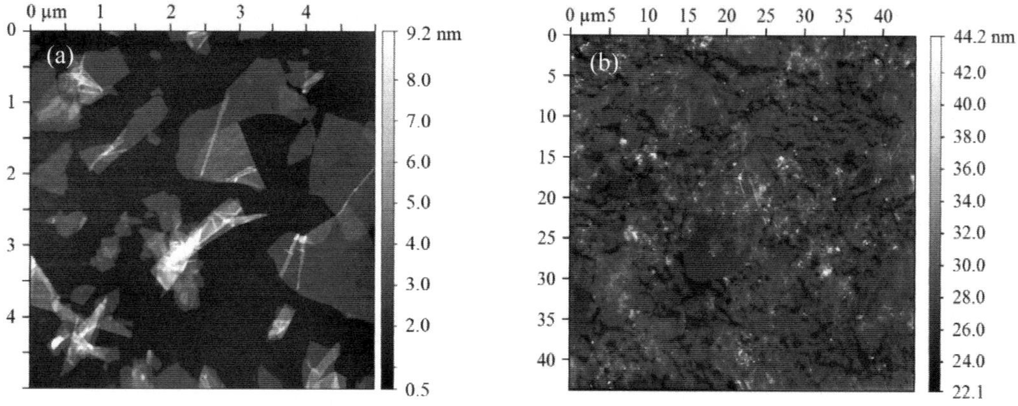

Figure 1. AFM images of graphene oxide (a) monolayer flakes; (b) a few-layer thin film on natural silicon oxide

Drying the suspension at room temperature was used for deposition of homogenous thick films up to 1.5 μm thick. Drying at increased temperatures (80–90°C) caused the appearance of the inhomogeneity and defects on the film surface. These graphene oxide films were locally treated under different conditions.

In air atmosphere, oxidation of carbon occurred with CO_2 formation due to the presence of oxygen. The lattice recovery of the carbon did not occur. Reduction of the background intensity in Raman spectrum after treatment (curves 1,2 in figure 2(a)) indicates removal of organic matter remained after the film deposition process. Intensities of the characteristic peaks of carbon I_D, I_{2D} and I_G in this case changed negligibly. The sheet resistance reduced from > 10 MOhm/□ to ~ 40 kOhm/□ due to the sintering processes between GO flakes. The Ar/Air partial pressure ratio changed in 10 % steps in the range 0–100 %. This experiment revealed that the most efficient reduction process of GO occurs at 90 % Ar and 10 % Air, that corresponds to 2 % of O_2.

Reducing the partial pressure of atmosphere oxygen in the chamber led to the shift of the process towards graphene oxide reduction: the removal of functional groups and restoration of the hexagonal lattice structure, which is confirmed by Raman spectra (figure 2).

Reduction of I_D and increase of I_{2D} peaks intensities occurred due to increase of argon partial pressure in the chamber (figure 2(b)). After argon pressure had excessed 90 % (at an oxygen content of < 2 %), several blue shifted peaks merged with I_G appeared, that can indicate the presence of amorphous carbon on the surface (curve 5 in figure 2(a)). Oxygen released during reduction was not sufficient for the removal of amorphous carbon from the film surface.

Amorphous films showed the highest conductivity, but they had lower structural perfection, low adhesion and they are not suitable for practical use as a functional material in micro- and nanoelectronics. Measurement of the sheet resistance of the material demonstrates its linear decrease to 101.17 Ohm/□ (figure 3).

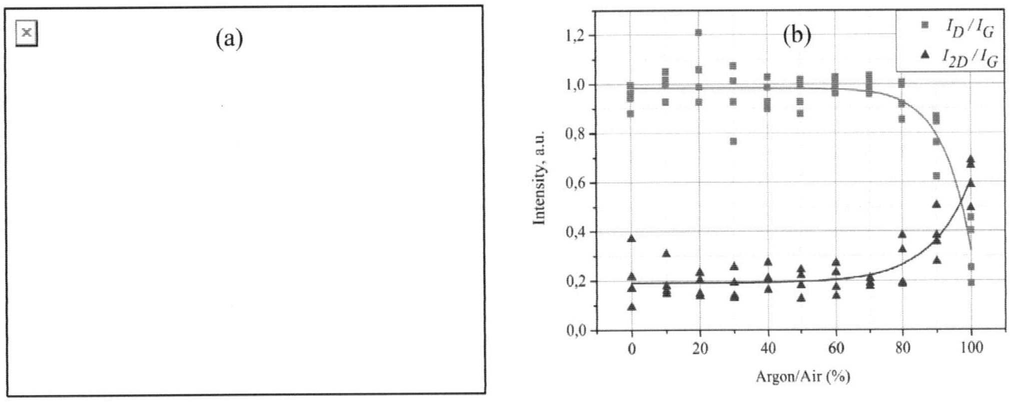

Figure 2. Raman spectra demonstrate the structural change of the material during processing (a); relative intensity of the Raman peaks I_D/I_G (square) and I_{2D}/I_G (triangle) as a function of the argon/air partial pressure ratio in the chamber (b).

We have experimentally obtained the dependence of sheet resistance and the I_{2D}/I_G, I_D/I_G ratio (which characterizes the structural perfection of films) on the laser irradiation exposure time. The exposure time of the laser beam on the sample was defined by the formula: $\tau=(d/V)\ D$, where d is the laser beam diameter, V is the laser movement speed, D is the control signal duty ratio. The exposure time varied from $4{\cdot}10^{-3}$ to $3{\cdot}10^{-1}$ s (figure 4).

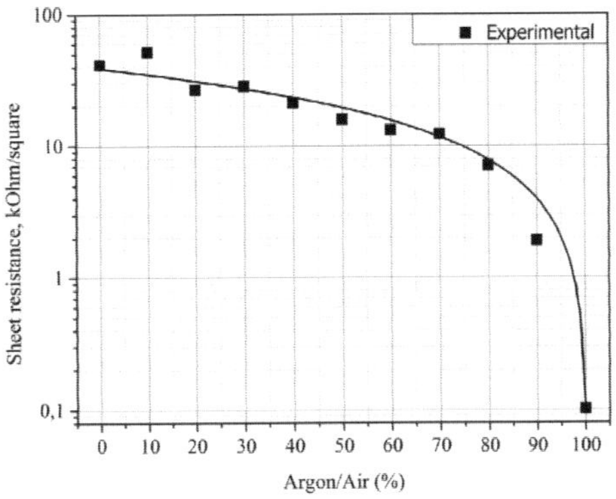

Figure 3. Sheet resistance as a function of the argon/air partial pressure ratio in the chamber. Recovery occurred at a constant exposure time of $4.29{\cdot}10^{-2}$ s.

The laser speed influence on reduction process was studied (within a range from 150 to 1500 µm/s, at 100% power) (figure 5). Existence of the critical exposure time of $\sim 4{\cdot}10^{-2}$ s is observed in all experiments. When we exceeded the critical time of exposure, the material reduction did not occur because the excess energy broke the carbon lattice bonds.

When the exposure time was less than the critical one, reduction of the material occurred that was confirmed by Raman spectra (increasing intensity of the I_{2D} peak, decreasing intensity of the I_D peak) and improving conductivity of the film. Further reduction of the exposure time resulted in a slight improvement of the material quality.

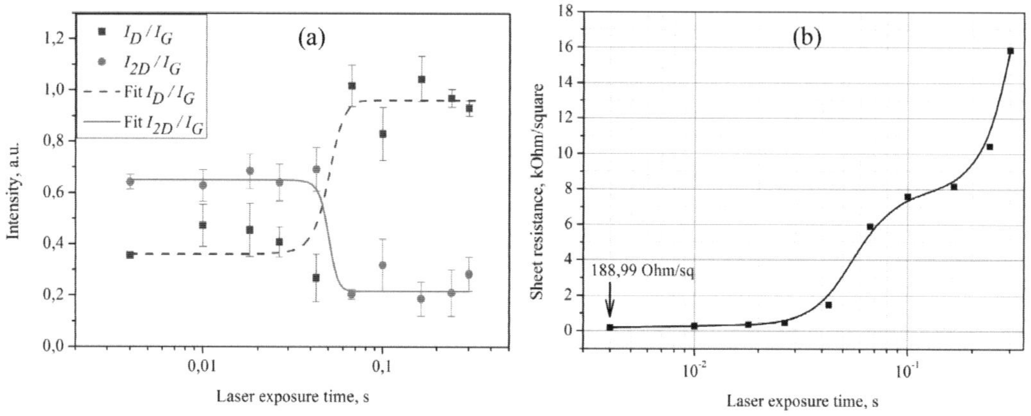

Figure 4. Relative intensity of Raman I_D/I_G and I_{2D}/I_G peaks (a); Sheet resistance (b) as a function of the laser irradiation exposure time.

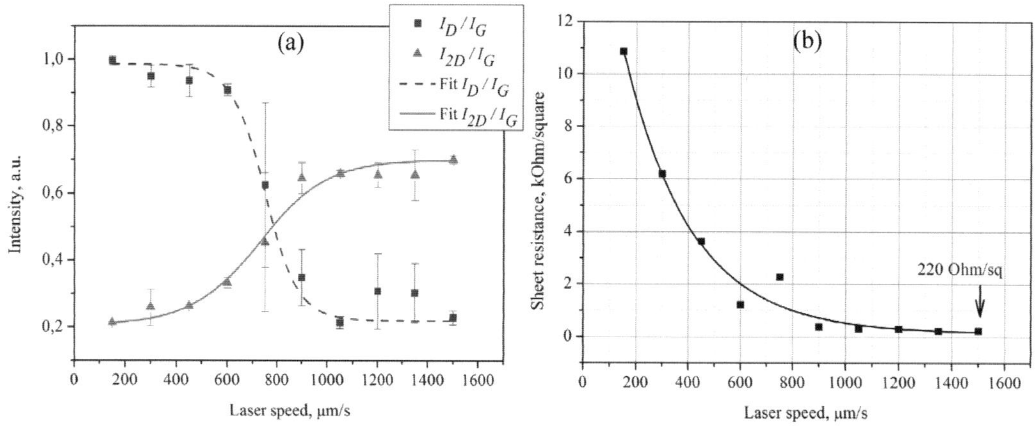

Figure 5. Relative intensity of Raman I_D/I_G and I_{2D}/I_G peaks (a); sheet resistance (b) as functions of the laser movement speed.

4. Conclusions

Processes of films and monolayers graphene oxide deposition on silicon and PEN substrates were practiced in this work. The effect of laser irradiation on the graphene oxide films was studied. We found the optimal ratio of argon to air (9:1) for the high-quality graphene films production. The dependence of material sheet resistance and its structural perfection on the laser irradiation exposure time and the laser movement speed was revealed. We found the critical exposure time required for graphene oxide reduction, which is $\sim 4{\cdot}10^{-2}$ s. We obtained the conductive traces with a sheet resistance of ~ 200 Ohm/□ and a thickness of 0.1 um, isolated by graphene oxide.

References

[1] Geim A K and Novoselov K S 2007 *Nature Materials* **6** 183
[2] Stankovich S, Dikin D A, Piner R D, Kohlhaas K A, Kleinhammes A, Jia Y, Wu Y, Nguyen S T and Ruoff R S 2007 *Carbon* **45** 1558

[3] Yung K C, Liem H, Choy H S, Chen Z C, Cheng K H and Cai Z X 2013 *Journ. Appl. Phys.* **113** 244903
[4] Pei S and Cheng H 2012 *Carbon* **50** 3210
[5] Chen W F, Yan L and Bangal P R 2010 *Carbon* **48** 1146
[6] Sinara D, Knopf G K, Nikumb S and Andrushchenko A 2014 *Proc. of SPIE* **8973** 89730K
[7] Zhang H and Miyamoto Y 2012 *Phys. Rev. B* **85** 033402
[8] Sokolov D A, Rouleau C M, Geohegan D B and Orlando T M 2013 *Carbon* **53** 81
[9] Tao Y, Varghese B, Jaiswal M, Wang S, Zhang Z, Oezyilmaz B, Loh K P, Tok E S and Sow C H 2012 *Appl. Phys. A* **106** 523
[10] Trusovas R, Ratautas K, Račiukaitis G, Barkauskas J, Stankevičiene I, Niaura G and Mažeikiene R 2013 *Carbon* **52** 574
[11] Ferrari A C 2007 *Solid State Communications* **143** 47
[12] Alexandrov G N, Smagulova S A, Kapitonov A N, Vasil'eva F D, Kurkina I I, Vinokurov P V, Timofee V B and Antonova I V 2014 *Nanotechnologies in Russia* **9** 363
[13] Alexenskii A E, Brunkov P N, Dideikin A T, Kirilenko D A, Kudashova Yu V, Sakseev D A, Sevryuk V A and Shestakov M S 2013 *Technical Physics. The Russian Journal of Applied Physics* **58** 1614

The ordering of Ge islands on a stepped Si(100) surface

M Yu Yesin[1], V A Timofeev[1], A R Tuktamyshev[1], A I Nikiforov[1,2] and I D Loshkarev[1]

[1] A.V. Rzhanov Institute of semiconductor Physics SB RAS, Novosibirsk 630090, Lavrentyev Avenue 13, Russia
[2] National Research Tomsk State University, 36, Lenina Avenue, Tomsk, 634050, Russia

E-mail: yesinm@isp.nsc.ru

Abstract. In the paper we investigate the Ge island nucleation on a Si(100) surface which was annealed to obtain the diatomic steps. It was observed that the islands tend to nucleate at the step edges.

1. Introduction

The self-organization in the Ge–Si system allows to obtain the nanometric size quantum dots with a density of 10^{10} – 10^{11} cm^{-2} [1]. The quantum dots system represents a great potential for nanoelectronics and photonics applications. The basic requirements of quantum dots are the following: the high island density, the size homogeneity and the nanoisland size providing manifestation of the dimensional quantization effect [2].

The past decade has seen rapid progress in research into high-performance Ge-on-Si photodetectors. Owing to their unique optoelectronic properties, which include high responsivity at near-infrared wavelengths, high bandwidths and compatibility with silicon complementary metal–oxide–semiconductor circuits, these devices can be monolithically integrated with silicon-based read-out circuits for applications such as high-performance photonic data links and IR detectors at low cost and low power consumption [3].

In paper [4], the clusters were found to preferably decorate the step edges in the form of a «necklace» indicating the possibility of ordered quantum dots creating. In paper [4], the author does not focus attention on the surface preparation for growth, the substrate disorientation specific characteristics and the step characteristics, that is necessary to know for studying the step influence on the Ge island growth. It is seen that dimer vacancy lines on the steps are directed perpendicularly to each other indicating that the surface has two-domains and, hence, the steps are monatomic.

At a certain temperature, the change in the two dimensional growth mechanism occurs. If the terrace width is larger than the average length of adatom migration, the film growth occurs due to the two-dimensional islands nucleation. If the terrace width is smaller than the average length of adatom migration, the film growth occurs due to the step movement. This is shown experimentally in [5] by the damping of specular beam intensity oscillations of reflection highenergy electron diffraction (RHEED). Also, the results of investigations of Ge island nucleation on the Si(100) surface are presented in [6, 7], where it was found that the temperature of growth mechanisms change is equal to 600 °C. Under heating of the surface in a Si flux, the steps transform from monoatomic to diatomic.

IOP Conf. Series: Journal of Physics: Conf. Series **816** (2017) 012015 doi:10.1088/1742-6596/816/1/012015

The control method consists in observation of disappearance of the reflexes belonging to one of two domains of the structure in the reflection high-energy electron diffraction (RHEED) pattern.

The goal of this work is to study the effect of Ge islands ordering on a Si(100) surface depending on the conditions of growth and surface preparation.

2. Experimental technique

The studies were carried out on a Si(100) substrate. Initially, chemical oxide SiO_2 was on the Si surface, meant for the surface protection from atmospheric contamination. Before the layer deposition, the Si substrate was annealed at a temperature of 800°C for SiO_2 layers desorption. The structures were obtained by molecular beam epitaxy (MBE). The investigations were carried out by reflection high-energy electron diffraction and atomic force microscopy (AFM). The AFM images were obtained in the contact mode.

3. Results and discussions

The samples with Ge islands on the Si(100) surface were obtained. In sample 1, the surface was annealed for 2 hours at a temperature of 600°C, then for 30 minutes at 700°C and for 10 minutes at 800°C. Upon the annealing, the Ge film with an effective thickness of 10 angstroms was grown at a temperature of 450°C. In sample 2, the silicon surface was annealed at a temperature of 1000°C during 10 minutes in a low Si flux of 10^{13} at.$cm^{-2}s^{-1}$. Upon the annealing, the Ge film with an effective thickness of 9 angstroms was grown at a temperature of 400°C.

Figure 1 shows the RHEED pattern in the azimuthal [100] direction from the Si(100) surface, where the surface was annealed for 2 hours at a temperature of 600°C, then for 30 minutes at 700°C and for 10 minutes at 800°C.

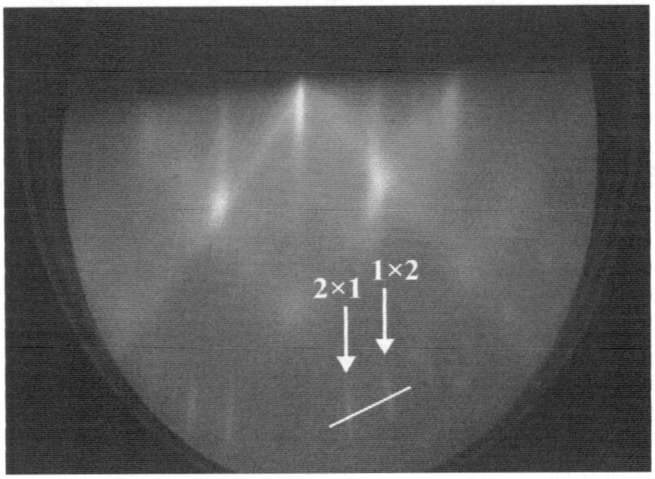

Figure 1. The RHEED pattern from the Si(100) 2×1+1×2 surface in the [100] direction in sample 1.

Figure 2 shows the RHEED pattern in the azimuthal [100] direction from the Si(100) surface in sample 2.

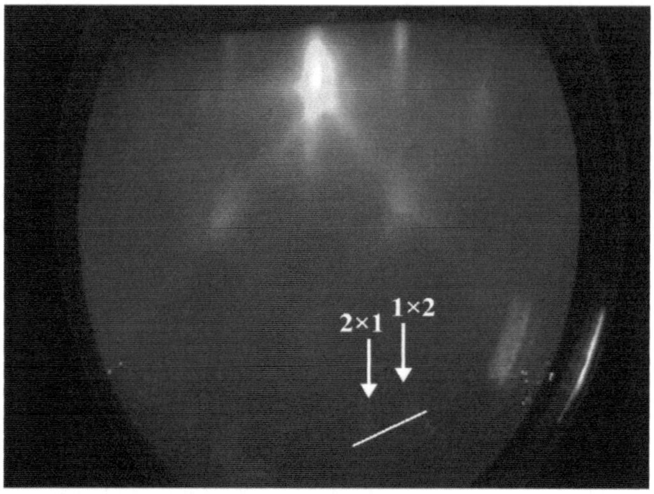

Figure 2. The RHEED pattern from the Si(100) 2×1+1×2 surface in in sample 2.

Figure 3 shows the RHEED pattern in the azimuthal [110] direction from the Si(100) surface in sample 1.

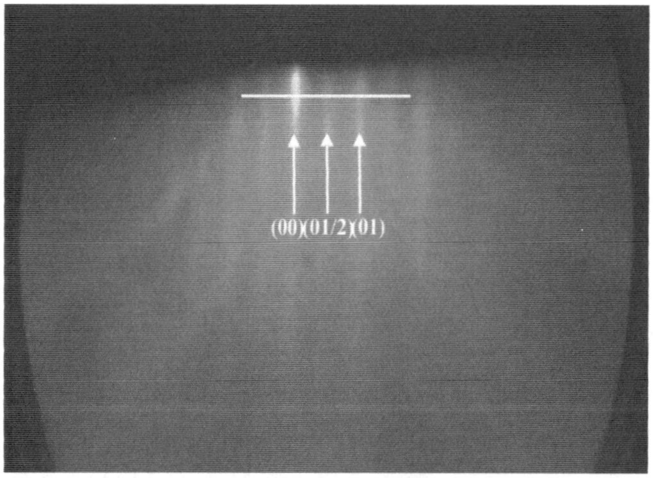

Figure 3. The RHEED pattern from the Si(100) 2×1+1×2 surface in the [110] direction in sample 1.

Figure 4 shows the RHEED pattern in the azimuthal [110] direction from the Si(100) surface in sample 2.

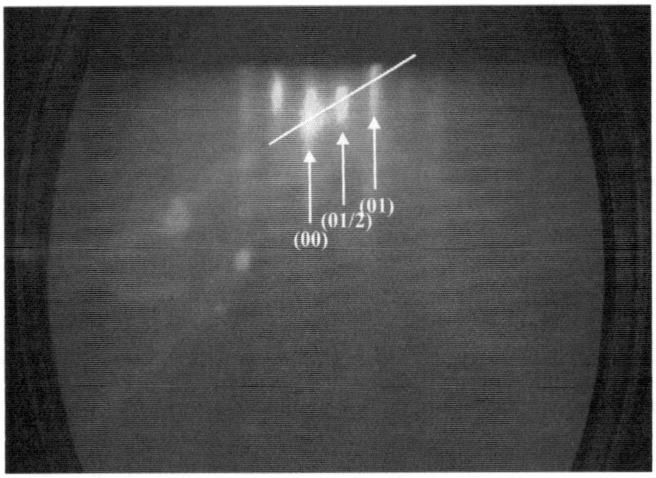

Figure 4. The RHEED pattern from the Si(100) 2×1+1×2 surface in the [110] direction in sample 2.

Figure 5 shows the profiles taken along the white line in figures 1 and 2.

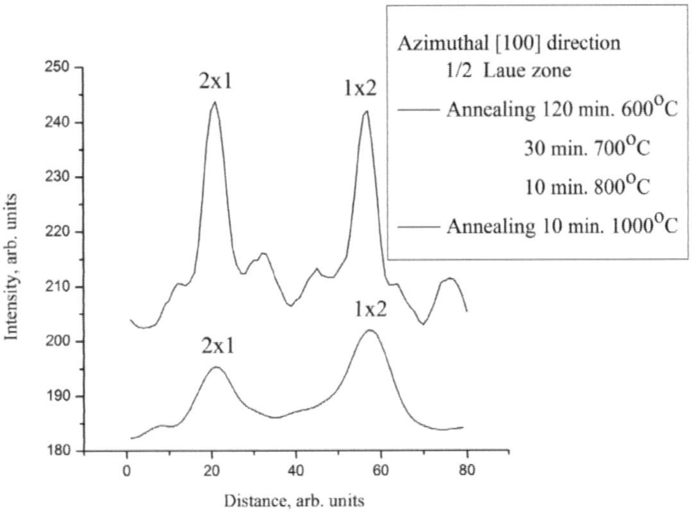

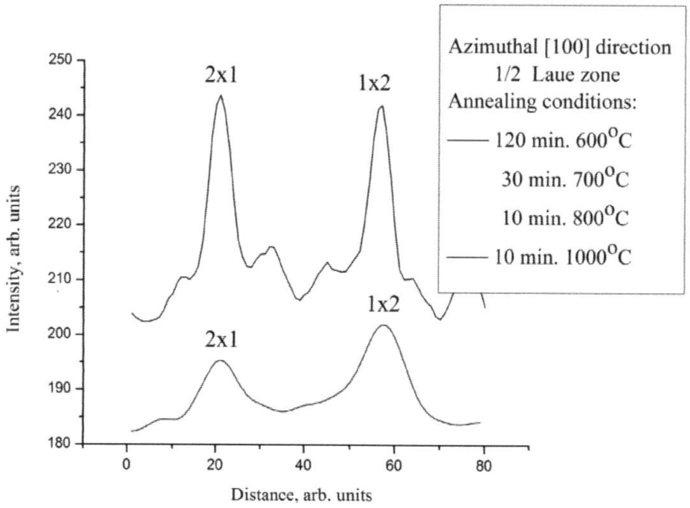

Figure 5. The profiles taken along the white line on the RHEED patterns in figures 1 and 2.

In the profile obtained from the RHEED patterns, it is seen that at higher annealing temperatures the reflex intensity from one sublattice of the two-domain surface becomes larger. Besides, the reflexes intensity generally falls and the background level decreases. This can be explained by the fact that the terrace length of one sublattice of the two-domain surface becomes wider. The reflex intensity decreasing can generally be related to the partial decay of the 2×1 reconstruction. The background decreasing can be related to the fact that the surface becomes less defective and atomically smoother.

Figure 6 shows the profile taken along the white line in figures 3 and 4.

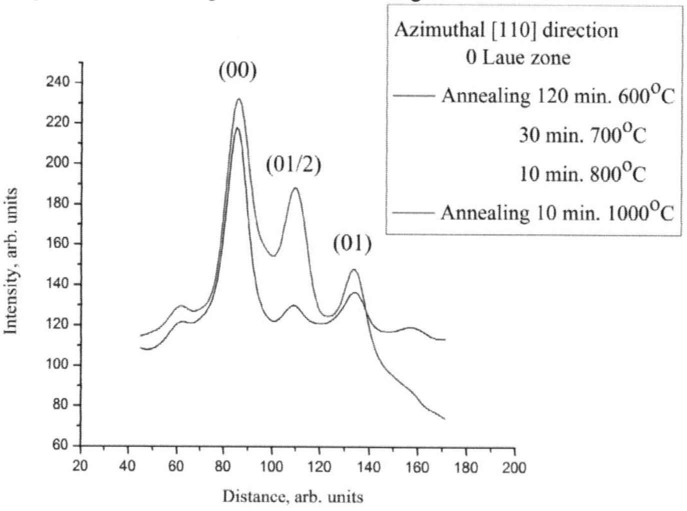

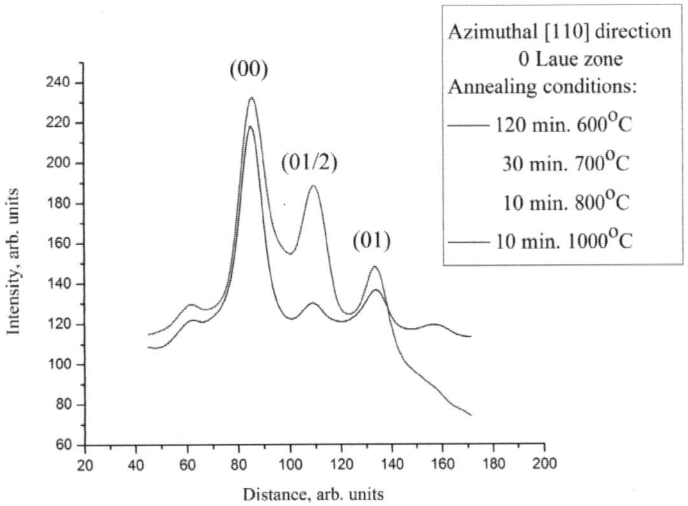

Figure 6. The profile taken along the white line on the RHEED pattern in figures 3 and 4.

The pattern analysis in the azimutal [110] direction shows that the intensity from the fractional order reflex increases with temperature increasing. This also indicates that the system departs from the equilibrium state and the terrace length for the two-domain surface changes.

Figure 7 shows the AFM image of the Ge island on the Si(100) surface of sample 1.

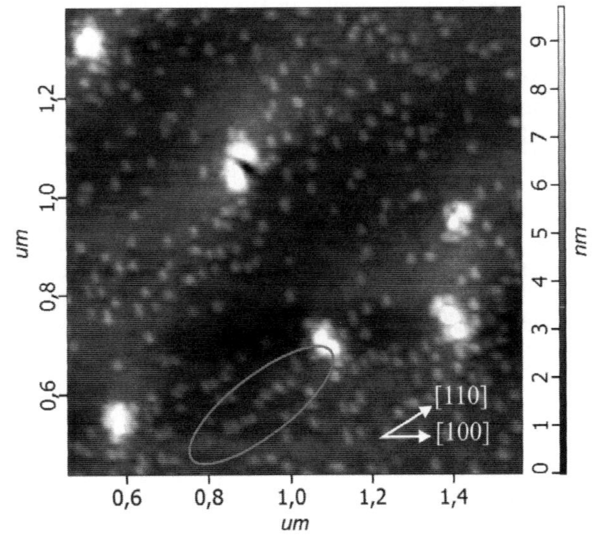

Figure 7. The Ge islands grown on the Si(100) surface of sample 1.

It is seen from the AFM image (figure 7) that some fraction of the island nucleates in parallel to the [110] direction. As it is known from the substrate characteristics, the steps are parallel to the [110] direction and the disorientation angle is 0,5°. The terrace widths approximately equal 16 nm and 31 nm for the monatomic and diatomic steps, respectively.

Figure 8 shows the AFM image of the Ge islands of sample 2.

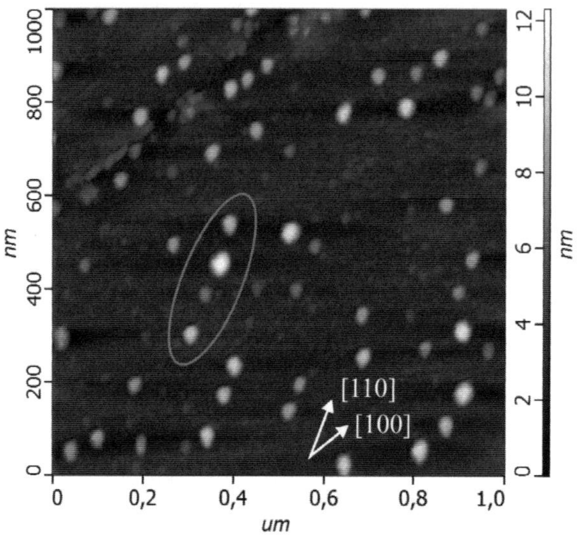

Figure 8. The Ge islands grown on the Si(100) surface of sample 2.

Analogously, it is seen from figure 8, that some fraction of the island nucleates is parallel to the [110] direction. But unlike sample 1, it is seen from the RHEED picture of sample 2 that the reflex intensities of superstructure on the zero-order Laue zone and the fractional-order Laue zone are not equal to each other. This indicates that the surface tends to a single-domain structure and, hence, to the diatomic steps.

Conclusions

As can be seen from the represented results, the diatomic steps have not been formed. However, it is seen from the RHEED picture that the reflexes from 2x1 and 1x2 domains of the reconstruction become not equal to each other. That is due to the different terrace widths. As it was observed on two such surfaces, the islands tend to nucleate on the step edges.

Acknowledgments

The work is supported by the Russian Foundations of Basic Research (Projects 14-29-07153, 16-29-03292, 16-32-60005, 16-32-00039), Russian Science Foundations (Projects 16-12-00023) and by The Tomsk State University Academician D.I. Mendeleev Foundation Program (Research grant No 8.2.10.2015).

References

[1] Pchelyakov O P, Bolkhovityanov Yu B, Dvurechenskii A V, Sokolov L V, Nikiforov A I, Yakimov A I and Voigtlander B 2000 *Semiconductors* **34** 1281
[2] Wang K, Cha D, Liu J and Chen C 2007*Proc. IEEE* **95** 1866
[3] Michel J, Liu J and Kimerling C 2010 *Nature photonics* **4** 527
[4] Goldfarb I. Hayden P T, Owen J H G and Briggs G A D 1997 *Phys. Rev. let.* **78** 3959
[5] Neave J H,. Dobson P J, Joyce B A and Zhang J 1985 *Appl. Phys. Lett.* **47** 100
[6] Cimalla V and Zekentes K 2000 *Appl. Phys. Lett.* **77** 1452
[7] Nikiforov A I, Cherepanov V A and Pchelyakov O P 2001 *Semiconductors* **35** 1032

Dipolar electron-hole liquid in a double-well SiGe/Si heterosystem

M A Akmaev[1,2] **and T M Burbaev**[1]

[1]Lebedev Physical Institute, Russian Academy of Sciences, Leninskiy prospect 53, Moscow 119991, Russia
[2]Moscow State University, Faculty of Physics, Leninskie Gory 1-2, Moscow 119991, Russia

E-mail: mark.akmaev@yandex.ru

Abstract. The transition from a dipolar to a spatially direct electron-hole liquid in two-dimensional layers of a type-II (*buffer* $Si_{1-y}Ge_y$)/*t*Si/*s*$Si_{1-x}Ge_x$/*t*Si/(*cap* $Si_{1-y}Ge_y$) heterostructure is investigated by photoluminescence spectroscopy at liquid-helium temperatures at high excitation levels. The transition takes place upon a reduction of the thickness of the *s*$Si_{1-x}Ge_x$ layer, which forms a quantum well for holes in the valence band and a barrier in the conduction band separating the electron quantum wells (*t*Si layers). The main characteristics of both types of electron-hole liquid are determined. The lifetime of dipolar excitons is determined from photoluminescence kinetics measurements.

1. Introduction

Interest in the investigation of multiparticle interactions in low-dimensional electron-hole systems of high density is stimulated by the discovery of new condensed phases with unusual properties in these systems, in particular an electron-hole liquid (EHL) [1] and a Bose-Einstein condensate (BEC) [2–5]. Theoretical analysis indicated that these phases should exhibit superconductivity and superfluidity [6, 7]. Long-range coherence of the BEC was observed experimentally [8, 9]. These studies were carried out in contravariant (type-I) GaAs/AlGaAs heterostructures with double or wide single quantum wells (QWs) for electrons and holes. The application of an external electric field to these heterostructures leads to the appearance of a spatially indirect energy gap. Then, at low temperatures, a system of nonequilibrium spatially indirect (dipolar) excitons forms under photoexcitation. At high excitation levels, various condensed phases emerge in this system. Photoluminescence (PL) spectroscopy is the main tool for the experimental investigation of the properties of these phases.

In this study, we use the PL spectroscopy to study covariant (type-II) Si/SiGe heterostructures, where the electron-hole system should exhibit equally interesting behavior [10–12]. In contrast to GaAs/AlGaAs structures, a spatially indirect electron-hole system can be created in SiGe structures without application of an external electric field. Lifetimes of nonequilibrium charge carriers in silicon, which is an indirect-gap material, are long (more than three orders of magnitude longer than those in GaAs), and this makes it possible to attain high concentrations of charge carriers and investigate their interactions leading to the formation of multiparticle states even at low excitation levels.

Here, we investigate the spectrum of multiparticle excited states in the exciton system in a double-QW SiGe/Si heterostructure in a wide range of excitation levels and temperatures. We study the transformation of the spectrum upon the transition from a dipolar to a spatially direct electron-hole system. We demonstrate that such a transition takes place as the thickness of the barrier layer separating the two QWs is reduced. We determine the main characteristics of both spatially direct and indirect EHL. Using time-resolved PL spectroscopy, we measure the lifetime of dipolar excitons.

2. Experimental techniques

A system of dipolar excitons was implemented in a $Si_{1-y}Ge_y$/*t*Si/*s*$Si_{1-x}Ge_x$/*t*Si/$Si_{1-y}Ge_y$ ($x > y$) heterostructure with two QWs for electrons, formed by tensile strained *t*Si layers separated by a compressively strained *s*$Si_{1-x}Ge_x$ layer acting as a barrier for electrons and a QW for holes. The heterostructure was grown by molecular-beam epitaxy on a single-crystal $Si_{0.92}Ge_{0.08}$ (001) substrate. The thickness of the pseudomorphic buffer layer of the same composition as the substrate was 100 nm. The thickness of the pseudomorphically grown strained *t*Si layers was 4 nm. The thickness d of the electron-barrier strained *s*SiGe was 4 or 2 nm (structures (4×4×4) and (4×2×4), respectively). A cap layer of the same composition as the substrate completed the design of this structure with two electron QWs.

The energy spectrum of the structures under study was investigated by PL spectroscopy in the near-infrared spectral range. Measurements were carried out at temperatures $T = 1.8$–60 K and excitation levels $P = 0.01$–$300\ \mathrm{W{\cdot}cm^{-2}}$. Quasi-continuous excitation of the structure was performed by a semiconductor laser emitting at a wavelength of $\lambda = 405$ nm. Recombination radiation was detected by a germanium *p-i-n* photodiode.

Analyzing the PL spectra of the structures with different barrier thicknesses ($d = 4$ or 2 nm), we can trace the transition from a dipolar to a spatially direct electron-hole system.

3. Energy-band diagram and the scheme of optical transitions

Figure 1 shows schematically the energy-band diagram of the heterostructures under study. In the first tensile-strained *t*Si layer grown on top of the unstrained $Si_{1-y}Ge_y$ (001) substrate, the sixfold degeneracy of the conduction-band bottom characteristic of unstrained Si is lifted. The bottom of the conduction band in this layer is formed by two Δ_2 valley located at the <001> axes of the reciprocal lattice that are perpendicular to the layer plane. The four Δ_4 minima for which the axes of the isoenergy surfaces are perpendicular to the growth direction appear at a higher energy. The compressively strained *s*$Si_{1-x}Ge_x$, grown on top of the first *t*Si layer, acts as a barrier for electrons and a well for holes. The degeneracy of the valence-band edge in the *s*$Si_{1-x}Ge_x$ layer is also lifted; the top of the valence band is formed by the quantum-confined states of heavy holes (*hh*) in the QW. The ground exciton state in such a system is formed by electrons from the Δ_2 valley of the Si conduction band, which occupy the quantum-confinement level in the QW formed by the *t*Si layer, and heavy holes occupying the quantum-confinement level in the QW formed in the valence band of the *s*$Si_{1-x}Ge_x$ layer. The second pseudomorphically grown *t*Si layer and the cap $Si_{1-y}Ge_y$ layer with the same composition as the substrate completes the layout of this structure with two electron QWs.

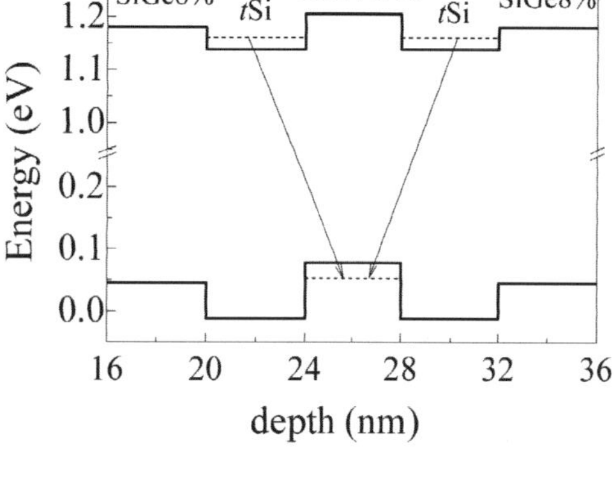

Figure 1. Energy-band diagram and the scheme of optical transitions.

4. PL spectra of the EHL and excitons in dipolar and spatially direct electron-hole systems

Figure 2 shows a series of the PL spectra of the (4×4×4) structure (the thickness of the *s*SiGe barrier layer $d = 4$ nm) measured at $T = 1.8$ K upon an increase in the excitation level P.

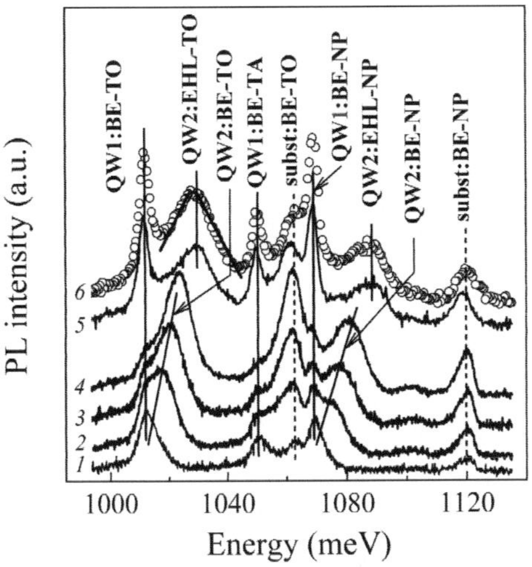

Figure 2. The formation of a dipolar EHL in the (4×4×4) double-QW structure with an increase in the pump level $P =$ (*1*) 0.7, (*2*) 1.0, (*3*) 4.5, (*4*) 8.0, (*5*) 15, and (*6*) 90 W·cm^{-2}; $T = 1.8$ K. The designations of the PL lines are explained in the text. For spectrum *6*, the solid line shows the calculated EHL line shape.

Dashed vertical lines mark the PL lines originating from the recombination of bound excitons (BE) in the substrate and the buffer and cap layers; the corresponding no-phonon line and its transverse-optical (TO) phonon replica are designated as (subst: BE-NP) and (subst: BE-TO), respectively. Solid vertical lines mark the PL lines related to the *t*Si – *s*SiGe transitions (figure 1). As the excitation level increases, the lines from the second QW (QW2:BE-NP and QW2:BE-TO) experience a blue shift as a result of Coulomb band bending caused by the accumulation of spatially indirect (dipolar) excitons collected from the substrate. Apparently, only the lower QW for electrons participates in the collection of excitons produced by photoexcitation from the substrate. This makes it possible to distinguish the radiative transitions from the *lower* *t*Si to the *s*SiGe layer (lines QW2, which experience a blue shift at high excitation levels) and from the *upper* *t*Si to the *s*SiGe layer (lines QW1: BE-NP, BE-TA (transverse-acoustic phonon replica), and BE-TO, which do not experience a blue shift). At low temperatures and high excitation levels, we observe in this sample a two-dimensional dipolar EHL with spatially separated electrons and holes, manifested in the emergence of the PL lines QW2: EHL-NP and EHL-TO. The shape of these lines does not vary with increasing excitation level, which is a signature of the EHL. In addition, we see that, for $P > 15$ W·cm^{-2}, the blue shift ceases. This takes place because the particle density in the EHL is constant and only the size and concentration of the EHL drops increase with increasing excitation level. We found that the critical density of the observed EHL is about 2 K.

Figure 3 shows a series of the PL spectra of the (4×2×4) structure (the thickness of the *s*SiGe barrier layer $d = 2$ nm) measured at $T = 1.8$ K upon an increase in the excitation level.

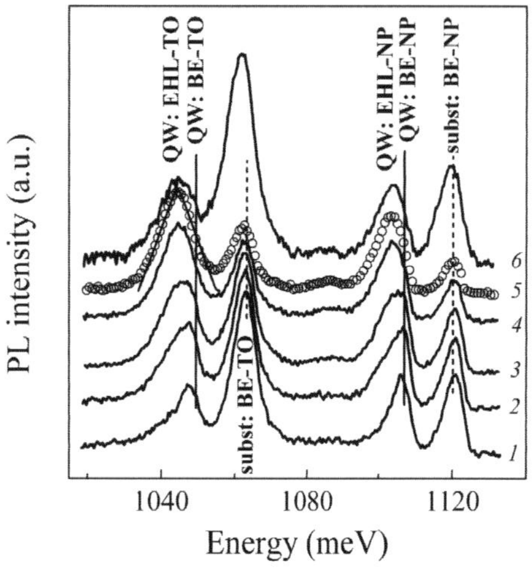

Figure 3. Formation of a spatially direct EHL in the (4×2×4) with an increase in the pump level P = (*1*) 0.09, (*2*) 0.4, (*3*) 0.75, (*4*) 1.0, (*5*) 8, and (*6*) 80 W·cm^{-2}, T=1.8 K.
The designations of the PL lines are explained in the text. For spectrum *5*, the solid line shows the calculated EHL line shape.

Dashed vertical lines mark the PL lines originating from the substrate and the buffer and cap layers. Solid vertical lines mark the PL lines related to the *t*Si – *s*SiGe transitions. In contrast to the previous case, no blue shift occurs in these spectra with increasing excitation level. The wave function of electrons spreads into the thin *s*SiGe barrier, so that the heterostructure becomes effectively type-I. The EHL line appears with increasing excitation level at the red tail of the exciton line. For $P > 1$ W·cm^{-2}, the line shape becomes independent of the excitation level, which gives evidence of the EHL formation. The EHL is formed by Δ_2 electrons and heavy holes. Thus, in this structure we observe PL from a spatially direct EHL. The critical temperature was found to be 6 K.

We note that the critical temperatures of the spatially direct and dipolar EHL in double-well heterostructures with tensile strained *t*Si layers is considerably lower than the corresponding critical temperatures of the EHL in Si/*s*SiGe/Si structures [10, 12]. We believe that the reduction in the critical temperatures of the EHL of both types is caused by a twofold decrease in the degree of degeneracy of the electron spectrum in *t*Si layers of $Si_{1-y}Ge_y$/*t*Si/*s*$Si_{1-x}Ge_x$/*t*Si/$Si_{1-y}Ge_y$ heterostructures as compared to Si layers in Si/*s*SiGe/Si(001) heterostructures.

In curve *6* (figure 2) and curve *5* (figure 3), circles show the experimental results and solid lines show the results of calculations of the EHL line shape. The calculations were carried out similarly to the analysis given in [13] for bulk semiconductors. In contrast to [13], we used an energy-independent density of states for electrons and holes. The two-dimensional density of holes in the EHL in the QW formed by the *s*SiGe layer obtained from the calculations is $1.2\cdot10^{12}$ and $2.2\cdot10^{12}$ cm^{-2} for the (4×2×4) and (4×4×4) structures, respectively.

One can see from figure 2 and figure 3 that the intensity of the no-phonon exciton and EHL recombination lines is comparable to or even higher than the intensity of their phonon replicas. Apparently, the high intensity of the NP components of the PL spectrum in the structures under study is related to the localization of dipolar excitons and electron-hole drops in lateral wells of the random potential profile resulting from the composition inhomogeneities in the SiGe layer and at heterointerfaces [14, 15].

4.1. Temporal characteristics of photoluminescence

Figure 4 shows the time evolution of the PL spectra of the dipolar exciton system in the (4×4×4) heterostructure measured at a temperature of 5 K (higher than the critical temperature of the dipolar EHL). The spectra in figure 4 are measured upon a stepwise increase in the delay time with respect to the excitation pulse. The evolution of the spectra with increasing delay time manifests the phase transition between the dipolar electron-hole plasma and the gas of dipolar excitons. From the PL decay kinetics, it was found that the lifetime of dipolar excitons in this structure is 6.3 μs. This is considerably longer than the lifetime of spatially direct excitons (<1 μs in Si/*s*SiGe/Si structures [16]).

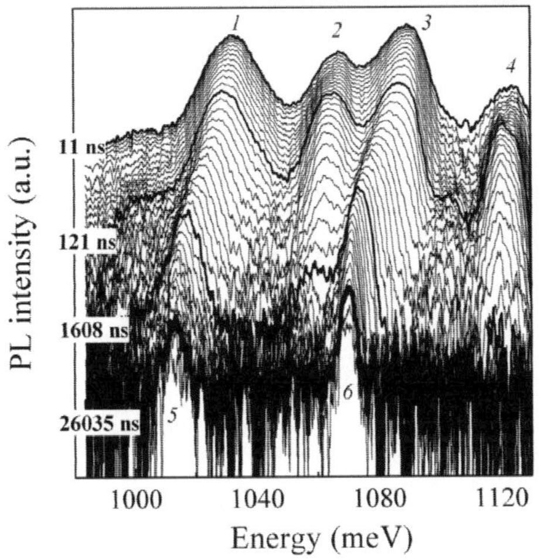

Figure 4. PL spectra of the (4×4×4) structure at $T = 5$ K at different delay times after the excitation pulse. Numbers indicate the PL lines (*1*) QW2:EHP-TO, (*2*) subst:BE-TO, (*3*) QW2:EHP-NP, (*4*) subst:BE-NP, (*5*) QW2:BE-TO, and (*6*) – QW2:BE-NP (see figure 2 and explanations in the text).

5. Conclusions

To summarize, the following results have been obtained.

$Si_{1-y}Ge_y$/*t*Si/*s*$Si_{1-x}Ge_x$/*t*Si/$Si_{1-y}Ge_y$ heterostructures with two QWs for electrons formed by tensile strained *t*Si separated by $Si_{1-x}Ge_x$ (acting as a barrier for electrons and a well for holes) have been designed and grown.

Using PL spectroscopy, we have observed the formation of two-dimensional EHL in these structures at liquid-helium temperatures and high excitation levels. Either a spatially direct or a dipolar EHL forms depending on the design of the structure. The two-dimensional density of holes is $2.2 \cdot 10^{12}$ and $1.2 \cdot 10^{12}$ cm^{-2} for dipolar and spatially direct EHL, respectively. The critical temperatures of spatially direct and dipolar EHL have been determined to be 6 and 2 K, respectively.

The transformation of the spectrum of the excited states upon the transition from a dipolar to a spatially direct electron-hole system has been investigated. This transition takes place as the thickness of the barrier layer separating the two QWs is reduced from 4 to 2 nm.

The lifetime of dipolar excitons has been determined to be 6.3 μs.

Acknowledgments

We are grateful to N.N. Sibeldin for useful discussions and valuable suggestions, to S.N. Nikolaev for performing time-resolved PL measurements, to D.N. Lobanov and A.V. Novikov for growing of structures.

This study was supported by grants of the Russian Foundation for Basic Research (project nos. 16-02-00986 and 16-29-03352).

References

[1] Timofeev V B, Larionov A V, Grassi-Alessi M, Capizzi M, Hvam J M 2000 *Phys. Rev.* B **61** 8420
[2] Larionov A V and Timofeev V B 2001 *JETP Lett.* **73** 301
[3] Butov L V, Gossard A C and Chemla D S 2002 *Nature* **418** 751
[4] Gorbunov A V and Timofeev V B 2004 *JETP Lett.* **80** 185
[5] Gorbunov A V and Timofeev V B 2006 *JETP Lett.* **84** 329

[6] Lozovik Yu E, Yudson V I 1975 *JETP Lett.* **22** 11
[7] Lozovik Yu E, Yudson V I 1976 *Sol. St. Comm.* **18** 628
[8] Butov L V 2004 *Journal of Physics: Condensed Matter* **16** 50 R1577
[9] Timofeev V B, Gorbunov A V, Larionov A V 2007 *Journal of Physics: Condensed Matter* **19** 29 295209
[10] Burbaev T M, Kozyrev D S, Sibeldin N N and Skorikov M L 2013 *JETP Lett.* **98** 823
[11] Orlov A, Levashova N, Burbaev T 2015 *Journal of Physics: Conference Series* **586** 012003
[12] Sibeldin N N *J. Exp. Theor. Phys.* 2016 **122** 3 587
[13] Pokrovsky Ya E Kaminsky A S and Svistunova K I 1970 *Proceedings of the Tenth International Conference on the Physics of Semiconductors (Cambridge, Mass.)*, (AEC Division of Technical Information, Springfield, Virginia) p 504
[14] Nayak D K, Usami N, Fukatsu S, and Shiraki Y 1993 *Appl. Phys. Lett.* **63** 3509
[15] Sheng S R, Rowell N L and McAlister S P 2003 *Appl. Phys. Lett.* **83** 857
[16] Bagaev V S, Zaitsev V V, Krivobok V S, Lobanov D N, Nikolaev S N, Novikov A V, Onishchenko E E 2008 *JETP* **107** 5 846

 doi:10.1088/1742-6596/816/1/01201

Photoluminescence in InGaAsSb/AlGaAsSb quantum wells: impact of nonradiative recombination

M Ya Vinnichenko[1], I S Makhov[1], A V Selivanov[1], D A Firsov[1], L E Vorobjev[1], L Shterengas[2] and G Belenky[2]

[1] Peter the Great St. Petersburg Polytechnic University, 29 Polytechnicheskaya str., St. Petersburg 195251, Russia

[2] Department of Electrical and Computer Engineering, State University of New York at Stony Brook, New York, 11794 USA

e-mail: mvin@spbstu.ru

Abstract. The photoluminescence in nanostructures with InGaAsSb/AlGaAsSb quantum wells of different widths is investigated experimentally under different pumping levels and theoretically. The experimentally determined dependencies of photoluminescence intensity on the optical pumping level are compared with the calculated dependences of photoluminescence intensity on the nonequilibrium carrier concentration. The presence of resonant nonradiative Auger recombination in one of the investigated samples is proved by the comparison between experiment and calculations.

1. Introduction

The InGaAsSb/AlGaAsSb quantum well (QW) nanostructures can be used as a basis for injection laser diodes that emit radiation in the wavelength range between 2 and 3 μm and above [1, 2]. Such laser diodes may be employed in many fields: infrared spectroscopy, safety systems, medicine, wireless communications, etc. Furthermore, the atmosphere is transparent in the spectral range 2 - 4 μm, thus an applicability area of these lasers expands. However, some problems of the development of effective laser diodes operating in this spectral range are not completely solved.

The efficiency and characteristics of the injection lasers are determined by the relation between radiative and nonradiative carrier lifetimes. It is well known, that at high injection levels the Auger recombination is one of the most important negative mechanisms of nonradiative recombination in QWs [3]. In some cases, Auger recombination can become resonant, which reduces the nonradiative lifetime by 2–3 orders of magnitude [4] and causes corresponding degradation of the laser parameters. For these reasons, the investigation of Auger recombination is undoubtedly of interest. The principal research method in our work is an analysis of interband photoluminescence spectra at different pump levels, which gives the essential information about the concentration and distribution function of the nonequilibrium charge carriers involved in the radiative recombination.

2. Sample and experiment

The structures for optical studies were MBE grown on GaSb substrates and contained ten InGaAsSb/AlGaAsSb QWs. The structures had different quantum well widths: 4, 5, 7 and 9 nm. The position of the energy levels in QWs was calculated in the context of Kane's model [5], in which the nonparabolic shape of subbands was taken into account. This approximation must be used since the electron energy in subbands is of the order of the band gap value. We used material parameters taken from the paper [6]. Compressive strain in QWs decreases the density of states in the heavy hole subband. Therefore, in energy spectrum calculations we used the heavy hole masses less than in a bulk material. The calculated energy gaps between the first electron ($e1$) and heavy hole ($hh1$) subbands, the first ($e1$) and second ($e2$) electron subbands, the first heavy hole ($hh1$) and the first split off by spin-orbit interaction ($so1$) subbands for different QW widths are shown in figure 1.

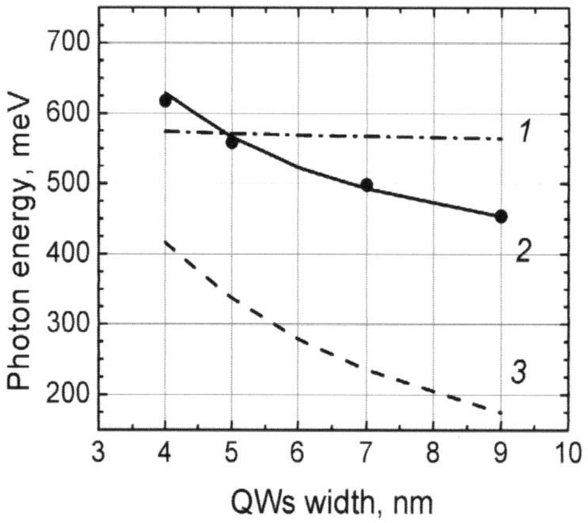

Figure 1. Experimental positions of the photoluminescence peaks related to the transitions $e1 \rightarrow hh1$ (circles) and calculated energy gaps (lines): $E(hh1)$–$E(so1)$, $E(e1)$–$E(hh1)$, $E(e2)$-$E(e1)$ marked as (1), (2) and (3), respectively. $T = 77$ K.

The interband photoluminescence spectra were studied using a vacuum Fourier transform infrared spectrometer Bruker Vertex 80v operating in a step-scan mode. For the experimental studies, the samples were mounted into an optical cryostat where the sample temperature can be varied from 80 K to 320 K. Electron-hole pairs were excited directly in QWs by a pulsed semiconductor laser with a photon energy of 1.17 eV. The optical excitation of the samples was obtained through a fused silica window. Photoluminescence emission of the samples was collected through a ZnSe window by an off-axis parabolic mirror of a spectrometer. A Ge filter was used in order to block the pumping light. An InSb photodetector was used as a detector of radiation.

Our goal was to investigate the influence of Auger processes on the carrier concentration. The condition of the resonance Auger recombination is the equality of the effective band gap and the energy separation between the first heavy hole level and the spin-orbit split-off level [3, 4] (see inset in figure 2):

$$E(e1) - E(hh1) \cong E(hh1) - E(so1). \qquad (1)$$

In our structures, this condition of resonant nonradiative Auger processes is satisfied only in the structure with 5 nm wide QWs at liquid nitrogen temperature ($T = 77$ K). So, in this structure the concentration of carriers taking part in the radiative recombination should be less than in other samples. It should be noted that the resonant Auger recombination in the structure with 5 nm wide QWs was early observed [3] when analyzing the photoluminescence dynamics measured by the up-conversion method. Here we present another experimental method.

3. Results and discussions

The interband photoluminescence spectra are presented in figure 2 for the maximum pumping level for all structures at $T = 77$ K. The spectral positions of the photoluminescence peaks for all structures are shown in figure 1 by solid circles. It is clear that photoluminescence intensity at the maximum of its spectral dependence (J^{PL}_{MAX}) is determined by the optical electron transitions $e1 \rightarrow hh1$. So, one can see a good agreement between experiment and calculations.

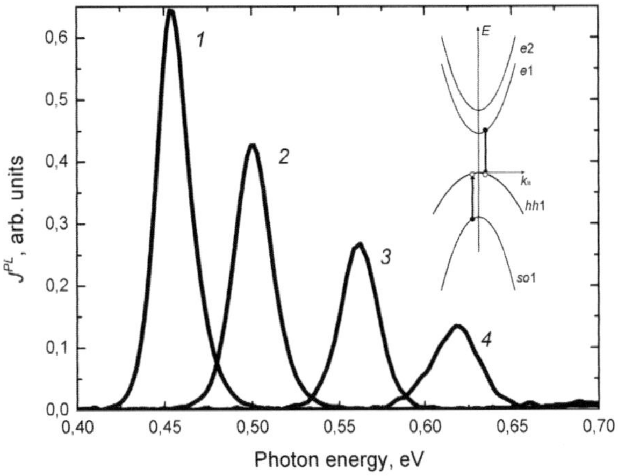

Figure 2. Photoluminescence spectra for the maximum optical pump level. Curves 1, 2, 3, 4 correspond to the samples with QW widths of 9, 7, 5, 4 nm. The lattice temperature $T = 77$ K. The inset shows the condition of resonant Auger recombination.

From the photoluminescence spectra measured at different pump levels, we received dependences of maximum photoluminescence intensity J^{PL}_{MAX} on the pump level J_{PUMP} (circles in figure 3). Concentration of excited charge carriers is proportional to the pump intensity. However, the photoluminescence intensity at a dedicated wavelength is not simply proportional to the concentration of carriers in the QWs. Using the method described in Ref. [7], we received the theoretical dependence of J^{PL}_{MAX} on carrier concentration (see lines in figure 3). For this purpose, we wrote down the number of photons $dq_{sp}^{i\to j}$ emitted per unit volume per unit time in the photon energy range from $h\nu$ to $h\nu + d(h\nu)$ due to the optical electron transitions from the electron level i to the hole level j:

$$dq_{sp}^{i\to j} = \frac{2}{\pi}\frac{n\, e^2}{\hbar\, m_0^2}\frac{m_{eh}}{L_{QW}}(h\nu)\frac{P^2}{c^3} I_{i\to j}(\mathbf{k}^2) f_e f_h d(h\nu), \tag{2}$$

where L_{QW} is the quantum well width, n is the refraction index, m_{eh} is the reduced effective mass, m_0 is the free electron mass, c is the light velocity, $\hbar$ is the reduced Planck's constant, $I_{i\to j}$ is the overlap integrals for the selected quasi-momentum calculated in [8], $\mathbf{k}$ is the wave vector, P is the Kane matrix element of the momentum operator written as for a bulk semiconductor:

$$P^2 = \frac{m_0}{2}\left(\frac{m_0}{m_e^*} - 1\right)\frac{E_g(E_g + \Delta_{so})}{(E_g + \frac{2}{3}\Delta_{so})}, \tag{3}$$

E_g is the bandgap energy, m_e^* is the electron effective mass, Δ_{so} is the energy of level split off by spin–orbit interaction, f_e, f_h is the Fermi-Dirac distribution functions of electrons and holes.

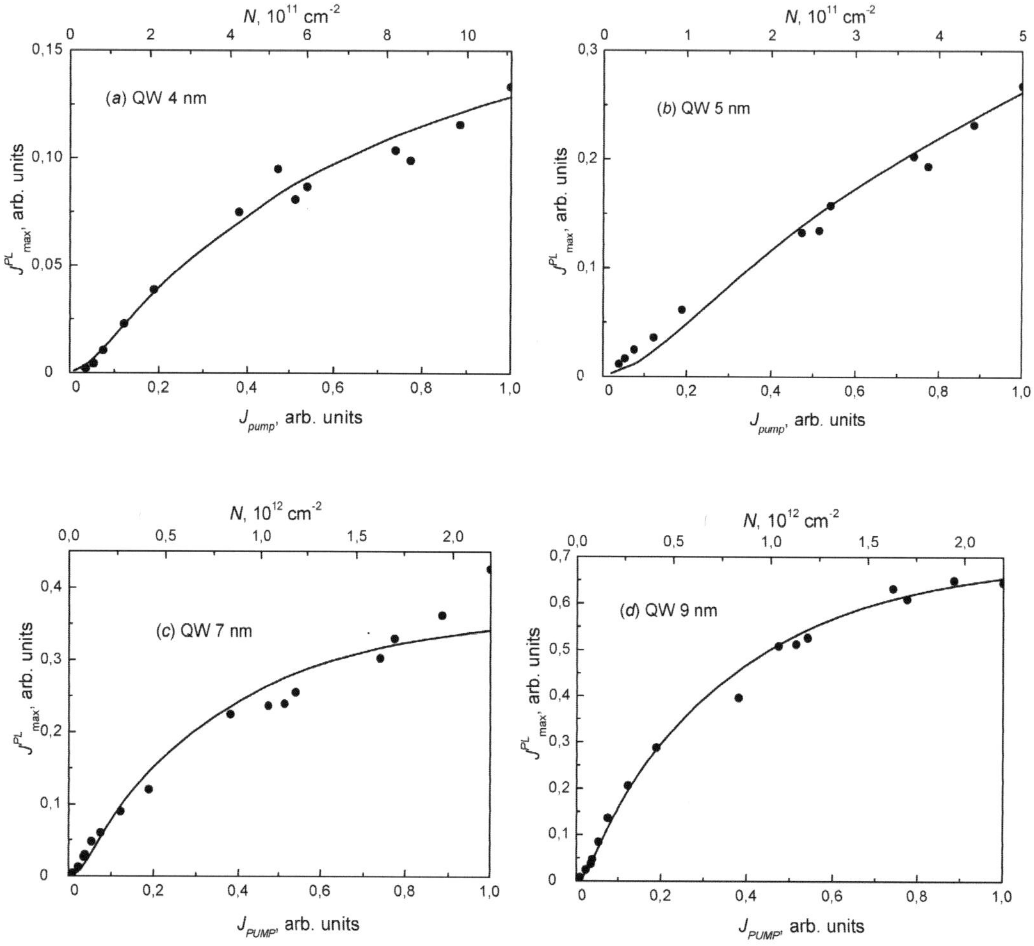

Figure 3. Dependencies of photoluminescence intensity measured at the maximum of spectra on the pump level (circles) and on the electron concentration (solid lines, calculations) for the samples with different QW widths at $T = 77$ K.

The theoretical results of the photoluminescence intensity maximum dependencies on the carrier concentration N for the samples with different QW widths are shown in figure 3 by circles. These results are in good agreement with the experimental dependencies of the photoluminescence intensity maximums on the pump level for all our samples (see figure 3). Comparison of theoretical and experimental dependencies gives us the information about the concentration of carriers taking part in the radioactive recombination. The different shapes of these curves can be explained by the following. At a low pump level, the photoluminescence intensity is proportional to the product of nonequilibrium electron and hole concentrations (we have neglected the equilibrium concentration). At a medium pump level, the electron gas becomes degenerated while holes are still non-degenerated due to the different effective masses of electrons and holes. As a result, the J^{PL}_{MAX} depends only on the concentration of excited holes. At a high pump level, the concentrations of electrons and holes become constant and $J^{PL}{}_{MAX}$ tends to the saturation. One can see that the concentration of carriers taking part in the radiative recombination is lower for the structure with 5 nm wide QWs (see the absence of saturation in figure 3 (b)).

A low carrier concentration in the 5 nm wide QW structure as compared to other structures with close parameters can be explained by the presence of an additional recombination channel. Based on the fulfillment of condition (1) for this structure, we suppose that this channel is related to resonant Auger recombination in accordance with the rate equation:

$$\frac{dN}{dt} = -AN - BN^2 - CN^3, \quad (3)$$

where the first term describes the Shockley–Read–Hall recombination, the second term describes the radiative recombination, and the third term describes the Auger recombination. The coefficient C reaches the maximum value in the 5 nm wide QW structure. In all structures except the 5 nm wide QWs, the concentration of carriers taking part in the radiative recombination is higher, that can be explained by the lack of resonant Auger recombination. These results confirm the assumption that the condition of resonant Auger recombination (1) can be satisfied only in the samples with 5 nm wide QWs. Similar findings were obtained in our previous study [3].

The shortwave tail of photoluminescence spectra describes the distribution function of carriers and can give information about carrier heating. Using the method described in [9], we determined the carrier temperature for all structures at different pump levels. The carrier temperature increase (ΔT_C) is approximately 10 K for 5 nm wide QWs for the maximum pump level. For other structures, $\Delta T_C \approx$ 20 K. It is associated with more frequent electron-electron collisions in structures without resonant nonradiative Auger recombination.

4. Conclusion

In the present work, the dependencies of the photoluminescence intensity at the maximum of the spectra on the pumping intensity and carrier concentration were experimentally and theoretically investigated in the nanostructures with different InGaAsSb/AlGaAsSb QW widths. The nonequilibrium carrier concentrations for the samples with 9, 7, 4 nm wide QWs are approximately the same, whereas the carrier concentration for the sample with 5 nm wide QWs is significantly less. This result can be explained by the mechanism of the resonant nonradiative Auger recombination, that is significant for this sample. The hot carrier temperature was estimated for all structures.

Acknowledgments
This work was supported by the Russian Foundation for Basic Research (grant 16-02-00863), Russian Federation President grant for young scientists No MK-4616.2016.2 and the Ministry of Education and Science of the Russian Federation (state assignment).

References

[1] Belenky G, Shterengas L, Kipshidze G and Hosoda T 2011 *IEEE J Sel Top Quantum Electron* **17** 1426
[2] Shterengas L, Belenky G, Hosoda T, Kipshidze G and Suchalkin 2008 S *Appl Phys Lett* **93** 011103
[3] Vinnichenko M Y, Firsov D A, Vorobjev L E, Mashko M O, Shterengas L and Belenky G 2013 *Semiconductors* **47** 146
[4] Danilov L V and Zegrya G G 2008 *Semiconductors* **42** 550
[5] Kane E 1957 *Journal of Physics and Chemistry of Solids, Elsevier* **1** 249
[6] Vurgaftman I, Meyer J R and Ram-Mohan L R 2001 *Journal of Applied Physics* **89** 5815

[7] Vorob'ev L E, Zerova V L, Borshchev K S, Sokolova Z N, Tarasov I S and Belenky G 2008 *Semiconductors* **42** 737
[8] Sokolova Z N and Khalfin V B 1989 *Soviet physics semiconductors – USSR* **23** 1117
[9] Vorobjev L E, Vinnichenko M Ya, Firsov D A, Zerova V L, Panevin V Yu, Sofronov A N, Thumrongsilapa P, Ustinov V M, Zhukov A E, Vasiljev A P, Shterengas L, Kipshidze G, Hosoda T and Belenky G 2010 *Semiconductors* **44** 1402

IOP Conf. Series: Journal of Physics: Conf. Series **816** (2017) 012018 doi:10.1088/1742-6596/816/1/012018

The binding energy of excitons in narrow quantum wells

P A Belov[1] and E S Khramtsov[2]
[1]Department of Computational Physics, St. Petersburg State University, Ulyanovskaya 1, 198504 St. Petersburg, Russia
[2]Spin Optics Laboratory, St. Petersburg State University, Ulyanovskaya 1, 198504 St. Petersburg, Russia

E-mail: pavelbelov@gmail.com

Abstract. The effects of discontinuities of the material parameters at the interfaces of narrow GaAs-based quantum wells as well as of the nonparabolicity of the conduction band on the exciton binding energy are studied. These effects are taken into account in the three-dimensional Schrödinger equation for the exciton. Accurate exciton ground state energies are obtained using the improved numerical scheme of Khramtsov *et al.* (2016 *J. Appl. Phys.* **119** 184301). The enhancement of the exciton binding energy for narrow quantum wells is observed and the contribution of each effect is estimated.

1. Introduction

Exciton states in quantum wells (QWs) have been experimentally and theoretically studied for several decades. The theoretical studies [1–12] were generally ahead of experimental ones due to the well-developed quantum theory and an obvious lag between a technological design and the practical realization of experimental techniques [13–19] for growth of high-quality heterostructures. The defects and roughness of interfaces of heterostructures, the segregation effect [20, 21] especially for narrow QWs, impair the accuracy of measurements of optical transitions between exciton states [22]. The high-quality heterostructures have only become available only recently and provide a good possibility for the experimental studies [23–26, 12, 27, 28]. Therefore, the precise theoretical results on the exciton states can significantly advance the ongoing experiments on heterostructures.

The binding energy of excitons in QWs is usually obtained from the solution of the two-body Schrödinger equation (SE) which is, in turn, complicated by impossibility to separate the center-of-mass motion. Furthermore, the degenerate valence band, in the simplest case, requires one to consider the diagonal terms of the Luttinger Hamiltonian [30] for the hole. These complications lead to the three-dimensional SE which can be solved numerically by the variational approach [4, 29, 5, 8, 10, 31], shooting [11] or perturbative methods [32, 33]. Recently, we proposed a new method [12] for direct numerical solution of this equation. The method is based on the finite-difference discretization and, since it is asymptotically exact, allows one to obtain accurate exciton energies for a wide range of QW widths and potential profiles [28]. Our method is superior to the variational approach because, in particular, it does not rely on a prescribed trial solution and, furthermore, makes it possible to calculate the excited exciton states.

In the present report, we extend the proposed numerical method [12] to the case when the discontinuities of the mass and dielectric constants are properly taken into account in the SE. These discontinuities considerably enhance the exciton binding energy for the narrow QWs. Moreover, we studied the effect of the conduction band nonparabolicity which results in further increase of the binding energy, but is rarely considered in theoretical calculations. We simulate the GaAs/Al$_x$Ga$_{1-x}$As heterostructures with a single rectangular QW and various Al concentrations, x, in order to provide a reliable reference for future experimental studies.

2. Theoretical model

We generally describe an exciton in a rectangular QW by the SE assuming the parabolic conduction and valence bands. The latter one is defined only by the diagonal terms of the Luttinger Hamiltonian, because for narrow QWs the nondiagonal terms contribute little to the energy states [34]. In other words, the heavy-hole–light-hole coupling is negligible. The derivation of the SE without material parameter mismatch is given in Ref. [12]. Here, we briefly outline the main points of the equation for the s-wave exciton states arising from the material discontinuities at the interfaces. This equation is given as [5]

$$\left(K+V_e^{pot}(z_e)+V_h^{pot}(z_h)+V_e^{self}(z_e)+V_h^{self}(z_h)+V^C(\rho,z_e,z_h)\right)\chi(z_e,z_h,\rho)=E_x\chi(z_e,z_h,\rho), \quad (1)$$

where the kinetic term K reads

$$K=-\frac{\hbar^2}{2}\frac{\partial}{\partial z_e}\frac{1}{m_{ez}(z_e)}\frac{\partial}{\partial z_e}-\frac{\hbar^2}{2}\frac{\partial}{\partial z_h}\frac{1}{m_{hz}(z_h)}\frac{\partial}{\partial z_h}-\frac{\hbar^2}{2}\frac{1}{\mu_{xy}}\left(\frac{\partial^2}{\partial\rho^2}-\frac{1}{\rho}\frac{\partial}{\partial\rho}+\frac{1}{\rho^2}\right).$$

In Eq. (1), indices e and h denote the electron and the hole, respectively. The potentials $V_{e,h}^{pot}(z_{e,h})$ are the finite rectangular confinement QW potentials. The self-interaction potentials $V_{e,h}^{self}(z_{e,h})$ come from the interaction of carriers with their images appeared in the materials with different dielectric constants [35, 36]. The Coulomb potential $V^C(\rho,z_e,z_h)$ also takes into account the dielectric constant mismatch. This mismatch gives rise to the electrostatic perturbation expansions over a small parameter $(\epsilon_w-\epsilon_b)/(\epsilon_w+\epsilon_b)$, where $\epsilon_{w,b}$ are the dielectric constants in the QW and in the barrier. These expansions, special corrections to avoid divergences at the interfaces, and results of variational calculations can be found in Refs. [29, 5, 10, 37]. In Eq. (1), the term $\mu_{xy}=m_{exy}m_{hxy}/[m_{exy}+m_{hxy}]$ is the reduced effective mass in the (x,y)-plane. We assume that the center-of-mass motion in the (x,y)-plane is separated. Other mass terms depend on the coordinate z of the confinement potential. In the case of nonparabolicity of the conduction band, the electron effective mass becomes dependent also on the confinement energy [38–41]. We use the linear nonparabolicity approximation: the electron mass linearly increases with the confinement energy.

The three-dimensional equation (1) cannot be solved analytically even for narrow QWs because of finite confinement potentials and complicated conditions at the QW interfaces. In our study, Eq. (1) is solved numerically and the binding energies for excitons in QWs of various widths and compositions of the QW layer and barriers are found.

3. Numerical method

Following Ref. [12], we numerically solved the boundary value problem for Eq. (1) and accurately obtained the exciton ground state energy, E_x. The exciton binding energy, E_b, is defined with respect to the quantum confinement energy of the electron, E_e, and the hole, E_h, in QW by the formula

$$E_b=E_e(\Delta m,\Delta\varepsilon)+E_h(\Delta m,\Delta\varepsilon)-E_x(\Delta m,\Delta\varepsilon).$$

Energies $E_e(\Delta m, \Delta\varepsilon)$ and $E_h(\Delta m, \Delta\varepsilon)$ are obtained from the solution of the corresponding one-dimensional SEs for the electron and the hole in QWs. These one-dimensional problems take into account the mass discontinuities, self-interaction potentials $V_{e,h}^{self}(z_{e,h})$ as well as the nonparabolicity of the conduction band for E_e. In the perturbation series of the electrostatic contributions to the potentials we kept the terms up to ten order of the expansion parameter.

The exponential decrease of the exciton wave function at large values of variables allows us to impose zero boundary conditions for the function χ at the boundary of some rectangular domain. For the discretization, we employ the second-order finite-difference (FD) approximation [42] of the partial derivatives in Eq. (1) on the equidistant grids over three variables. We use the central second-order FD formula for approximation of the terms with the second partial derivative over z with the discontinuity at the interface [43, 44]:

$$-\frac{\hbar^2}{2\Delta_z^2}\left(\frac{2}{m_{i-1}+m_i}\chi(z_{i-1}) - \left[\frac{2}{m_{i-1}+m_i} + \frac{2}{m_i+m_{i+1}}\right]\chi(z_i) + \frac{2}{m_i+m_{i+1}}\chi(z_{i+1})\right). \quad (2)$$

The grid steps over each variable have been taken to be the same, $\Delta = \Delta_{z_e} = \Delta_{z_h} = \Delta_\rho$. Eq. (2) defines the theoretical uncertainty of the numerical solution of order of Δ^2 as $\Delta \to 0$. The discontinuities of the rectangular confinement potential are treated in a similar way.

The nonzero solution of the homogeneous equation (1) with trivial boundary conditions can be found by a diagonalization of the matrix constructed from this equation. A small part of the matrix spectrum is obtained by the Arnoldi algorithm [45]. As a result, we have calculated the lowest eigenvalue of the matrix and the corresponding eigenvector. After the extrapolation to the limit $\Delta = 0$ [12], the accurate asymptotic results are obtained. Thus, we found the energy $E_x(\Delta m, \Delta\varepsilon)$ for various QW widths.

The calculations of the binding energies as a function of the QW widths were performed for the GaAs/$Al_xGa_{1-x}As$ model heterostructures with various concentrations of aluminum. Material and energy gap parameters used for solving the eigenvalue problem (1) are based on the data from Refs. [8, 46]. In particular, the difference of the gap energies, as a function of x, is modelled by the formula $\Delta E_g = 1087x + 438x^2$ meV. A ratio of potential barriers is taken to be $V_e/V_h = 65/35$. The Luttinger parameters used in calculations are $\gamma_1 = 6.85$, $\gamma_2 = 2.10$ for GaAs and $\gamma_1 = 3.76$, $\gamma_2 = 0.82$ for AlAs; the dielectric constants are 12.53 and 10.06, respectively. Masses and dielectric constants for the ternary alloys are obtained by a linear interpolation on x. The nonparabolicity parameters are taken to be the same for GaAs and AlAs from Ref. [41].

It is also worth noting that, despite the robustness of the theoretical model and the numerical method, the underlying envelope function approximation used for derivation of the SE becomes inappropriate for very thin QWs, namely for widths less than 2 nm that is a few monolayers of GaAs [12]. For such cases, an application of the interface short-range corrections [47] or solution of the SE in the nodal representation [48] can be considered as more reliable approaches.

4. Results of calculations

First of all, we studied the one-dimensional problems for $E_{e,h}(\Delta m \neq 0, \Delta\varepsilon \neq 0)$ with respect to $E_{e,h}$. The former defines the binding energy obtained with the material mismatches, the latter — without ones ($\Delta m = 0, \Delta\varepsilon = 0$). The differences of the studied values are shown in Figure 1. It is seen that the dominant discrepancy comes from the mass mismatch at the interfaces, whereas the difference between $E_{e,h}(\Delta m \neq 0, \Delta\varepsilon \neq 0)$ and $E_{e,h}(\Delta m \neq 0, \Delta\varepsilon = 0)$ is less than 0.06 meV, and the corresponding curves are indistinguishable in the figure. Therefore, the self-potential terms $V_{e,h}^{self}(z_{e,h})$ in such one-dimensional problems contribute little to the energy shifts. From the one-dimensional result we can conclude that the change of the confinement energy in Eq. (1) due to the mass mismatch takes place only for narrow QWs. For QWs of widths larger than 10 nm, this change is negligible. Taking into account the nonparabolicity of the conduction band

IOP Conf. Series: Journal of Physics: Conf. Series **816** (2017) 012018 doi:10.1088/1742-6596/816/1/012018

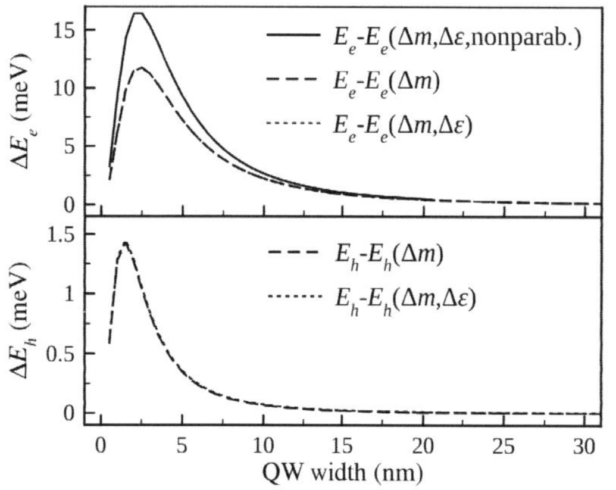

Figure 1. The difference of energies $E_{e,h}$ and $E_{e,h}(\Delta m, \Delta\varepsilon)$ for the one-dimensional problems in $GaAs/Al_{0.3}Ga_{0.7}As$ QWs. The quantity $E_{e,h}(\Delta m)$ denotes the energy $E_{e,h}(\Delta m \neq 0, \Delta\varepsilon = 0)$. The solid curve shows the energy obtained when the nonparabolicty is additionally taken into account. Dashed curves at each panel visually overlap.

results in an even greater reduction of the confinement energy for widths less than 10 nm. One can see in Figure 1 that the difference becomes more profound.

Figure 2 shows the heavy-hole exciton binding energy for three cases: without mismatches, with the mass discontinuity only, and with the mass and dielectric constant mismatches. The uncertainty of the calculated energy is 0.05 meV. One can see that the mass mismatch affects the binding energies only for QW widths of less than 10 nm. Nevertheless, this shift for narrow

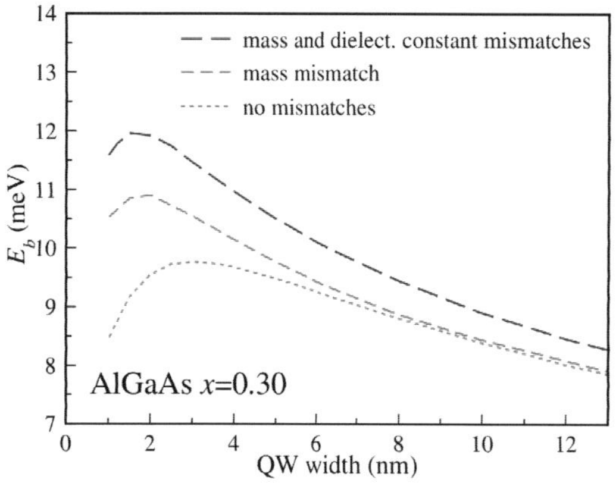

Figure 2. The binding energy of the exciton in $GaAs/Al_{0.3}Ga_{0.7}As$ QWs obtained by solving the three-dimensional problem (1). Effect of the mass mismatch, both mass and dielectric constant mismatches are shown separately.

QWs is considerable. It enhances the binding energy up to almost 25%. The dielectric constant discontinuity further increases the binding energy by about 10% for the whole studied range of QW widths. That comes from the Coulomb potential term $V^C(\rho, z_e, z_h)$ including the effect of the image charges. Our results may be compared with the calculations of Gerlach *et al.* [8]. In contrast to that paper, according to our calculations, the mass and dielectric constant mismatches both result in an increase of the binding energy. This contradiction may be caused by a truncation of the perturbative series for the electrostatic contribution after the first order term in Ref. [8].

The exciton binding energy in studied heterostructures with QW is believed to grow with increase of the Al concentration. In Figure 3 the binding energy for heavy-hole excitons in $GaAs/Al_xGa_{1-x}As$ heterostructures for various concentrations of Al, $x = 0.15, 0.3, 0.4$, is shown. The growth of the concentration increases the relative shift of the binding energy. For $x = 0.30$,

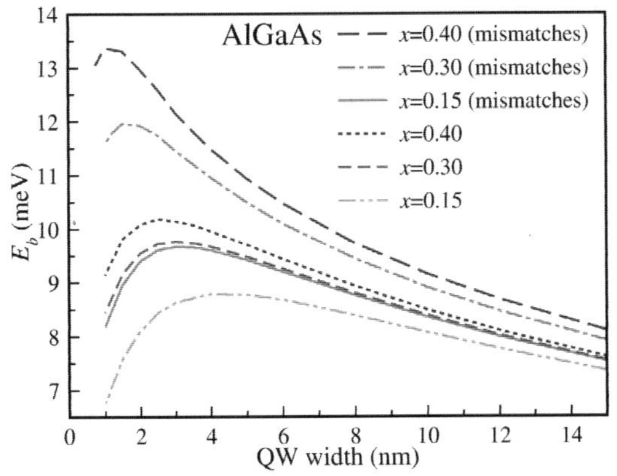

Figure 3. The binding energy of the exciton in GaAs/Al_xGa_{1-x}As QWs obtained by solving the three-dimensional problem (1) for various concentrations, x, of the alloy. The results of calculations with the material parameter mismatches as well as without ones are presented.

the relative shift of about 25% is slightly larger than that estimated preliminary in Ref. [12]. The maximum of the binding energy is achieved for smaller QW widths as the concentration increases. For $x = 0.40$, the binding energy close to 13.4 meV is reached for the width of 1 nm. We compared the obtained energy for $x = 0.40$ with the results of Shuvayev *et al.* [10], who used the same Luttinger parameters, dielectric constants and the perturbation series up to several orders of the expansion parameter, but employed a different method of calculation. Both results well agree when material parameters in the QW and in the barrier are different as well as when they are identical, see Figure 4. The calculated data of Shuvayev for GaAs/$Al_{0.4}Ga_{0.6}$As differ from our results by less than 0.2 meV. This comparison additionally confirms the correctness of our method.

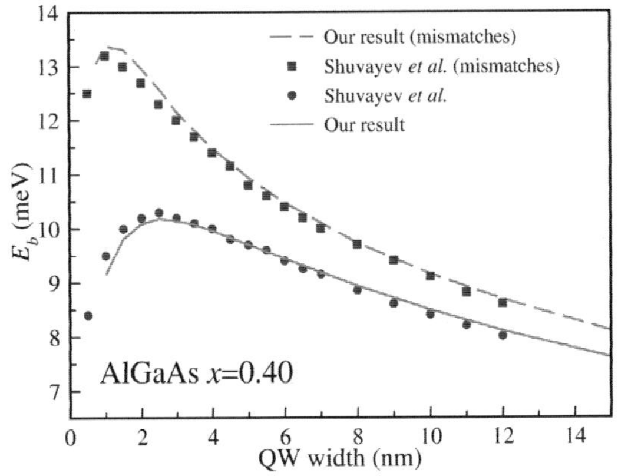

Figure 4. The comparison of the binding energy of the exciton in GaAs/$Al_{0.4}Ga_{0.6}$As QWs obtained by our method with the results of Shuvayev *et al.* [10]. The results of calculations with the material parameter mismatches as well as without ones are shown.

The nonparabolicity of the conduction band can be taken into account by changing the electron effective masses in the z-direction and in the (x, y) plane [38–41]. We use the linear approximation of the effective mass dependence on energy, E, which is valid for QW widths larger than 3 nm. One can see in Figure 5 that due to nonparabolicity of the conduction band, the binding energy increases by about 10%, that is comparable with the case of dielectric constant mismatch accounting. The binding energy grows as the QW width diminishes. The larger concentration of the alloy causes larger enhancement of the binding energy. Therefore, a proper theoretical treatment of the binding energy in narrow QWs should take into account not only discontinuities of material parameters but also the nonparabolicity of the conduction band.

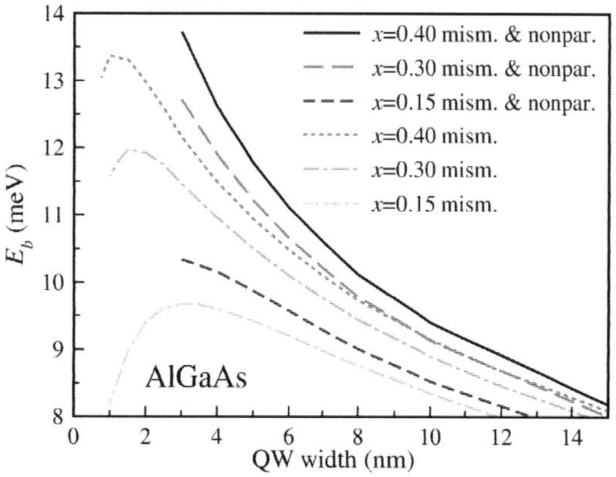

Figure 5. The binding energy of exciton in GaAs/Al_xGa_{1-x}As QWs for different Al concentrations and taking into account the mismatches of the material parameters. Effect of the nonparabolicity of the conduction band is shown separately for not very narrow QWs.

5. Conclusions

In summary, we have calculated the exciton binding energies for narrow GaAs/Al_xGa_{1-x}As QWs taking into account the discontinuities of the material parameters at the QW interfaces and the nonparabolicity of the conduction band. The results of calculations show that the mass and dielectric constant mismatches considerably enhance the binding energies for narrow QWs. The exciton binding energy increases up to 25% and 10%, respectively. Moreover, the nonparabolicity of the conduction band causes further increase of the binding energy, comparable with the case of dielectric constant mismatch. We hope that our results will provide a reliable reference for ongoing and future experimental studies.

Acknowledgments

Financial support from SPbU (grants No. 11.38.213.2014 and No. 11.38.241.2015), RFBR (grants No. 16-02-00245 and No. 17-01-00221) and DFG in the frame of Project ICRC TRR 160 is acknowledged. The calculations were carried out using the facilities of the SPbU Resource Center "Computational Center of SPbU".

[1] Keldysh L V 1979 *JETP Lett.* **29** 658 [*Pis'ma Zh. Eksp. Teor. Fiz.* **29** 716]
[2] Miller R C, Kleinman D A, Tsang W T and Gossard A C 1981 *Phys. Rev. B* **24** 1134
[3] Bastard G, Mendez E E, Chang L L and Esaki L 1982 *Phys. Rev. B* **26** 1974
[4] Greene R L, Bajaj K K and Phelps D E 1984 *Phys. Rev. B* **29** 1807
[5] Thoai D B T, Zimmermann R, Grundmann M and Bimberg D 1990 *Phys. Rev. B* **42** 5906
[6] Andreani L C and Pasquarello A 1990 *Phys. Rev. B* **42** 8928
[7] Petrov A G and Shik A Ya 1993 *Fiz. Tech. Polupr.* **27** 1047
[8] Gerlach B, Wüsthoff J, Dzero M O and Smondyrev M A 1998 *Phys. Rev. B* **58** 10568
[9] Ivchenko E L 2005 *Optical spectroscopy of semiconductor nanostructures* (Harrow: Alpha Science Int.)
[10] Shuvayev V A, Deych L I, Ponomarev I V and Lisyansky A A 2006 *Superlatt. Microstr.* **40** 77
[11] Sivalertporn K, Mouchliadis L, Ivanov A L, Philp R and Muljarov E A 2012 *Phys. Rev. B* **85** 045207
[12] Khramtsov E S, Belov P A, Grigoryev P S *et al.* 2016 *J. Appl. Phys.* **119** 184301
[13] Voliotis V, Grousson R, Lavallard P and Planel R 1995 *Phys. Rev. B* **52** 10725
[14] Evstigneev S V, Imamov R M, Lomov A A *et al.* 2000 *Semiconductors* **34** 693
[15] Orlov L K, Ivina N L, Romanov Yu A and Rubtsova R A 2000 *Phys. Solid State* **42** 548 [*Fiz. Tverd. Tela* **42** 537]
[16] Evtikhiev V P, Konstantinov O V, Matveentsev A V and Romanov A E 2002 *Semiconductors* **36** 74
[17] Astakhov G V, Yakovlev D R, Kochereshko V P *et al.* 2002 *Phys. Rev. B* **65** 165335

[18] Toropov A A, Kaibyshev V K, Terent'ev Y V *et al.* 2011 *Semiconductors* **45** 208
[19] Agekyan V F, Borisov E V, Serov A Y *et al.* 2015 *Phys. Solid State* **57** 1831 [*Fiz. Tverd. Tela* **57** 1784]
[20] Muraki K, Fukatsu S, Shiraki Y and Ito R 1992 *Appl. Phys. Lett.* **61** 557
[21] Avakyants L P, Bokov P Yu, Galiev G B *et al.* 2015 *Semiconductors* **49** 1202
[22] Seisyan R P 2012 *Semicond. Sci. Technol.* **27** 053001
[23] Poltavtsev S V and Stroganov B V 2010 *Phys. Solid State* **52** 1899 [*Fiz. Tverd. Tela* **52** 1769]
[24] Poltavtsev S V, Efimov Yu P, Dolgikh Yu K *et al.* 2014 *Solid State Comm.* **199** 47
[25] Loginov D K, Trifonov A V and Ignatiev I V 2014 *Phys. Rev. B* **90** 075306
[26] Trifonov A V, Korotan S N, Kurdyubov A S *et al.* 2015 *Phys. Rev. B* **91** 115307
[27] Grigoryev P S, Yugov O A, Eliseev S A *et al.* 2016 *Phys. Rev. B* **93** 205425
[28] Grigoryev P S, Kurdyubov A S, Kuznetsova M S *et al.* 2016 *Superlatt. Microstr.* **97** 452
[29] Kumagai M and Takagahara T 1989 *Phys. Rev. B* **40** 12359
[30] Luttinger J M 1956 *Phys. Rev.* **102** 1030
[31] Peter A J and Lee C W 2012 *Phys. Scr.* **85** 015704
[32] Lee Y C, Mei W N and Liu K C 1982 *J. Phys. C* **15** L469
[33] Jiang T F 1984 *Solid State Comm.* **50** 589
[34] Baldereschi A and Lipari N O 1971 *Phys. Rev. B* **2** 439
[35] Grigoryan G B, Rodina A V and Efros Al L 1990 *Sov. Phys. Solid State* **32** 2037 [*Fiz. Tverd. Tela* **32** 3512]
[36] Wendler L and Hartwig B 1991 *J. Phys.: Condens. Matter* **3** 9907
[37] Onodera Ch, Yoshida M, Shoji T and Taguchi T 2009 *Jpn. J. Appl. Phys.* **48** 121101
[38] Braun M and Rössler U 1985 *J. Phys. C* **18** 3365
[39] Ekenberg U 1989 *Phys. Rev. B* **40** 7714
[40] Milanović V, Radovanović J and Ramović S 2009 *Phys. Lett. A* **373** 3071
[41] Niculescu E C 2013 *J. Lumin.* **143** 254
[42] Samarskii A A 1989 *The theory of difference schemes* (Moscow: Nauka)
[43] Demikhovskii V Ya and Savinskii S S 1992 *Sov. Phys. Solid State* **34** 1276 [*Fiz. Tverd. Tela* **34** 2382]
[44] Glutsch S 2004 *Excitons in low-dimensional semiconductors* (Berlin: Springer)
[45] Sorensen D C, Lehoucq R B and Vu P 1995 *ARPACK: an implementation of the Implicitly Restarted Arnoldi iteration that computes some of the eigenvalues and eigenvectors of a large sparse matrix*
[46] Vurgaftman I, Meyer J R and Ram-Mohan L R 2001 *J. Appl. Phys.* **89** 5815
[47] Glinskii G F, Lakisov V A, Dolmatov A G and Kravchenko K O 2000 *Nanotechnology* **11** 233
[48] Glinskii G F and Mironova M S 2014 *Semiconductors* **48** 1324

Optical properties of GaN/AlGaN nanostructures in the terahertz frequency range

A I Galimov[1], V A Shalygin[1], M D Moldavskaya[1], G A Melentev[1], M Ya Vinnichenko[1], A A Artemyev[1], D A Firsov[1], L E Vorobjev[1], A V Sakharov[2], E E Zavarin[2], E Yu Lundina[2] and W V Lundin[2]

[1] Department of Physics of Semiconductors & Nanoelectronics, Peter the Great St.Petersburg Polytechnic University, Polytechnicheskaya 29, St.Petersburg 195251, Russia
[2] Ioffe Institute, 26 Polytechnicheskaya str., St. Petersburg 194021, Russia

E-mail: shalygin@rphf.spbstu.ru

Abstract. Optical transmission and reflection for a single GaN/AlGaN heterojunction grown on sapphire have been investigated with high and low spectral resolution in the spectral range of 8–47 meV. For comparison, the same spectra have been measured for the sapphire substrate. Refractive index dispersion has been determined for sapphire from the spectra measured with high resolution. Then the data on 2D electron absorption in the GaN/AlGaN heterojunction were obtained from the low resolution spectra. The spectra of terahertz emission from the GaN/AlGaN heterojunction under 2D electron heating in strong electric field have been measured for the first time.

1. Motivation

In the last decade, intensive studies of GaN based structures are conducted with the aim of application in terahertz (THz) photonics [1–6]. Development of THz emitters and detectors based on GaN/AlGaN nanostructures calls for reliable experimental data on their optical properties in the terahertz frequency range. In the present work, we first carried out a series of far infrared optical investigations of a single GaN/AlGaN heterojunction grown on a sapphire substrate including measurements of THz reflection, transmission and emission spectra. The THz reflection and transmission were studied in equilibrium conditions at room and cryogenic temperatures. The THz emission spectra were studied under conditions of 2D electron heating by lateral electric field.

2. Samples

The heterostructure was grown by metal-organic vapor phase epitaxy (MOVPE) on a (0001) sapphire substrate with a thickness of 430 μm. The heterostructure consists of a 2.8 μm GaN buffer layer, a 1 nm AlN interface layer, a 31.7 nm $Al_{0.214}Ga_{0.786}N$ barrier layer and a 4.4 nm GaN cap layer. No intentional doping was performed during MOVPE growth. Two dimensional electron gas (2DEG) is formed at the GaN/AlN interface due to piezoelectric and spontaneous polarization induced effects in the heterostructure [7]. In accordance with the Hall effect and conductivity measurements, 2DEG with a concentration N_e of 1.0×10^{13} cm^{-2} and a low-field mobility of 1500 and 8000 cm^2/V·s at temperatures of 300 and 77 K, respectively, was formed in the structure under investigation. For the optical reflection and transmission measurements, the samples with an area of about 1cm^2 were cut from a wafer. For investigation of the THz emission, two Ti/Ni/Al/Au electrical contacts of 7 mm length with a distance of 3 mm were fabricated on the sample surface. By applying a voltage to the contacts, we produced 2D electron heating in the GaN/AlGaN heterojunction.

3. Results and discussion

The optical transmission spectra of the GaN/AlGaN nanostructures were experimentally studied in the THz spectral range using a Fourier spectrometer Bruker Vertex 80v operating in a rapid-scan mode. The measurement pressure was about 4 mbar. A globar was used as a source of THz radiation and a Mylar beam-splitter was applied. The intensity of THz radiation was measured by a pyroelectric detector. Measurements were carried out at room and cryogenic (77 K) temperatures. The experimental transmission spectra for the case of normal incidence of radiation (incidence angle $\theta = 0°$, angular aperture of the beam $\Delta\theta = 16°$) are presented in figure 1.

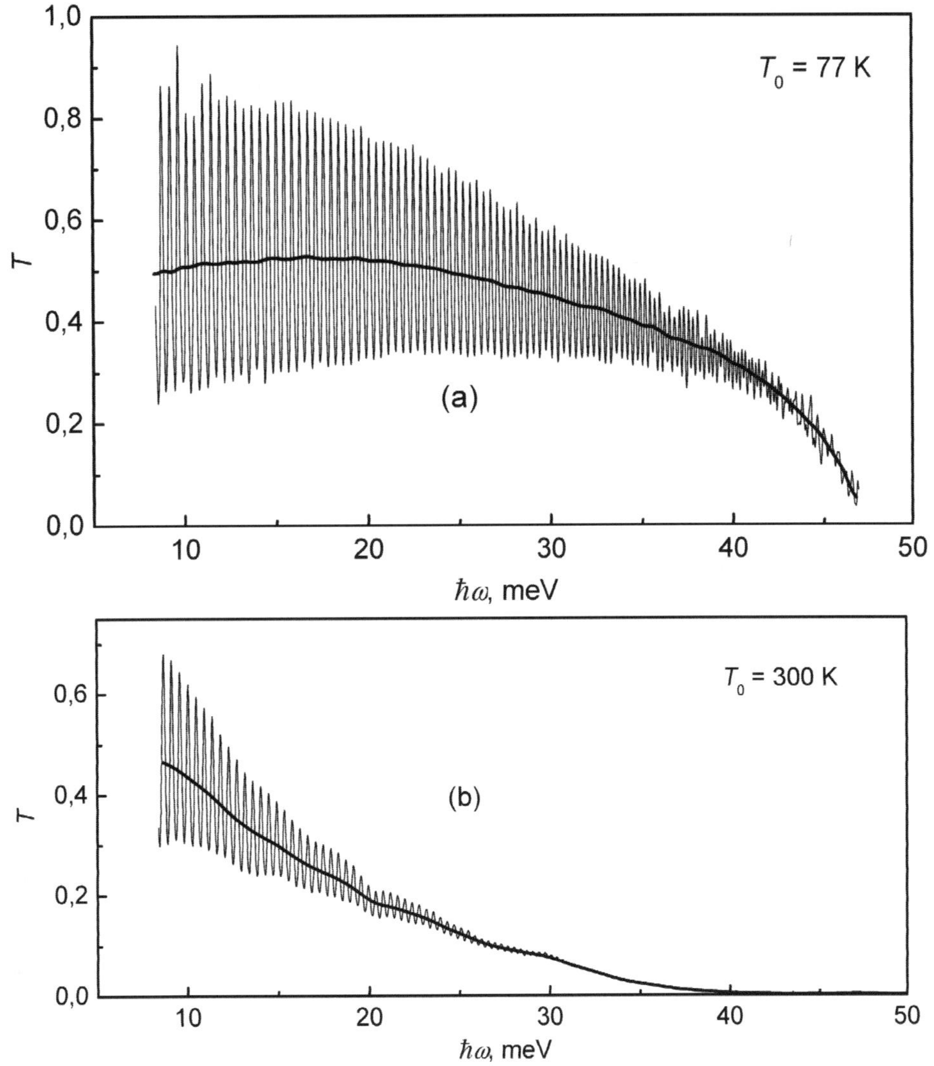

Figure 1. Experimental spectra of optical transmission of the GaN/AlGaN nanostructure at temperatures of 77 K (a) and 300 K (b). Thin and thick lines correspond to the spectral resolutions of 0.06 and 1 meV, respectively.

IOP Conf. Series: Journal of Physics: Conf. Series **816** (2017) 012019 doi:10.1088/1742-6596/816/1/012019

The transmission spectra of the GaN/AlGaN nanostructure measured at high spectral resolution (0.06 meV) demonstrate oscillations caused by the radiation interference under multiple reflections on the outer facets of the structure. In the case of low enough resolution, when the radiation coherence length becomes less than the structure thickness, these oscillations disappear (see thick lines in figure 1).

Similar transmission spectra were measured also for the sapphire substrate before nanostructure growth. Analyzing interference oscillations in the spectra, we determined the refractive index for the sapphire with the help of the following formula:

$$n_S = \frac{\pi\hbar c}{\hbar(\Delta\omega)_{2\pi} d_S}, \tag{1}$$

where $\hbar(\Delta\omega)_{2\pi}$ is a spectral period of the transmission curve (the period monotonically decreases with the photon energy $\hbar\omega$), d_S is the thickness of the sapphire substrate, and c is the light velocity in vacuum. The spectral dependencies of the refractive index n_S obtained for temperatures of 77 K and 300 K are presented in figure 2. The both dependencies coincide within the limits of experimental accuracy. A monotonic increase of the refractive index n_S in the spectral range of 8–42 meV is associated with the low-frequency phonon mode [8]. These data give us an opportunity to find the reflection coefficient r_S for the vacuum/sapphire and sapphire/vacuum interfaces in accordance with the relationship:

$$r_S = \frac{(n_S - 1)^2}{(n_S + 1)^2}, \tag{2}$$

which is true for the case of normal incidence of radiation in approximation $n_S >> k_S$ (here k_S is the extinction coefficient for sapphire).

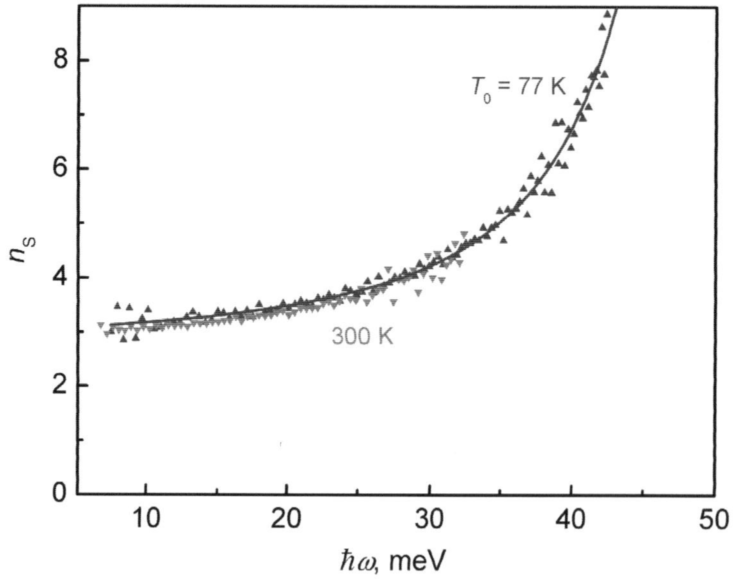

Figure 2. Spectral dependencies of the refractive index n_S of the sapphire substrate at two temperatures T_0. Triangles show experimental points, the solid line represents the Lorentzian fit to the experimental data at T_0 = 77 K.

The value of the optical transmission of the sapphire substrate measured at low spectral resolution (when interference oscillations are completely smoothed) can be described by the formula for incoherent radiation taking into account multiple reflections:

$$T_S = \frac{(1-r_S)^2 V_S}{1-r_S^2 V_S^2}, \tag{3}$$

where $V_S = \exp(-\alpha_S d_S)$ and α_S is the absorption coefficient which is related to the extinction coefficient by the equation:

$$\alpha_S = 2\frac{\omega}{c}k_S. \tag{4}$$

Analyzing the low-resolution transmission curve for the sapphire substrate we determined the spectral dependence of the optical absorption in sapphire with the help of equations (3) and (2). Results of this analysis for room temperature are shown in figure 3 by the dotted red curve representing the values of $\alpha_S d_S = -\ln V_S$. The same analysis for the temperature $T_0 = 77$ K has shown that, in this case, the sapphire substrate has negligible optical absorption ($\alpha_S d_S < 0.03$) in the spectral range of 8–42 meV (it is not plotted in figure 3).

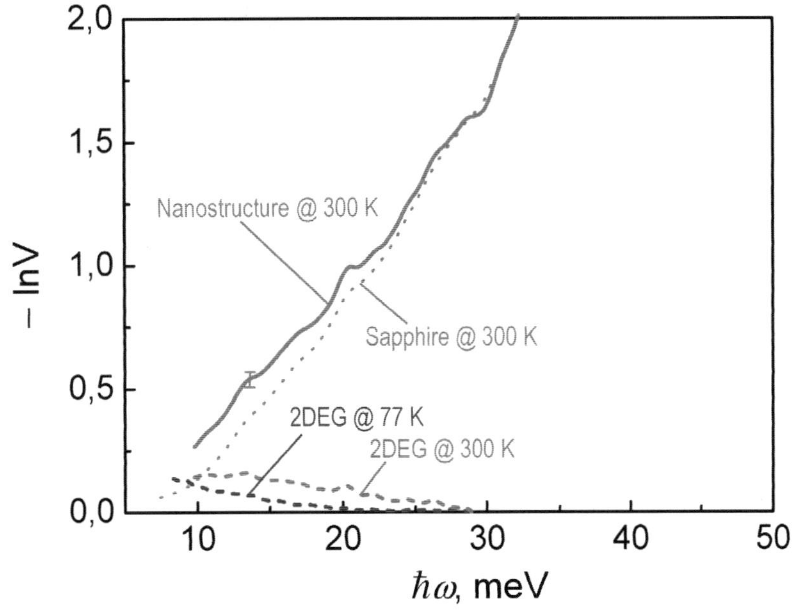

Figure 3. Optical absorption spectra for the GaN/AlGaN nanostructure (solid red line) and sapphire substrate (dotted red line) at room temperature. Contribution of 2D electrons to the absorption is shown by the dashed red line (at $T_0 = 300$ K) and the dashed blue line (at $T_0 = 77$ K).

The main goal of the optical transmission study of the GaN/AlGaN nanostructure was to determine 2D electron contribution to the optical absorption. For this purpose, the transmission spectra with totally smoothed interference effects are more appropriate (thick curves in figure 1). In the first approximation, these spectra can be described by the formula for incoherent radiation taking into account multiple reflections:

$$T_N = \frac{(1-r_S)(1-r_{GaN})V_S V_{2DEG}}{1 - r_S r_{GaN} V_S^2 V_{2DEG}^2}, \quad (5)$$

where r_{GaN} is the reflection coefficient for the interface vacuum/GaN and V_{2DEG} is the optical transmittance of 2D electron gas at the GaN/AlGaN interface (we assume the absence of THz absorption in intentionally undoped GaN, AlN and AlGaN layers of the nanostructure and negligible reflection at the GaN/sapphire interface).

Let us first analyse the experimental results on the nanostructure transmission for room temperature (thick curve in figure 1 (b)). Using equation (5) and previously determined data on the optical properties of sapphire, namely $n_S(\hbar\omega)$, $r_S(\hbar\omega)$ and $V_S(\hbar\omega)$, we extracted the experimental dependence $V_{2DEG}(\hbar\omega)$. The esulting data for room-temperature 2DEG absorption are presented in figure 3 by the dashed red line. One can see that 2D electrons give a significant contribution to the total nanostructure absorption at photon energies less than 30 meV. The magnitude of this contribution monotonically rises with photon energy decrease.

Then we analyzed nanostructure transmission data for T_0 = 77 K (thick curve in figure 1 (a)) using previously obtained data on $n_S(\hbar\omega)$ and $r_S(\hbar\omega)$ for this temperature. We took into account that $V_S(77\ \text{K}) \approx 1$ in the spectral range $\hbar\omega < 30$ meV. The resulting data on 2D electron absorption at T_0 = 77 K are presented in figure 3 by the dashed blue line. One can see that 2D electron absorption decreases with temperature decrease from 300 to 77 K. In accordance with the Drude model, it can be explained by the increase of 2D electron mobility under sample cooling. Nevertheless, 2D absorption is rather significant for photon energies $\hbar\omega < 20$ meV.

By means of the same experimental technique, the optical reflection of the nanostructure was studied as well. Reflectivity measurements were carried out under oblique incidence of a radiation beam at an angle $\theta = 11°$. An aluminum mirror was used as a reflectivity reference. The reflectivity spectrum of the nanostructure at room temperature is shown in figure 4. For comparison, the spectrum of the sapphire substrate is also presented.

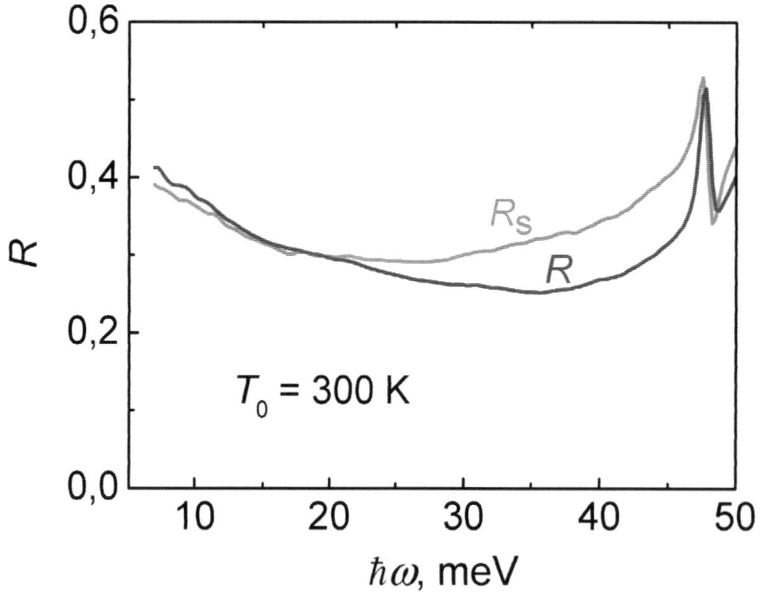

Figure 4. Optical reflectivity spectra of the GaN/AlGaN nanostructure (R) and sapphire substrate (R_S) at room temperature.

Both spectral curves demonstrate a sharp reflection maximum at the photon energy of 47.5 meV which is a fingerprint of the low-frequency phonon resonance in sapphire [8]. At lower photon energies, the both spectra represent non-monotonic dependencies with a minimum. Increase of the reflectivity at the lowest photon energies is caused by multiple reflections inside the sample which arise when the sample becomes partially transparent. It is clearly seen from comparison of figure 4 and figure 1. It should be emphasized that an arbitrary increase of the reflectivity at the long-wavelength edge of the spectrum for the nanostructure is significantly larger than for the sapphire substrate. This feature indicates a substantial contribution of 2D electrons to the reflectivity of the nanostructure.

Applying lateral electric field to the GaN/AlGaN nanostructure, we heated 2DEG and studied emission of THz radiation. The sample under investigation was placed into an optical closed cycle cryostat and cooled to the temperature of about 9 K. The voltage pulses with a duration of about 1 μs and a repetition frequency of 77 Hz were applied via the sample contacts. The THz radiation was collected in the direction perpendicular to the sample surface within the angular aperture of 16°. The emission spectra were investigated using a Fourier spectrometer operating in a step-scan mode. A liquid-helium-cooled silicon bolometer was used as a detector. The signal from the detector was measured at the frequency of 77 Hz by means of a lock-in amplifier.

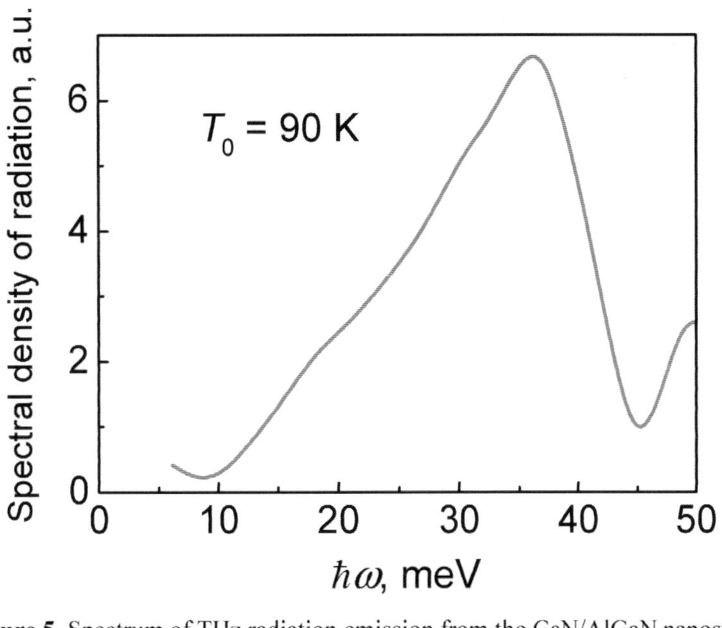

Figure 5. Spectrum of THz radiation emission from the GaN/AlGaN nanostructure under 2D electron heating in electric field. Electric field E = 1000 V/cm. The effective temperature of hot electrons $T_e \sim 280$ K.

Before the emission study, a study of the current-voltage characteristics was performed. Electrical measurements were carried out at different sample temperatures (T_0 = 4.2–300 K) in a wide range of electric fields (E = 30–3000 V/cm). Single voltage pulses of different duration (1–4 μs) were applied. The field dependencies of the 2D electron mobility at different values of T_0 have been determined. On the basis of these dependencies, the effective temperature of hot 2D electrons T_e has been found as a function of two parameters: T_0 and E (by the method described in [5]). The THz emission was investigated under rather high heating electric field (E = 1000 V/cm). It should be noted that due to periodic sample excitation, the stationary value of the time-averaged sample temperature T_0 was about 90 K (much higher than the initial value of 9 K). During the voltage pulse we observed 2D electron heating up to $T_e \sim 280$ K. The amplitude of the electron temperature modulation $\Delta T_e \sim 190$ K was determined with the help of the above mentioned dependence $T_e(T_0, E)$. It was found that the temperature of the sapphire substrate is also slightly modulated with a frequency of 77 Hz.

From the measured spectrum of the photoresponse signal, we restored the spectral density of THz radiation from the nanostructure as a function of photon energy (figure 5). The presented curve is obtained taking into account the spectra of the bolometer detectivity and optical path transmission. As far as we know, this is the first experimental result on the spectral dependence of THz emission from a GaN/AlGaN nanostructure under 2D electron heating. The preliminary analysis of the THz emission spectrum shows that in addition to THz emission by hot 2D electrons which we are interested in, there is some contribution caused by a weak temperature modulation of the sapphire substrate.

4. Conclusions

We have investigated optical properties of the GaN/AlGaN nanostructure with 2D electron gas. The measurements of the transmittivity, reflectivity and spectral density of THz radiation emission have been performed in the spectral range of 8–47 meV. For comparison, the transmittivity and reflectivity spectra have been measured for the sapphire substrate as well. The contribution of 2D electrons to the optical absorption has been determined. At a temperature of 77 K, it is significant in the spectral range $\hbar\omega < 20$ meV. At room temperature, optical absorption of 2D electrons becomes observable also at higher photon energies ($\hbar\omega \sim 30$ meV). THz radiation emission has been investigated under conditions of 2DEG heating by lateral electric field. The spectrum of THz radiation emission from the GaN/AlGaN nanostructure under 2D electron heating has been experimentally determined for the first time.

The results of our studies can be applied for the development of portable sources of THz radiation operating under electric pumping, in particular in the field of THz plasmonics.

Acknowledgements
The authors are grateful to V.Yu. Panevin and I.S. Makhov for the help in the experiments. This work was supported by the Russian Foundation for Basic Research (grants 16-32-60085, 16-02-00863, 17-02-01191) and the Ministry of Education and Science of the Russian Federation (state assignment).

References

[1] Hao Y, Yang L-A and Zhang J-C 2008 *Terahertz Science and Technology* **1** 51
[2] Terashima W and Hirayama H 2011 *Phys. Stat. Sol. (c)* **8** 2302
[3] El Fatimy A, Dyakonova N, Meziani Y, Otsuji T, Knap W, Vandenbrouk S, Madjour K, Théron D, Gaquiere C, Poisson M A, Delage S, Prystawko P and Skierbiszewski C 2010 *J. Appl. Phys.* **107** 024504
[4] Starikov E, Shiktorov P, Gružinskis V, Varani L, Palermo C, Millithaler J-F, and Reggiani L 2008 *J. Phys.: Condens. Matter* **20** 384209
[5] Shalygin V A, Vorobjev L E, Firsov D A, Sofronov A N, Melentyev G A, Lundin W V, Nikolaev A E, Sakharov A V and Tsatsulnikov A F 2011 *J. Appl. Phys.* **109** 073108
[6] Melentev G A, Shalygin V A, Vorobjev L E, Panevin V Yu, Firsov D A, Riuttanen L, Suihkonen S, Korotyeyev V V, Lyaschuk Yu M, Kochelap V A and Poroshin V N 2016 *J. Appl. Phys.* **119** 093104
[7] Ambacher O, *et al* 2000 *J. Appl. Phys.* **87** 334
[8] Barker A S Jr 1963 *Phys. Rev.* **132** 1474

Sn influence on MBE growth of GeSiSn/Si MQW

A R Tuktamyshev[1], V A Timofeev[1], A I Nikiforov[1,2], V I Mashanov[1], A K Gutakovskii[1] and N A Baydakova[3]

[1] Rzhanov Institute of Semiconductor Physics SB RAS, 630090 Novosibirsk, Russia
[2] National Research Tomsk State University, 634050 Tomsk, Russia
[3] Institute for Physics of Microstructures RAS, 607680 Nizhny Novgorod, Russia

E-mail: tuktamyshev@isp.nsc.ru

Abstract. Temperature and composition dependencies of the critical thickness of transition from two-dimensional to three-dimensional growth for GeSiSn films on Si(100) with a lattice mismatch of 1 – 5% were experimentally determined. To understand the Sn influence on growth of SiGeSn/Si multi-quantum wells, the phase diagram of surface superstructures during the growth of pure Sn on Si(100) was created. A possibility of synthesizing multilayer structures by molecular beam epitaxy was shown, and the crystal lattice constants were determined using high-resolution transmission electron microscopy. We obtained GeSiSn/Si MQW structures which demonstrated photoluminescence for the Sn content in GeSiSn layers of up to 6%.

1. Introduction

The interest in GeSiSn materials is associated with the ability to create a direct bandgap semiconductor, as a double compound GeSn [1], and a ternary compound GeSiSn [2]. These compounds can be grown directly on a silicon wafer, providing the ability to create silicon photonics and optoelectronics devices operating in the IR spectrum. In recent years, a lot of articles on the creation of emitting devices and photodetectors, based on GeSiSn materials, have been published [3-6]. The structures, which were obtained in these studies, were grown on a silicon wafer using a 1 μm thick Ge buffer layer.

In the present work, we suggest to use pseudomorphic elastic-strained GeSiSn films grown directly on Si rather than relaxed layers. The principal advantage of pseudomorphic films against thick relaxed layers is that they are free of dislocations and coherent to the substrate. GeSiSn films are more thermostable than GeSn [7], their lattice constant and bandgap can be individually controlled as dependent on the composition.

In this work, GeSiSn multi-quantum wells (MQW) structures were grown directly on Si(100) substrates by molecular-beam epitaxy (MBE). The data about the initial growth stages of ternary compounds Ge-Si-Sn on Si(100) have been earlier obtained [8]. Using the reflection high-energy electron diffraction (RHEED) method [9], the kinetic diagrams of growth of $Ge_{1-x-y}Si_xSn_y$ on Si(100) in a wide temperature range (150 – 500°C) and at different lattice mismatch (1 – 5%) between a GeSiSn film and a Si substrate were determined (figure 1). Based on the temperature dependencies of 2D-3D transition, the growth temperature and thickness of GeSiSn layers were chosen to achieve a pseudomorphic growth. The GeSiSn layer thickness was 2 – 3 nm and the growth temperature of GeSiSn layers was 150 – 200°C. GeSiSn layers act as quantum wells which are covered by a Si layer

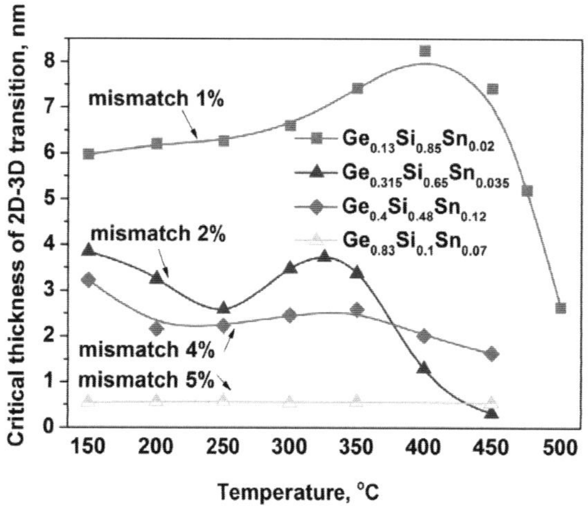

Figure 1. Kinetic diagram of GeSiSn film growth at different lattice mismatch. Generally, an increase in the critical thickness of 2D-3D transition is observed with a decrease of the lattice mismatch. Sn content is varied from 2 to 12%.

at higher growth temperatures (400 – 500°C) for smoothing the surface and also to create a potential barrier. The number of periods of a GeSiSn/Si heterostructure ranged up to 15 repetitions.

2. Results and discussion

Using the *in situ* RHEED technique during the growth of GeSiSn/Si MQW, a variety of superstructures such as (2x1), (2xN), (4x4), (6x4), (5x1) were observed on the growth surface. Similar surface structures were observed during the growth of pure Sn on Si(100) [10]. Thus, the appearance of various surface structures demonstrates the Sn surface segregation.

To prove the Sn segregation and to study the Sn influence on growth of GeSiSn/Si MQW, the phase diagram of surface superstructures during the growth of pure Sn on Si(100) was created (figure 2). Sn films were grown at a temperature of 100°C and then structures were annealed at a higher temperature up to 800°C. The Sn layer thickness reached 1.7 monolayer (MN). Different

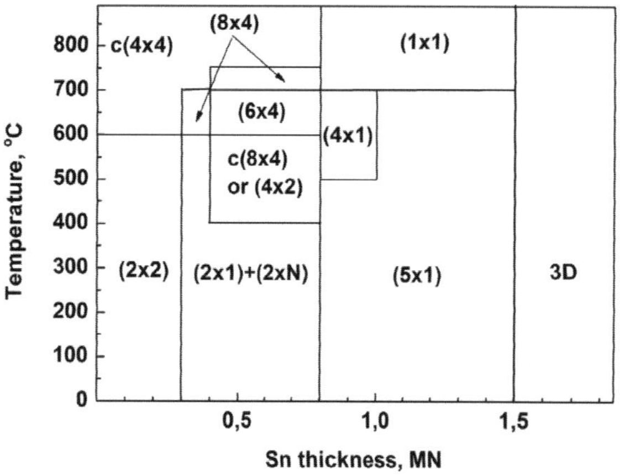

Figure 2. Phase diagram of surface superstructures during the growth of pure Sn on Si(100).

superstructures were observed at various annealing temperatures and growth thicknesses. It is clear that the appearance of (5x1) superstructure indicates a higher Sn content on the surface. Therefore, the appearance and disappearance of various surface reconstructions during the growth of SiGeSn/Si MQW show the surface quality and can assist in selecting the optimal growth parameters such as the growth temperature, a composition of the ternary compound and the number of periods of MQW structure.

High-resolution transmission electron microscopy (TEM) was used to characterize the crystal structure of the obtained samples (figure 3(a)). Inspection of the TEM data allows to conclude that the structures are free of threading dislocations or their concentration is negligible. The crystal lattice distortion and deformation fields were visualized and measured in TEM images of the multilayer GeSiSn/Si structure using the method of geometric phase [11] (figure 3(b)). The experimentally measured interplanar spacing (d_{002} = 0.29 nm, d_{111} = 0.32 nm, d_{220} = 0.192 nm), with the sample presented in figure 3 (c), are to show that the tetragonal lattice with constants a = 0.543 nm and c = 0.58 nm is characteristic of GeSiSn layers. This implies that GeSiSn layers feature either a non-strained strained tetragonal or a cubic crystal lattice that is elastically deformed and coherently conjugated conjugated with the silicon crystal lattice at the heteroboundary, i.e. a pseudomorphic GeSiSn film is formed.

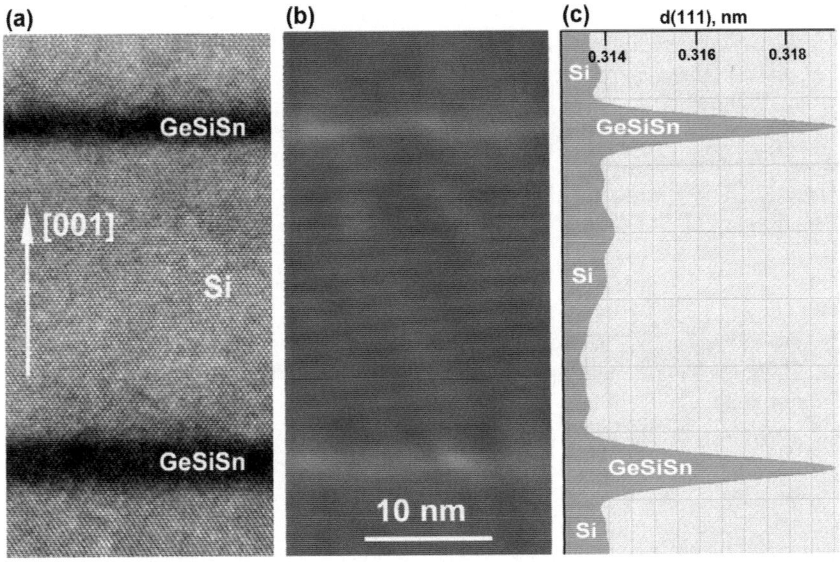

Figure 3. (a) Fragment of a high-resolution multilayer structure including Si-covered GeSiSn layers; (b) the map of interplanar (111) distances; (c) the profile of interplanar (111) distances versus the thickness of the film.

The optical properties of multilayer structures were studied using photoluminescence (PL). The PL signal was excited by a Nd:YAG laser (532 nm). Figure 4 shows the spectra of PL at a temperature of 77 K. PL is observed in the range of 0.6 eV to 0.85 eV. With increasing Sn concentration in the composition of a ternary compound GeSiSn, the maximum of PL intensity decreases from 0.77 eV to 0.65 eV for $Ge_{0.315}Si_{0.65}Sn_{0.035}$ and $Ge_{0.4}Si_{0.54}Sn_{0.06}$, respectively. With decreasing growth temperature, and by increasing the thickness of the GeSiSn layer, PL signal reduces, which may be caused by an increase in point defects in the crystal structure. Progress to wavelengths longer than 2 μm requires an increase in the Sn content in GeSiSn layers higher than 10%.

3. Conclusion

Regularities of the MBE formation of multilayer structures on quantum wells comprising pseudomorphic GeSiSn layers without relaxed buffer layers but creating the structures directly on Si were studied for the first time. The obtained TEM data proved the crystal perfection of the samples under study. To investigate the Sn influence on growth of SiGeSn/Si MQW, the phase diagram of surface superstructures during the growth of pure Sn on Si(100) was created. The appearance and disappearance of various surface reconstructions during the growth of SiGeSn/Si MQW show the surface quality and can assist in selecting the optimal growth parameters. Nanostructures based on GeSiSn layers have demonstrated the photoluminescence at 0.6-0.85 eV. With increasing Sn content in the GeSiSn composition, the maximum of PL signal appears in the long-wave IR region.

Acknowledgments

The work is supported by the Russian Foundations of Basic Research (Projects 16-32-00039, 14-29-07153, 16-32-60005, 16-29-03292) and by The Tomsk State University Academician D.I. Mendeleev Foundation Program (Research Grant no.8.2.10.2015). HRTEM studies were performed using the equipment of CCU "Nanostructures". The part of the work carried out with the support of RSCF (Project no.14- 22-00143).

References

[1] Kostecki K, Oehme M, Koerner R, Widmann D, Gollhofer M, Bechler S, Mussler G, Buca D, Kasper E and Schulze J 2014 *ECS Trans.* **64** 811
[2] Gallagher J D, Xu C, Jiang L, Kouvetakis J and Menendez J 2013 *Appl. Phys. Lett.* **103** 202104
[3] Schwartz B, Saring P, Arguirov T, Oehme M, Kostecki K, Kasper E, Schulze J and Kittler M 2015 *Solid State Phenomena* **242** 361
[4] Fischer I A, Wendav T, Augel L, Jitpakdeebodin S, Oliveira F, Benedetti A, Stefanov S, Chiussi S, Capellini G, Busch K and Schulze J 2015 *Optics Express* **23** 25048
[5] Wirths S, Geiger R, von den Driesch N, Mussler G, Stoica T, Mantl S, Ikonic Z, Luysberg M, Chiussi S, Hartmann J M, Sigg H, Faist J, Buca D and Grutzmacher D 2015 *Nature Photonics* **9** 88
[6] Stange D, von den Driesch N, Rainko D, Schulte-Braucks C, Wirths S, Mussler G, Tiedemann A T, Stoica T, Hartmann J M, Ikonic Z, Mantl S, Grutzmacher D and Buca D 2016 *Optics Express* **24** 1358
[7] Xie J, Tolle J, D'Costa V, Chizmeshya A, Menendez J, Kouvetakis J 2009 *Appl.Phys. Lett.* **95** 181909
[8] Tuktamyshev A R, Mashanov V I, Timofeev V A, Nikiforov A I and Teys S A 2015 *Semiconductors* **49** 1582
[9] Nikiforov A I, Cherepanov V A and Pchelyakov O P 2002 *Mater. Sci. Eng. B* **89** 180
[10] Ueda K and Kinoshita K 1984 *Surface Science* **145** 261
[11] Gutakovskii A K, Chuvilin A L and Song S A 2007 *Bull.Russ.Acad.Sci.Phys.* **71** 1426

Photoluminescence and infrared spectroscopy studies of thin $Ge_{1-x}Sn_x$ heterostructures

V S Krivobok[1], Yu A Aleshchenko[1,2], V M Martovitsky[1], A V Klekovkin[1], M A Bazalevsky[1] and A B Mehiya[3]

[1] *P.N. Lebedev Physical Institute, Russian Academy of Sciences, Moscow, 119991 Russia*

[2] National Research Nuclear University MEPhI, Moscow, 115409 Russia

[3] *Moscow Institute of Physics and Technology, Dolgoprudny, 141700, Russia*

e-mail: aklekovkinbox@gmail.com

Abstract. Elastically strained metastable GeSn layers with mole fraction of tin up to 0.15 was grown on (001) Si substrates with different misorientation. Photoluminescence spectra (PL) at room temperature and infrared (IR) spectroscopy spectra at helium temperature were measured. The direct edge of intrinsic absorption in the region of 0.71-0.72 eV was observed in grown structures with a tin content of about 12-13%.

1. Introduction

Alloys in the Si–Ge–Sn material system hold promise for fabrication of monolithic optoelectronic systems based entirely on Group-IVA and working in the near- and mid-infrared regions. Previously, direct optical transitions have been demonstrated in momentum space in the photoluminescence spectra up to room temperature [1, 2]. Modulators, light emitting diodes (LEDs), and photodetectors based on $Ge_{1-x}Sn_x$ are promising for using in silicon photonics [3-6]. In [7] the possibility of obtaining laser generation in $Ge_{0.88}Sn_{0.12}$ layers grown on a silicon wafer with a germanium buffer layer was shown.

With increasing Sn content in a $Ge_{1-x}Sn_x$ alloy, the energy of the Γ7 conduction-band minimum decreases more rapidly than that of the L6 minimum. Theoretical calculation of band structures of germanium and gray tin (α-Sn), which is a semimetal with a band gap of – 0.40 eV, predicts that an unstrained $Ge_{1-x}Sn_x$ alloy should have a direct band gap at a Sn mole fraction $x \approx 0.6$. However, calculations show [8, 9] that the biaxial-tension strains, including heteroepitaxial ones, also result in a band-gap decrease; the effect of strain on the width of the direct band gap is much stronger. The transition from an indirect to a direct-gap material can be observed at a relatively small tin mole fraction $x = 0.02$. Available experimental data give a value $x \approx 0.09$ for the implementation of this transition [10, 11].

While the material system under discussion obviously holds much promise for both fundamental studies and applications, there are a number of factors hindering the fabrication of experimental samples with high crystalline perfection. At thermodynamic equilibrium, the solubility limit in germanium of α-Sn, which has a diamond-like crystal lattice and exists only at temperatures lower than 13.5°C, does not exceed 0.5%, and the solubility limit in silicon is 1.2%. The Sn atom radius (0.158 nm) is larger than that of Si and Ge (0.133 and 0.139 nm, respectively). This, along with a low value of the surface free energy of Sn, leads to Sn segregation during the epitaxial growth. Furthermore, the mismatch of the lattice-constant of α-Sn and Ge and Si is 14.7% and 19.7%, respectively, that inevitably leads to problems associated with the formation of defects during the epitaxial growth of GeSn on Si or Ge wafers.

The epitaxial growth of Si–Ge–Sn structures has been conventionally performed by low-temperature (~300–400°C) low-pressure gas-phase epitaxy [12]. Molecular-beam epitaxy (MBE) has also been used for this purpose [13]. The main problem encountered in MBE is the strong tendency toward the surface segregation of tin under ultrahigh-vacuum conditions. For suppression of this effect, the use of even lower growth temperatures (around 150°C) is required. Epitaxy at such low temperatures results in a high density of intrinsic point defects, which adversely affect the photoluminescence efficiency. In this paper, we present structural and optical characterization of MBE grown $Ge_{1-x}Sn_x$ for group IV photonic devices.

2. Experimental

$Ge_{1-x}Sn_x$ heterostructures were grown on a Ge-buffered Si (100) wafer (Ge-VS) with up to 4° misorientation towards [110]. A set of $Ge_{1-x}Sn_x$ structures with Sn concentrations ranging between 0.09 and 0.15 have been grown and covered by a Ge cap-layer [14].

X-ray diffraction (XRD) has been used to study the crystalline quality and the strain relaxation mechanisms in the grown layers. The optical quality and the light emission were investigated by room temperature photoluminescence (PL) and infrared (IR) transmission spectroscopy in a wide temperature range.

Wafer misorientation leads to an improvement of conditions for the incorporation of tin and germanium adatoms into steps with a height of one monolayer which separate the terraces on the growing layer surface and represent drains for adatoms. This is especially important for low-temperature growth when characteristic migration length of adatoms is small. Surface segregation during epitaxial growth also depends on the efficiency of incorporation of tin adatoms into the GeSn crystal lattice.

3. Results and discussion

Recently, in work [14] the results of XRD for the $Ge_{1-x}Sn_x$ structures and the description of phase decay of metastable $Ge_{1-x}Sn_x$ layers have been demonstrated. Figure 1 shows the XRD rocking curves near the (004) reflection for samples C and D.

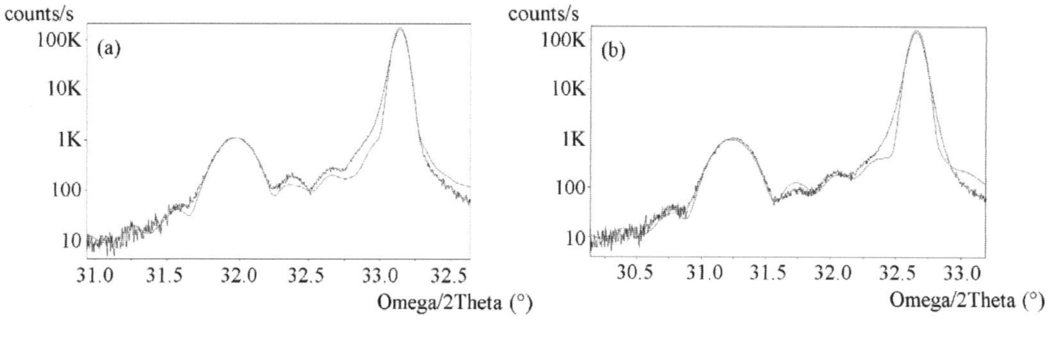

Figure 1. Experimental rocking curves (004) of samples C (a), D (b).

Despite the metastable character of $Ge_{1-x}Sn_x$ layers, they maintain huge elastic deformation (>2%) of the lattice without signs of plastic relaxation. The tin mole fraction in the samples with unrelaxed GeSn layers is as large as $x = 0.153$. The main parameters of the grown layers are summarized in Table 1.

Table 1. Parameters of the $Ge_{1-x}Sn_x$ structures grown by MBE

Number	Substrate	Orientation	T_{Sn}, ^0C	Sn concentration, %	Total thickness, nm	Ge cap-layer, nm
A	Si (001)	4° to (110)	980	9.5	17,8	–
B	Si (001)	0.5°	980	9.7	18,8	5
C	Si (001)	0.5°	1000	12.8	18	14
D	Si (001)	4° to (110)	1000	15.3%	16	4

It is worthwhile to say that, according to the calculations [15], $Ge_{1-x}Sn_x$ solid alloys under large compressive strain are indirect-gap semiconductors. The transition to the direct-gap semiconductors is expected at compressive strain of about -1.05%. The remarkable property of $Ge_{1-x}Sn_x$ grown on Ge-VS is that solid alloys with the direct and indirect fundamental absorption edges can be obtained by changing the built-in strain.

In some theoretical works [16, 17], it was shown that for relaxed $Ge_{1-x}Sn_x$ solid alloys with a tin content x over 0.06, the bottom of the conduction band is the G-point of the Brillouin zone. As the valence band top, regardless of the composition, is also at the G-point, relaxed $Ge_{1-x}Sn_x$ with $x > 6\%$ should be a semiconductor with a direct edge of intrinsic absorption.

For strained $Ge_{1-x}Sn_x$ epitaxial layers grown on Ge-VS, the difference between a local minimum at the G point and lowest local minima at the L-points is further increased due to compressive strain. Therefore, the transition to the direct self-absorption edge for strained $Ge_{1-x}Sn_x$ epitaxial layers is achieved at $x = 10\%$. Despite the indirect absorption edge in $Ge_{1-x}Sn_x$ solid alloys with x of about 4% or more, it is possible to observe phononless luminescence that is caused by the recombination of electrons from the G-point with holes also located near the G-point. The emergence of electrons in the neighborhood of a local minimum in the G-point is due to the thermal filling of the respective states. Therefore, photoluminescence (PL) measuring at relatively high temperatures (around room temperature) is crucial to determine the spectral position of the direct self-absorption edge. The IR transmission spectra was also used to evaluate the spectral position of the direct self-absorption edge.

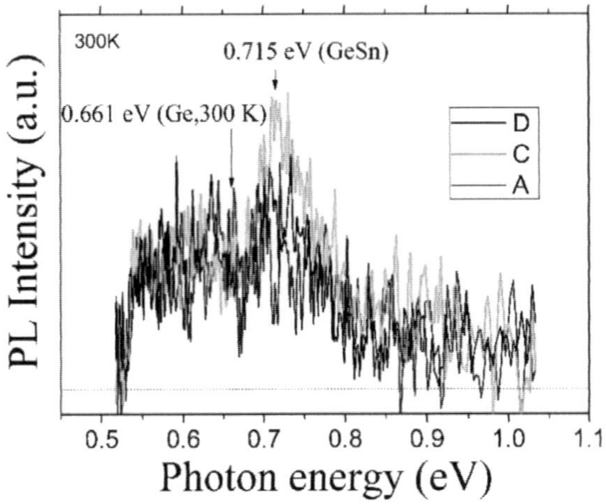

Figure 2. PL spectra at room temperature for structures D, A and C.

PL spectra were measured at 300K at excitation power density on the sample of about 5 W/cm^2. The radiation was registered by the method of synchronous detection with using a suitably selected cooled InGaAs diode (with the red edge of the sensitivity at a wavelength of 2.4 μm), equipped with a low-noise preamplifier with a cooled load. An asymmetrical PL peak with a maximum at 0.715 eV for film C was registered (figure 2). The FWHM band (40 meV) was of the order of the thermal energy at room temperature. Thus, this band has the characteristic features of radiation of low-density electron-hole plasma at a temperature of about 300 K. The spectral position of the band maximum is significantly above the indirect absorption edge of Ge (0.66 eV) and significantly lower than the direct self-absorption edge of GeSn (0.75 eV). Thus, the peak near 0.715 eV should be attributed to the radiation from a GeSn layer, which corresponds to the recombination of electrons and holes located near the G point.

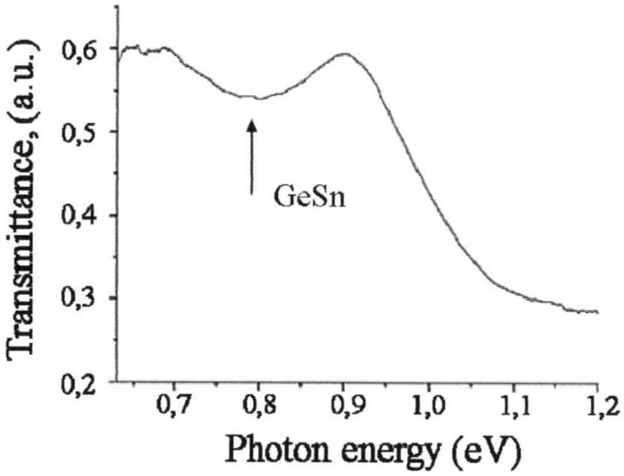

Figure 3. IR transmission spectrum of structure D at 5 K.

IR transmission spectra at helium temperature were measured to confirm the interpretation of the emission line. Figure 3 shows IR transmission spectrum for structure D at 5 K. This structure is characterized by a higher tin content, and has a smaller thickness, which is important to look for features associated with the direct edge. The spectrum distinctly shows a valley at about 0.79 eV. This value is significantly above the indirect self-absorption edge of Ge and considerably lower than the direct self-absorption edge of Ge at 5 K. Therefore, the valley around 0.79 eV can be associated with the direct self-absorption edge in the GeSn layer.

4. Conclusion

In conclusion, taking into account the temperature dependence of the GeSn band gap and the fact that the concentration of tin is larger in structure D compared to C, the IR transmission spectrum in figure 3 confirms our suggestion. As a matter of record, it is suggested that at room temperature the investigated structures with a tin content of about 12-13% have the direct self-absorption edge located in the region of 0.71-0.72 eV. This value corresponds to a tin concentration of about 5% in the $Ge_{1-x}Sn_x$ layer with a greater thickness, according to available literature data. This contradiction can be explained by the fact that the tin concentration is affected by the GeSn layer imperfection and the magnitude of the strain. Obviously, these characteristics may be radically different for thick (90 nm) GeSn layers [6].

This work was supported by the Russian Foundation for Basic Research (project no. 16-02-00986).

References

[1] R Ragan, H A Atwater 2000 *Appl. Phys. Lett.* **77** 3418
[2] G Grzybowski, R T Beeler, L Jiang, D J Smith, J Kouvetakis, J Menendez 2012 *Appl. Phys. Lett.* **101** 072105
[3] B R Conley, A Mosleh, S A Ghetmiri, W Du, R A Soref, G Sun, J Margetis, J Tolle, H A Naseem and S-Q Yu 2014 *Opt. Express* **22** 15639
[4] M Oehme, J Werner, M Gollhofer, M Schmid, M Kaschel, E Kasper and J Schulze 2011 *IEEE Photonics Technol. Lett.* **23** 1751

[5] E Kasper, M Kittler, M Oehme and T Arguirov 2013 *Photonics Res.* **1** 69
[6] W Du, Y Zhou, S A Ghetmiri, A Mosleh, B R Conley, A Nazzal, R A Soref, G Sun, J Tolle, J Margetis, H A Naseem and S-Q Yu 2014 *Appl. Phys. Lett.* **104** 241110
[7] S Wirthst, R Geiger, N von den Driesch, G Mussler, T Stoica, S Mantl, Z Ikonic,M Luysberg, S Chiussi, J M Hartmann, H Sigg, J Faist, D Buca, D Grützmacher 2015 *Nature photonics* **9** 88
[8] R A Sofer, L Friedman 1993 *Superlattices and Microstructures* **14** 189
[9] O Gurdal, P Desjardins, J R A Carlsson, N Taylor, H H Radamson, J-E Sundgren, J E Greene J 1998 *Appl. Phys.* **83** 162
[10] G He, H A Atwater 1997 *Phys. Rev. Lett.* **79** 1937
[11] J Mathews, R T Beeler, J Tolle, C Xu, R Roucka, J Kouvetakis, J Menéndez 2010 *Appl. Phys. Lett.* **97** 221912
[12] M R Bauer, C S Cook, P Aella, J Tolle, J Kouvetakis, P A Crozier, A V G Chizmeshya, D J Smith, S Zollner 2003 *Appl. Phys. Lett.* **83** 3489
[13] H Lin, R Chen, W Lu, Y Huo, T I Kamins, J S Harris 2012 *Appl. Phys. Lett.* **100** 102109
[14] Yu G Sadofyev, V P Martovitsky, A V Klekovkin, V V Saraykin, I S Vasil'evskii 2015 *Semiconductors* **49** 1564
[15] D Stange, S Wirths, N von den Driesch, G Mussler, T Stoica, Z Ikonic, J M Hartmann, S Mantl, D Grützmacher and D Buca 2015 *ACS Photonics* **2** 124
[16] V R D'Costa, Y-Y Fang, J Tolle, J Kouvetakis and J Menéndez 2010 *Thin Solid Films* **518** 2531
[17] N Bhargava, J P Gupta, T Adam and J Kolodzey 2014 *J. Electron. Mater.* **43** 931

Electrochemical capacitance–voltage measurements and modeling of GaAs nanostructures with delta-doped layers

L Shestakova, G Yakovlev and V Zubkov

Department of micro- and nanoelectronics, St. Petersburg State Electrotechnical University "LETI", Prof. Popov str. 5, St. Petersburg 197376, Russia

E-mail: L94Shes@yandex.ru

Abstract. The paper presents the results of electrochemical capacitance-voltage profiling and simulation of quantum-sized semiconductor structures with quantum wells and delta-doped layers based on gallium arsenide. The experimental ECV data were obtained by superposition of measured capacitance-voltage characteristics during the gradual etching of the nanostructure. As a result of simulation, the concentration distribution and energy lineups for structures with delta-layers and quantum wells in gallium arsenide were calculated. The results of simulation are in qualitative agreement with the experimental results and data found in literature.

1. Introduction

Devices based on gallium arsenide have almost achieved their limit in characteristics, but the needs of optics and high frequency electronics lead to necessity of further enhancement of their properties [1]. The alteration of the depth of the potential wells and the concentration of impurities in the barriers in order to improve the device parameters have been brought to perfection, so the actual question is the use of artificially fabricated structures such as delta-doped layers [2].

Delta-doping is the introducing a few atomic layers of very high concentrations in the matrix of semiconductor [3]. Since the electron wave functions are localized in a wider area than the thickness of a delta layer, the charge carriers move in the undoped region, thereby the scattering by ionized atoms is reduced. Often delta-doped layers are used as a source of charge carriers for quantum wells. In this case, it can significantly increase the concentration of two-dimensional electron gas in a quantum well. When producing, designing or investigating such devices, it is useful to evaluate and predict their characteristics. To do this, we have carried out a set of measurements of electrochemical capacitance-voltage (ECV) characteristics together with the self-consistent solution of the Schrödinger and Poisson equations. While the traditional capacitance–voltage (C-V) technique has a common limitation in depth of profiling due to the relatively high charge carrier concentration, the ECV method gives the possibility to overcome the problem and to obtain the carrier concentration distribution at any necessary depth. Further, to minimize the errors due to the influence of series resistance of lightly doped layers on the measured capacitance, the experimental results were obtained by the superposition of measured capacitance–voltage characteristics across the etching depth. For the simulation of concentration profiles of samples with a delta-doped layer, we chose 1DPoisson software. The results of simulation are then compared to the electrochemical carrier concentration profiles of the structures.

2. Samples and experimental technique

Two samples based on GaAs substrates were investigated, their layer sequences are shown in figure 1. Sample #1 (figure 1(a)) has a single carbon delta layer and sample #2 (figure 1(b)) has a InGaAs/GaAs quantum well (QW) and a single carbon delta layer. All the samples were grown by MOCVD at 620 °C growth process. Nominally, the undoped regions have the impurity concentration of about 10^{16} cm^{-3}. The composition of the quantum well was $In_{0.22}Ga_{0.78}As$, the impurity concentration in the delta layer was about 10^{19} cm^{-3}, and the concentration in the GaAs substrate was about 10^{18} cm^{-3}. In the buffer layer, the concentration increases from the value of "undoped" region to the value of the substrate.

(a)
GaAs cap layer (undoped), 140 nm
GaAs spacer (undoped), 13 nm
δC
GaAs spacer (undoped), 240 nm
GaAs (C) buffer, 480 nm
p-GaAs substrate

(b)
GaAs cap layer (undoped), 140 nm
InGaAs QW (undoped), 10 nm
GaAs spacer (undoped), 13 nm
δC
GaAs spacer (undoped), 240 nm
GaAs (C) buffer, 480 nm
p-GaAs substrate

Figure 1. Layer sequences for the investigated samples: sample #1 (a) and sample #2 (b)

Measurements of the concentration of the majority charge carriers in the depth of the structures were carried out at room temperature using an ECVPro profiler (Nanometrics) with a 200 W UV Hg-Xe lamp. This profiler allows etching almost all semiconductors, particularly both *p*- and *n*-type wide bandgap ones such as GaN and SiC [2, 3]. The solution of 0.1 M tyron ($C_6H_2(OH)_2(SO_3Na)_2$) was used to form an electrolyte barrier with a nominal contact area of 0.1 cm^2. This etch solution is a polishing one for GaAs and provides precise etching with an appropriate speed. The frequency and amplitude of the ac test signal were 300 Hz and 100 mV, respectively. The etching was performed with a 1 nm step.

For calculation of concentration profiles and energy diagrams with quantum levels we used 1DPoisson free software [4]. This software is based on self-consistent solution of the Schrödinger and Poisson equations using the difference method with some additional changes: a simple matrix transformation that will preserve the symmetry of the result matrix while allowing a variable mesh size. If the optimal nonuniform mesh is used, this method will provide a computationally efficient solution of the band profile and the electron density distributed over a large spatial dimension.

3. Experimental and calculation results and discussion

The results of modeling of charge carrier concentration profiles and the corresponding ECV-profiling for the two investigated samples are presented in figure 2. Here, the grey lines illustrate experimental results and the black lines,modeling ones. The experimental results were obtained by the gradual superposition of measuring capacitance-voltage characteristics across the etching depth. This approach allows getting the carrier distribution over a broad distance into the depth of the semiconductor and does minimize the errors due to the influence of series resistance of lightly doped layers on the measured capacitance. Experimental and calculated charge carrier profiles showed quite good coincidence, especially for the structure with a single delta-layer (sample #1). The differences between the simulations and the measurements relate to the fact that the calculations do not take into account the diffusion effects and the broadening caused by the inherent peculiarities of the ECV-technique. Also one should take into account that the recommended growth temperature is 550 °C [3], but in our case the growth temperature was 620 °C in order to grow the sharp delta-layers. So, it can be an additional reason of sharpness for the modelled profiles.

Figure 2(b) presents the measurement and calculation of concentration profiles for the structure with a delta-doped layer and a quantum well. As for sample #1, the calculation did not take into account the diffusion effects and the apparatus function of the ECV-measurement method.

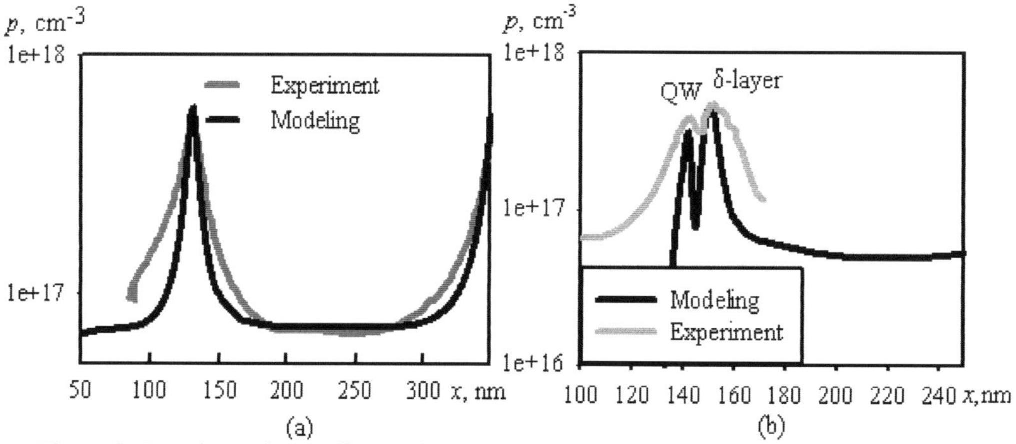

Figure 2. Experimental (gray line) and modeling (black line) charge carrier concentration profiles for sample #1 (a) and sample #2 (b)

Figures 3(a) and 3(b) show the results of calculation of energy diagram and the quantization levels for light and heavy holes for the two investigated samples. It also shows how much the energy structure of the quantum well is deformed by the presence of a delta-doped layer with a high carrier concentration.

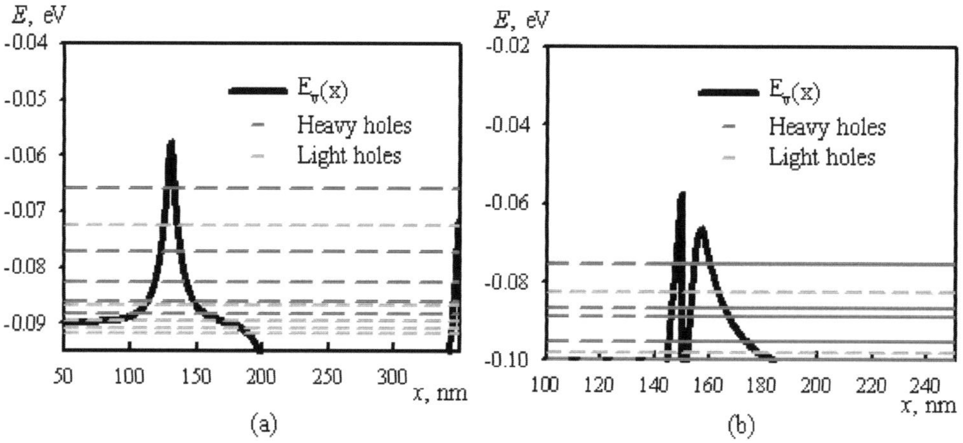

Figure 3. Energy diagram and the quantization levels for light and heavy holes for sample #1 (a) and sample #2 (b). In both cases the position of Fermi level is taken as zero.

The results of calculation of the energy diagram are presented in figure 3(b). The first five quantization levels for light holes and first five levels for heavy holes are shown. It can be concluded that heavy holes make the main contribution to the total charge carrier concentration. Indeed, the most part of carrier concentration falls in the first energy subband. As it could be seen, the first level is the heavy hole energy level [7].

If to compare figure 3(a) and 3(b) one can notice that the energy levels within the quantum size structure shift down in energy in the presence of a delta-layer. This is due to the fact that the delta-layer serves as a reservoir filling the quantum well with additional carriers, which makes the energy levels in a quantum well deeper relative to the Fermi level.

4. Conclusion

ECV-measurements give precise information about the charge carrier distribution along the nanostructure, overcoming the limitations of the common CV-technique. On the other hand, the computer modeling can clear up the fundamental properties of the energy band structure and electronic spectrum of such objects. In this study, we reported the results of ECV-measurements and simulation of concentration profiles for nanostructures with delta-layers and a single quantum well, fabricated on gallium arsenide substrates. The experimental data were obtained by the superposition of measured capacitance–voltage characteristics during the gradual etching of the nanostructure. This approach allows getting the carrier distribution over a broad distance in the depth of the semiconductor. The experimental and simulated data are in quite good agreement; the reasons for the noticed difference in results are the diffusion effects and the apparatus function of the ECV-measurement method, which could not be taken into account in the simulations.

Acknowledgments

This work was supported by the Ministry of Education and Science of the Russian Federation (Agreement № 16.1307.2014/K).

References

[1] Baca A G and Ashby C I 2005 *Fabrication of GaAs Device* (London: The Institution of Engineering and Technology)
[2] Lukashin V M *et al* 2014 *Semiconductors* **48** 666
[3] Schubert E F 1994 *Semiconductors and Semimetals* **40** 1
[4] Zubkov V, Kucherova O, Frolov D and Zubkova A 2013 *Phys. Stat. Sol.* C **10** 342
[5] Yakovlev G E, Zubkov V I, Frolov D S, Zubkova A V, Sterlyadkin O K and Sorokin S A 2016 *Semiconductors* **50(3)** 320
[6] Tan I-H, Snider G L and Hu E L 1990 *J. Appl. Phys.* **68** 4071
[7] Zubkov V I 2007 *Semiconductors* **41** 320

IOP Conf. Series: Journal of Physics: Conf. Series **816** (2017) 012023 doi:10.1088/1742-6596/816/1/012023

Persistent spin helices in 2D electron systems

A S Kozulin, A I Malyshev and A A Konakov

Department of Theoretical Physics, National Research Lobachevsky State University of Nizhni Novgorod, 23 Gagarin Avenue, Nizhni Novgorod, 603950, Russia

E-mail: SashaKozulin@yandex.ru

Abstract. We present a theoretical investigation of persistent spin helices in two-dimensional electron systems with spin-orbit coupling. For this purpose, we consider a single-particle effective mass Hamiltonian with a generalized linear-in-***k*** spin-orbit coupling term corresponding to a quantum well grown in an arbitrary crystallographic direction, and derive the general condition for the formation of the persistent spin helix. This condition applied for the Hamiltonians describing quantum wells with different growth directions indicates the possibility of existence of the persistent spin helix in a wide class of 2D systems apart from the [001] model with equal Rashba and Dresselhaus spin-orbit coupling strengths and the [110] Dresselhaus model.

1. Introduction

Although properties of two-dimensional electron gas (2DEG) with spin-orbit coupling (SOC) have been studied intensively during several decades [1], some of its fascinating features were established relatively recently. In 2006 Bernevig et al. [2] discovered a new type of SU(2) spin rotation symmetry for two well-known special cases of 2DEG with SOC: the [001] model with equal Rashba and Dresselhaus SOC strengths (ReD model) and the Dresselhaus [110] model. They predicted the persistent spin helix (PSH) which is a special spin precession pattern with the precession angle depending only on the net displacement in specific directions (±[110] for the ReD model and $\pm[1\bar{1}0]$ for the Dresselhaus [110] model). In 2009, PSHs were experimentally observed in [001]-grown GaAs/AlGaAs quantum wells (QWs) [3] and since then many theoretical [4-8] and experimental [9-15] investigations have been devoted to the studies of different manifestations of the PSH. However, almost all of them primarily focus on two above-mentioned cases and do not cover the other types of 2D electron systems with SOC. The main goal of the present paper is to derive the general condition of the PSH formation in 2DEG contained in a QW grown in an arbitrary crystallographic direction and outline some specific materials that are good candidates for realization of the PSH state.

2. The generalized SOC-Hamiltonian

Motivated by our main goal, we consider the following generalized form of a single-particle effective mass Hamiltonian with linear-in-***k*** SOC:

$$\hat{H} = \frac{\hbar^2}{2m}\left(\hat{k}_x^2 + \hat{k}_y^2\right)\hat{\sigma}_0 + \left(\alpha_{11}\hat{\sigma}_x + \alpha_{12}\hat{\sigma}_y + \alpha_{13}\hat{\sigma}_z\right)\hat{k}_x + \left(\alpha_{21}\hat{\sigma}_x + \alpha_{22}\hat{\sigma}_y + \alpha_{23}\hat{\sigma}_z\right)\hat{k}_y, \quad (1)$$

where we use Cartesian coordinates with z axis perpendicular to the plane of 2DEG, m is the effective electron mass, six parameters α_{ij} define the asymmetry induced SOC, $\hat{\sigma}_0$ and $\hat{\sigma}_x$, $\hat{\sigma}_y$, $\hat{\sigma}_z$ are the 2×2 unit matrix and the Pauli matrices, respectively.

In [6, 16] it was shown that the Rashba and Dresselhaus SOC in two dimensions can be regarded as a Yang-Mills non-Abelian gauge field. Following the approach developed in [6], we introduce the SOC gauge

$$\hat{\gamma}^{SO} = \{\hat{\gamma}_x, \hat{\gamma}_y\} = \frac{m}{\hbar}\left\{(\vec{\alpha}_1 \cdot \hat{\vec{\sigma}}), (\vec{\alpha}_2 \cdot \hat{\vec{\sigma}})\right\}, \tag{2}$$

and rewrite the Hamiltonian (1) in the form:

$$\hat{H} = \frac{1}{2m}\left((\hat{p}_x\hat{\sigma}_0 + \hat{\gamma}_x)^2 + (\hat{p}_y\hat{\sigma}_0 + \hat{\gamma}_y)^2\right) - V\hat{\sigma}_0, \tag{3}$$

with the constant potential $V = \left(|\vec{\alpha}_1|^2 + |\vec{\alpha}_2|^2\right)/2\hbar^2$, where $\vec{\alpha}_1 = \{\alpha_{11}, \alpha_{12}, \alpha_{13}\}$ and $\vec{\alpha}_2 = \{\alpha_{21}, \alpha_{22}, \alpha_{23}\}$ are two symbolic vectors constructed from the SOC parameters. In general, the components of the SO gauge do not commute,

$$[\hat{\gamma}_x, \hat{\gamma}_y] = \mathrm{i}\frac{2m^2}{\hbar^2}\left(\hat{\vec{\sigma}} \cdot [\vec{\alpha}_1 \times \vec{\alpha}_2]\right). \tag{4}$$

However, if the cross product of $\vec{\alpha}_1$ and $\vec{\alpha}_2$ is a null vector, the commutator (4) vanishes. In such situations, the operator of the spin projection on the direction $\vec{\alpha}_1$ commutes with the Hamiltonian (1) and defines an extra conserved physical quantity. Moreover, in these exceptional symmetric cases, the effective SOC-induced magnetic field with components $\{\alpha_{11}\hat{k}_x + \alpha_{21}\hat{k}_y, \alpha_{12}\hat{k}_x + \alpha_{22}\hat{k}_y, \alpha_{13}\hat{k}_x + \alpha_{23}\hat{k}_y\}$ becomes unidirectional, i.e. its direction does not depend on the components of the wave vector and is defined by the unit vector $\vec{b} = \vec{\alpha}_2/|\vec{\alpha}_2|$, and the Hamiltonian (1) can be written in the form as $\hat{H}_{SO} = (q\hat{k}_x + \hat{k}_y)(\vec{\alpha}_2 \cdot \hat{\vec{\sigma}})$, where a real constant q is determined from the relation $\vec{\alpha}_1 = q\vec{\alpha}_2$.

Performing the global spin rotation generated by the operator

$$\hat{U} = \hat{\sigma}_0 \cos\frac{\chi}{2} - \mathrm{i}(n_x\hat{\sigma}_x + n_y\hat{\sigma}_y)\sin\frac{\chi}{2}, \tag{5}$$

where $\vec{n} = [\vec{b} \times \vec{e}_z] = \left\{\alpha_{22}/\sqrt{\alpha_{21}^2 + \alpha_{22}^2}, -\alpha_{21}/\sqrt{\alpha_{21}^2 + \alpha_{22}^2}, 0\right\}$ with $\vec{e}_z = \{0, 0, 1\}$, and χ is the angle between $\vec{e}_z$ and $\vec{b}$ (see figure 1), we transform the Hamiltonian (1) to the diagonal form:

$$\hat{H}_1 = \frac{\hbar^2}{2m}\left(\hat{k}_x^2 + \hat{k}_y^2\right)\hat{\sigma}_0 + |\vec{\alpha}_2|\left(q\hat{k}_x + \hat{k}_y\right)\hat{\sigma}_z. \tag{6}$$

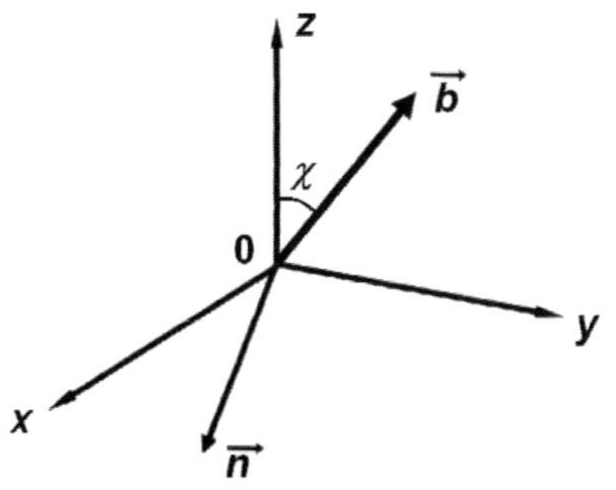

Figure 1. Global transformation generated by the operator (5) is a rotation on the angle χ around the $\vec{n}$ axis which is perpendicular to the direction of the unidirectional effective magnetic field and z axis.

Physically, it means that the direction of the effective magnetic field is chosen as the spin quantization axis and the quantum states are characterized by the projection of the electron spin on this axis.

The energy spectrum of the Hamiltonian (6)

$$E_\lambda(k_x,k_y)=\frac{\hbar^2}{2m}\left(k_x^2+k_y^2\right)+\lambda|\vec{\alpha}_2|\left(qk_x+k_y\right) \tag{7}$$

with $\lambda=\pm1$ which numerates the energy bands, has an important shifting property

$$E_{-1}(k_x+Q_x,k_y+Q_y)=E_1(k_x,k_y) \quad \Leftrightarrow \quad E_{-1}(\vec{k}+\vec{Q})=E_1(\vec{k}), \tag{8}$$

where $\vec{Q}=\frac{2m|\vec{\alpha}_2|}{\hbar^2}\{q,\ 1\}$ is the so-called "magic" vector [2]. Accordingly, the eigenfunctions of the Hamiltonian (6) corresponding to quantum numbers $\lambda=\pm1$ have the following simple form:

$$\psi_{\vec{k},1}=\exp\left(\mathrm{i}(\vec{k}\cdot\vec{r})\right)\begin{Vmatrix}1\\0\end{Vmatrix}, \quad \psi_{\vec{k},-1}=\exp\left(\mathrm{i}(\vec{k}\cdot\vec{r})\right)\begin{Vmatrix}0\\1\end{Vmatrix}. \tag{9}$$

The ReD and [110] Dresselhaus models being particular cases of the Hamiltonian (6) are characterized by the exact SU(2) symmetry which is introduced in [2] in terms of creation and annihilation operators corresponding to the many-particle problem. Below we demonstrate that the Hamiltonian (6) has the same SU(2) symmetry with generators which are the single-particle counterparts of the generators presented in [2]. Namely, two single-particle operators

$$\begin{gathered}\hat{C}_{\vec{Q}}^{+}=\frac{\exp\left(-\mathrm{i}(\vec{Q}\cdot\vec{r})\right)}{2}(\hat{\sigma}_x+\mathrm{i}\hat{\sigma}_y)=\exp\left(-\mathrm{i}(\vec{Q}\cdot\vec{r})\right)\begin{Vmatrix}0&1\\0&0\end{Vmatrix},\\ \hat{C}_{\vec{Q}}^{-}=\frac{\exp\left(\mathrm{i}(\vec{Q}\cdot\vec{r})\right)}{2}(\hat{\sigma}_x-\mathrm{i}\hat{\sigma}_y)=\exp\left(\mathrm{i}(\vec{Q}\cdot\vec{r})\right)\begin{Vmatrix}0&0\\1&0\end{Vmatrix},\end{gathered} \tag{10}$$

that act on the eigenfunctions (9) in the following way:

$$\hat{C}_{\vec{Q}}^{+}\psi_{\vec{k},-1}=\psi_{\vec{k}-\vec{Q},1},\quad \hat{C}_{\vec{Q}}^{+}\psi_{\vec{k},1}=0,\quad \hat{C}_{\vec{Q}}^{-}\psi_{\vec{k},-1}=0,\quad \hat{C}_{\vec{Q}}^{-}\psi_{\vec{k},1}=\psi_{\vec{k}+\vec{Q},-1}, \tag{11}$$

together with $\hat{\sigma}_z$ obey the commutation relation for the angular momentum operator

$$\left[\hat{\sigma}_z,\hat{C}_{\vec{Q}}^{+}\right]=2\hat{C}_{\vec{Q}}^{+},\quad \left[\hat{\sigma}_z,\hat{C}_{\vec{Q}}^{-}\right]=-2\hat{C}_{\vec{Q}}^{-},\quad \left[\hat{C}_{\vec{Q}}^{+},\hat{C}_{\vec{Q}}^{-}\right]=\hat{\sigma}_z. \tag{12}$$

In addition, due to the shifting property (8) $\hat{C}_{\vec{Q}}^{+}$ and $\hat{C}_{\vec{Q}}^{-}$ commute with the Hamiltonian (6):

$$\left[\hat{H}_1,\hat{C}_{\vec{Q}}^{+}\right]=\left(E_1(\vec{k}-\vec{Q})-E_{-1}(\vec{k})\right)\hat{C}_{\vec{Q}}^{+}=\hat{0},\quad \left[\hat{H}_1,\hat{C}_{\vec{Q}}^{-}\right]=\left(E_{-1}(\vec{k}+\vec{Q})-E_1(\vec{k})\right)\hat{C}_{\vec{Q}}^{-}=\hat{0}. \tag{13}$$

It is also obvious that $\left[\hat{H}_1,\hat{\sigma}_z\right]=\hat{0}$. The latter relation together with (13) show that single-particles operators $\hat{C}_{\vec{Q}}^{+}$, $\hat{C}_{\vec{Q}}^{-}$ and $\hat{\sigma}_z$ are generators of the exact SU(2) symmetry responsible for formation of the PSH state. Concrete examples of 2D electron systems, where the PSH patterns are expected to appear, are given in the next section.

3. What 2D electron structures are good candidates for realization of the PSH state?

The obtained condition $[\vec{\alpha}_1\times\vec{\alpha}_2]=\vec{0}$, in fact, defines the possibility of formation of the PSH patterns in arbitrary 2D electron systems with SOC. Nevertheless, the most attractive structures for experimental realization of the PSH state are expected to be found in specific QWs due to opportunities of the SOC parameters manipulation via (a) external electric field and (b) fitting of QW characteristics (type of material and geometrical parameters) for achieving the above-mentioned relation. In this section, we examine a wide class of semiconductor QWs and outline among them some good candidates for realization of the PSH patterns. In addition, relations between the SOC parameters that should be satisfied in order to achieve the PSH state will be obtained.

We begin the analysis with the zinc-blende type QWs. In such structures, the absence of inversion symmetry in the bulk material leads to the Dresselhaus SOC-Hamiltonian:

$$\hat{H}_D = \beta_{ij} \hat{k}_i \hat{\sigma}_j , \tag{14}$$

with six parameters β_{ij} ($i = x, y, j = x, y, z$). Exact expression of the Dresselhaus term depends on the QW growth direction. The other contribution to the SOC-Hamiltonian is connected with the structure inversion asymmetry and described by the Rashba term:

$$\hat{H}_R = \alpha_R \left(\hat{k}_x \hat{\sigma}_y - \hat{k}_y \hat{\sigma}_x \right). \tag{15}$$

Hereafter, we divide all zinc-blende QWs into symmetric (the corresponding Hamiltonian contains only the Dresselhaus term) and asymmetric (both Dresselhaus and Rashba terms are present) types. SOC-parts of the electron effective mass Hamiltonians corresponding to different QWs are presented in table 1. Application of the criterion $[\vec{\alpha}_1 \times \vec{\alpha}_2] = \vec{0}$ for every Hamiltonian from this table allows to conclude whether the PSH state is realized in a specific QW or not.

Table 1. Formation of the PSH in symmetric (S) and asymmetric (A) QWs with different growth direction. All SOC Hamiltonians are taken from [17] and correspond to materials with zinc-blende structure.

Growth direction	SOC-Hamiltonian	$\vec{\alpha}_1$ and $\vec{\alpha}_2$	Existence of the PSH	Relation between the SOC parameters when the PSH state is realized
[001]	S: $\hat{H} = \beta(\hat{k}_x\hat{\sigma}_x - \hat{k}_y\hat{\sigma}_y)$	$\vec{\alpha}_1 = \{\beta, 0, 0\}$, $\vec{\alpha}_2 = \{0, -\beta, 0\}$	No	–
	A: $\hat{H} = \beta(\hat{k}_x\hat{\sigma}_x - \hat{k}_y\hat{\sigma}_y) + \alpha(\hat{k}_y\hat{\sigma}_x - \hat{k}_x\hat{\sigma}_y)$	$\vec{\alpha}_1 = \{\beta, -\alpha, 0\}$, $\vec{\alpha}_2 = \{\alpha, -\beta, 0\}$	Yes	$\lvert\alpha\rvert = \lvert\beta\rvert$
[110]	S: $\hat{H} = \beta \hat{k}_x \hat{\sigma}_z$	$\vec{\alpha}_1 = \{0, 0, \beta\}$, $\vec{\alpha}_2 = \{0, 0, 0\}$	Yes	$\beta \neq 0$
	A: $\hat{H} = (\beta_1 + \alpha)\hat{k}_y\hat{\sigma}_x + (\beta_2 - \alpha)\hat{k}_x\hat{\sigma}_y + \beta_3 \hat{k}_x \hat{\sigma}_z$	$\vec{\alpha}_1 = \{0, \beta_2 - \alpha, \beta_3\}$, $\vec{\alpha}_2 = \{\beta_1 + \alpha, 0, 0\}$	Yes	$\beta_1 + \alpha = 0$ or $(\beta_2 - \alpha)^2 + \beta_3^2 = 0$
[111]	S: $\hat{H} = \beta(\hat{k}_y\hat{\sigma}_x - \hat{k}_x\hat{\sigma}_y)$	$\vec{\alpha}_1 = \{0, -\beta, 0\}$, $\vec{\alpha}_2 = \{\beta, 0, 0\}$	No	–
	A: $\hat{H} = (\alpha + \beta)(\hat{k}_y\hat{\sigma}_x - \hat{k}_x\hat{\sigma}_y)$	$\vec{\alpha}_1 = \{0, -(\alpha + \beta), 0\}$, $\vec{\alpha}_2 = \{\alpha + \beta, 0, 0\}$	No	–
[113]	S: $\hat{H} = \beta_1\hat{k}_y\hat{\sigma}_x + \beta_2\hat{k}_x\hat{\sigma}_y + \beta_3\hat{k}_x\hat{\sigma}_z$	$\vec{\alpha}_1 = \{0, \beta_2, \beta_3\}$, $\vec{\alpha}_2 = \{\beta_1, 0, 0\}$	Yes	$\beta_1 = 0$ or $\beta_2 = \beta_3 = 0$
	A: $\hat{H} = (\beta_1 + \alpha)\hat{k}_y\hat{\sigma}_x + (\beta_2 - \alpha)\hat{k}_x\hat{\sigma}_y + \beta_3\hat{k}_x\hat{\sigma}_z$	$\vec{\alpha}_1 = \{0, \beta_2 - \alpha, \beta_3\}$, $\vec{\alpha}_2 = \{\beta_1 + \alpha, 0, 0\}$	Yes	$\beta_1 + \alpha = 0$ or $(\beta_2 - \alpha)^2 + \beta_3^2 = 0$

Firstly, in the most studied case of [001] QWs, the helices are not formed when a well is symmetric and appear in asymmetric wells only if modules of the Rashba and Dresselhaus SOC strengths are equal (ReD model). This result agrees with previous investigations [2-4, 6, 11-14]. Secondly, the PSH structures exist in the symmetric [110] QW for any value of the Dresselhaus SOC strength. This case is known as the Dresselhaus [110] model. In order to guarantee the formation of the PSH it is necessary just to prepare a symmetric well. Therefore, such kind of QWs seems very attractive for realization of the PSH state.

Application of the general condition allows to find all possible 2D systems apart from the ReD and Dresselhaus [110] models which provide the formation of the PSHs. In particular, our calculations show that the PSH patterns can also be realized for some combinations of the SOC parameters in [110]-asymmetric and [113] QWs. Apparently, they are expected to appear in [112] and miscut [001] QWs.

For QWs prepared on [013]-oriented substrates belonging to the trivial point group C_1, there are no restrictions on the relation between the spin and wavevector components arising from the symmetry requirements. Hence, the SOC-Hamiltonian for such systems has the most general form and coincides with the spin-orbit part of the Hamiltonian (1). We pay attention to this case because of the fact that [013]-oriented substrates are used for growth of HgTe-based 2D topological insulators [17].

In our analysis we neglected the interface inversion asymmetry which may yields extra linear-in-$\boldsymbol{k}$ terms in the Hamiltonian (1) caused by noninversion symmetric bonding of atoms at a heterostructure interface [18-20]. Although such contributions in the SOC have the same form as the Dresselhaus term in [001] QWs made from III-V materials, their inclusion may be important for formation of the PSH state in [110] QWs [21] and essential for SiGe QWs [22]. Therefore, effects of the interface inversion asymmetry should be taken into account in more rigorous investigations in the above-mentioned systems.

In summary of this section, our analysis points on the possibility of the PSH patterns formation in a wide class of 2D electron systems. In a specific QW, the PSH regime can be achieved through manipulation of the Rashba SOC be means of external electric field or (and) adjustment of the Dresselhaus SOC parameters by fitting of the QW width and other details of its design.

4. Conclusion

In conclusion, we have theoretically studied the spin precession in 2D electron systems with SOC and have derived the general condition of the PSH state realization. This condition applied for the Hamiltonians describing QWs with different growth directions indicates the possibility of existence of the PSH patterns in a wide class of 2D systems including the well-studied [001] model with equal the Rashba and Dresselhaus SOC strengths and the [110] Dresselhaus model. The latter statement requires experimental verification and we hope that the appropriate studies will be realized soon.

Acknowledgements

The authors are grateful to G.M. Maksimova for valuable discussions. We also would like to thank D.V. Khomitsky for technical assistance. The work was in part supported by RFBR (grants no. 15-42-02254, 16-07-01102, 16-32-00683, 16-32-00712 and 16-57-51045) and by the Russian Ministry of Education and Science (project no. 3.285.2014/K).

References

[1] Winkler R. 2003 *Spin-Orbit Coupling Effects in Two-Dimensional Electron and Hole Systems* (Berlin: Springer-Verlag, Heildelberg)
[2] Bernevig B A, Orenstein J, Zhang S-C 2006 *Phys. Rev. Lett.* **97** 236601
[3] Koralek J D, Weber C P, Orenstein J, Bernevig B A, Zhang S-C, Mack S and Awschalom D D 2008 *Nature (London)* **458** 610
[4] Liu M-H, Chen K-W, Chen S-H and Chang C-R 2006 *Phys. Rev. B* **74** 235322
[5] Liu M-H, Chang C-R 2006 *Journal of Magnetism and Magnetic Materials* **304** 97
[6] Chen S-H and Chang C-R 2008 *Phys. Rev. B* **77** 045324
[7] Lüffe M C, Kailasvuori J and Nunner T S 2011 *Phys. Rev B* **84** 075326

[8] Liu X and Sinova J 2012 *Phys Rev B.* **86** 174301
[9] Sacksteder IV V E and Bernevig B A 2014 *Phys. Rev. B* **89** 161307(R)
[10] Kohda M, Lechner V, Kunihashi Y, Dollinger T, Olbrich P, Schönhuber C, Caspers I, Belkov V V, Golub L E, Weiss D *et al* 2012 *Phys. Rev. B* **86** 081306(R)
[11] Walser M P, Reichl C, Wegscheider W and Salis G 2012 *Nature Physics* **8** 757
[12] Altmann P, Walser M P, Reichl C, Wegscheider W and Salis G 2014 *Phys. Rev. B* **90** 201306(R)
[13] Schönhuber C, Walser M P, Salis G, Reichl C, Wegscheider W, Korn T and Schüller C 2014 *Phys. Rev. B* **89** 085406
[14] Salis G, Walser M P, Altmann P, Reichl C and Wegscheider W 2014 *Phys Rev B* **89** 045304
[15] Sasaki A, Nonaka S, Kunihashi Y, Kohda M, Bauernfeind T, Dollinger T, Richter K and Nitta J 2014 *Nature Nanotechnology* **9** 703
[16] Hatano N, Shirasaki R and Nakamura H 2007 *Phys. Rev. A* **75** 032107
[17] Ganichev S D and Golub L E 2014 *Phys. Status Solidi B* **251** 1801
[18] Krebs O and Voisin P 1997 *Phys. Rev. Lett.* **77** 1829
[19] Krebs O, Seidl W, Andre J P, Bertho D, Jonani C and Voisin P 1997 *Semicond. Sci. Techol.* **12** 938
[20] Vervoort L and Voisin P 1997 *Phys. Rev. B* **56** 12744
[21] Nestoklon M O, Tarasenko S A, Jancu J-M and Voisin P 2012 *Phys. Rev. B* **85** 205307
[22] Nestoklon M O, Golub L E and Ivchenko E L 2006 *Phys. Rev. B* **73** 235334

InAs quantum dots grown by MOCVD in GaAs and metamorphic InGaAs matrixes

R A Salii[1], N A Kalyuzhnyy[1,2], N V Kryzhanovskaya[2], M V Maximov[2], S A Mintairov[1,2,3], A M Nadtochiy[2,3], V N Nevedomskiy[1] and A E Zhukov[2]

[1] Ioffe Institute, Saint Petersburg, Russia
[2] St Petersburg Academic University, St Petersburg, Russia
[3] Solar Dots Ltd, St Petersburg, Russia

e-mail: r.saliy@mail.ioffe.ru

Abstract. We have studied MOCVD-grown structures with InAs layers in GaAs and metamorphic InGaAs matrices. Deposition of 2 ML of InAs in a GaAs or $In_{0.24}Ga_{0.76}As$ matrix results in quantum dots formation with photoluminescence (PL) peaks at 1240 and 1380 nm, respectively. In case of deposition of 2 ML of InAs in a $In_{0.30}Ga_{0.70}As$ metamorphic matrix, formation of a corrugated quantum well emitting at 1450 nm has been revealed. The integrated PL intensity of the metamorphic structures is much higher than that for the GaAs based structure. For the metamorphic structures, the thermal escape of carriers from InAs quantum dots (a quantum well) results in high intensity of the PL line due to InGaAs matrix at room temperature.

1. Introduction

Quantum size objects such as quantum wells (QWs), quantum wires and quantum dots (QDs) provide significant improvement of the performance of semiconductor devices [1]. Using InAs QDs with pseudomorphic QW capping in a GaAs matrix has allowed obtaining the lowest laser threshold currents [2-4].The QWs and QDs have also found an application in photovoltaic converters (PVC) providing improvements of their parameters [5-8]. Among other advantages, QWs and QDs enable achieving a longer emission wavelength as compared to bulk materials. This becomes especially apparent in case of QDs. For instance, the use of InAs/InGaAs QDs in a GaAs matrix has provided low threshold lasing in the range of 1250 - 1350 nm.

One of the promising ways of further wavelength increase, e.g. towards a 1.55 μm telecom window, is the use of a metamorphic InGaAs matrix for epitaxial structures on GaAs substrates [9]. MBE-grown lasers emitting in the 1.3 - 1.55 μm interval have been created using InAs QDs in the InGaAs matrix grown on a GaAs substrate via a metamorphic buffer [10, 11]. Metamorphic buffers have also been successfully used for PVC [12-14]. However, most of the research in the field of QDs has been carried out using molecular beam epitaxy (MBE), whereas at the moment the metal-organic chemical vapor deposition (MOCVD) has become the main production technology for epitaxy of various semiconductor devices.

In the present work, we study InAs QD formation by MOCVD in a lattice matched GaAs matrix as well as in a metamorphic InGaAs matrix.

2. Experimental details

Structures under study were grown using a horizontal MOCVD reactor under 100 mbar on GaAs (100) substrates. Threemethylgallium, threemethylaluminum and threemethylindium were used as III-group atom precursors. Arsine was used as a source of As atoms.

We have grown and studied three structures (figure 1). In all the structures, the active region was formed by the deposition of 2 monolayers (ML) of InAs. The InAs layer was centered in the 500 nm-thick $In_xGa_{1-x}As$ matrix. $(Al_{0.5}Ga_{0.5})_{1-x}In_xAs$ wide-bandgap layers were used as barriers preventing carrier diffusion from the matrix to the substrate and structure surface. Structure N1 was made lattice matched to the GaAs substrate ($x = 0$) (figure 1(a)), whereas for structure N2 $x = 24\%$ (figure 1(b)) and for structure N3 $x = 30\%$ (figure 1(c)). The details of the metamorphic buffer design are presented elsewhere [13].

Transmission electron microscopy (TEM) study was carried out with a JEOL JEM 2100F microscope at 200 kV accelerating voltage. The samples were prepared by preliminary mechanical thinning followed by final ion beam milling (3.5 keVAr+ ions).

Photoluminescence (PL) spectra were measured using a 532nm Nd:YAG laser with an output power of 350mW. PL was focused on a monochromator. All measurements were carried out by means of a cooled Ge photodetector using a standard lock-in amplifier.

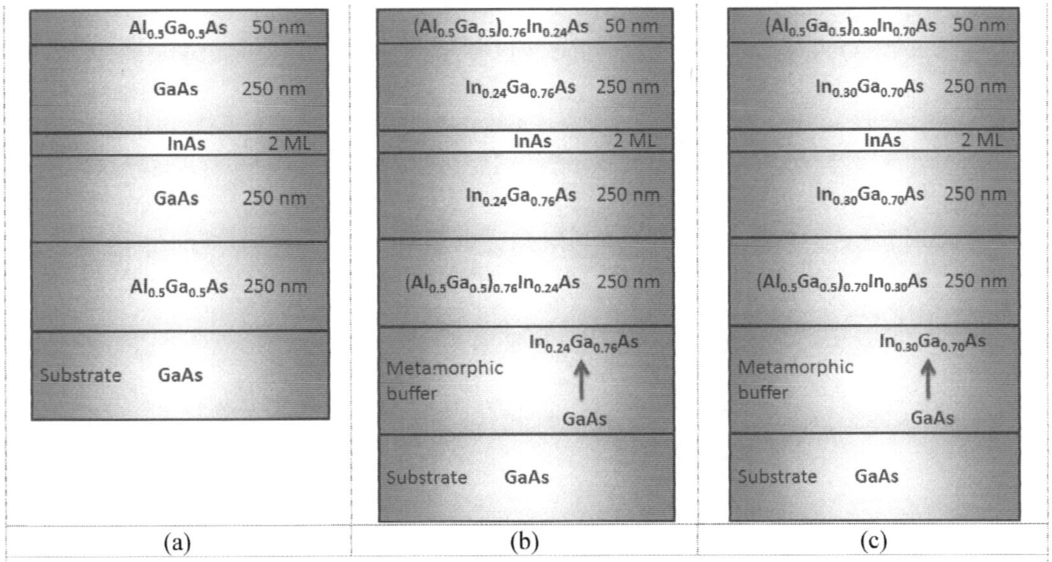

Figure 1. Schematic of the structure N1 (a), structure N2 (b) and structure N3 (c).

3. Results and discussion

TEM studies revealed formation of self-assembled Stransky–Krastanov (SK) QDs for structure N1 and structure N2 (figure 2(a), (b)), whereas for structure N3 only a corrugated thin InAs layer was observed (figure 2(c)). The latter result can be explained by the insufficient InAs thickness in case of using the $In_{0.30}Ga_{0.70}As$ matrix. It has been shown [8], that for InAs QDs grown by MOCVD in a GaAs matrix, the SK QDs with sufficiently high crystal quality are formed in the 1.4 – 2.4 ML range. The critical thickness necessary for coherent QDs formation due to elastic strain relaxation is 1.4 ML. The thickness of QDs coalescence with formation of large defected InAs islands is 2.4 ML. Increasing the In content in the metamorphic matrix leads to a reduction of the mismatch of lattice constants of the matrix and InAs. As a result, the critical thickness increases. Thus, one can conclude that the critical thickness exceeds 2ML in case of the $In_{0.30}Ga_{0.70}As$ matrix. The surface density of the SK QDs grown in the $In_{0.24}Ga_{0.76}As$ is ~10^{11} cm^{-2} (figure 2(b)) and seems to be higher than that for the GaAs matrix (figure 2(a)).

For structure N1 the PL spectrum under low excitation power at 300 K shows two peaks at 1140 and 1240 nm which correspond to QDs of different sizes or ground-state and excited-state optical transitions (figure 3(a)). The bi-modality of the MOCVD grown InAs QDs in GaAs matrix has been discussed elsewhere [8, 15]. For structure N2 a broad peak centered at 1380 nm is observed (figure 3(a)). We conclude that this peak corresponds to a recombination via SK QDs shown in figure 1(b). Besides this peak, a narrower peak at 1140 nm equal to the bandgap of the $In_{0.24}Ga_{0.76}As$ matrix is observed. Although the QDs were not observed in the TEM pictures of structure N3 (figure 1(c)), its PL spectrum looks similar to that for structure N2 (figure 3(a)). The long wavelength PL peak corresponding to recombination via the InAs layer is observed at 1450 nm. The large spectral width of this line can be explained by the strong corrugation and non-uniformity of the InAs layer. The spectral position of the second narrower peak (1240 nm) is equal to the E_g of $In_{0.30}Ga_{0.70}As$.

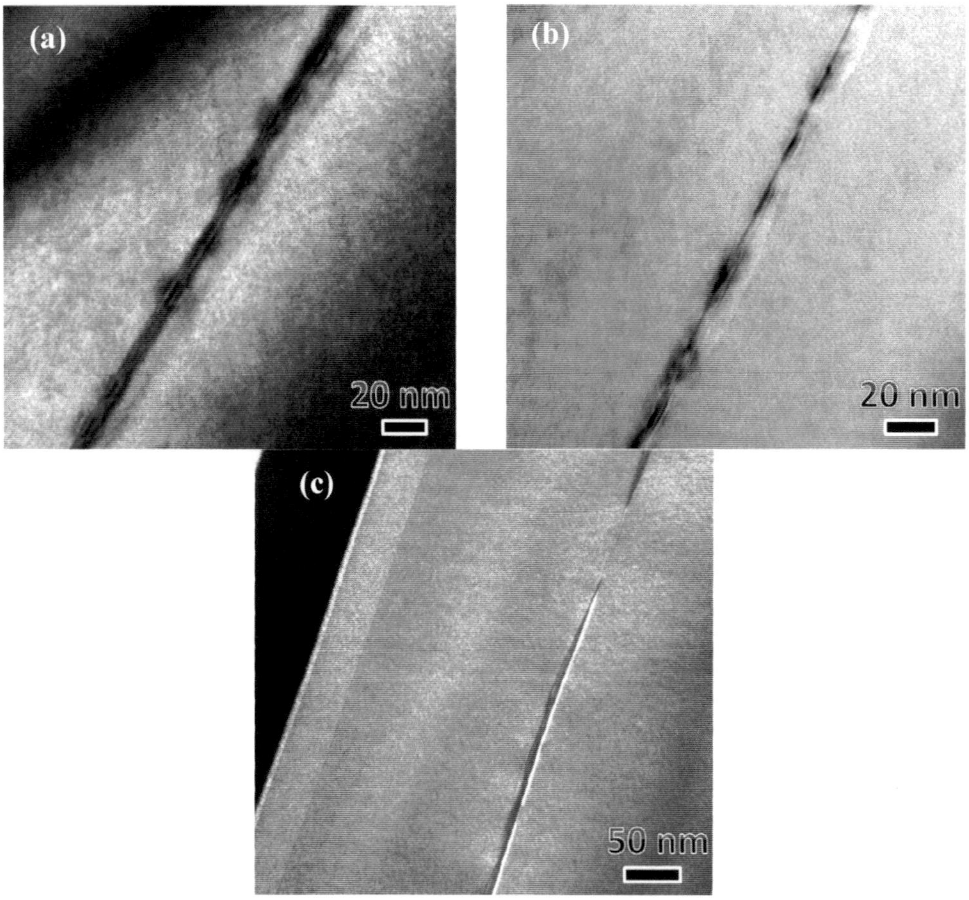

Figure 2. Bright-field cross-section TEM micrographs of QD regions for structure N1 (a), structure N2 (b) and structure N3 (c).

The presence of the matrix peak at the PL spectra for structure N2 and structure N3 can be explained by enhanced thermal escape of the carriers from InAs into the matrix. Indeed, the difference between the energies of the QDs and the matrix bandgaps for structure N2 (~ 180 meV) is two times lower than that for structure N1 (~ 440 meV). For structure N3, this difference is ~ 140 meV. This explanation is also confirmed by the PL data taken at 77 K (figure 3(b)), which do not reveal matrix peaks for all the structures.

At 77 K, the integrated intensities of the PL spectra for metamorphic structures N2 and N3 are several times higher in comparison with GaAs-based structure N1, whereas at 300 K the intensities of all three structures are comparable. This can be explained by non-radiative recombination on defects in metamorphic structures. As the defects capture cross section decreases with decreasing temperature, their impact on the PL spectra at 77 K becomes negligible.

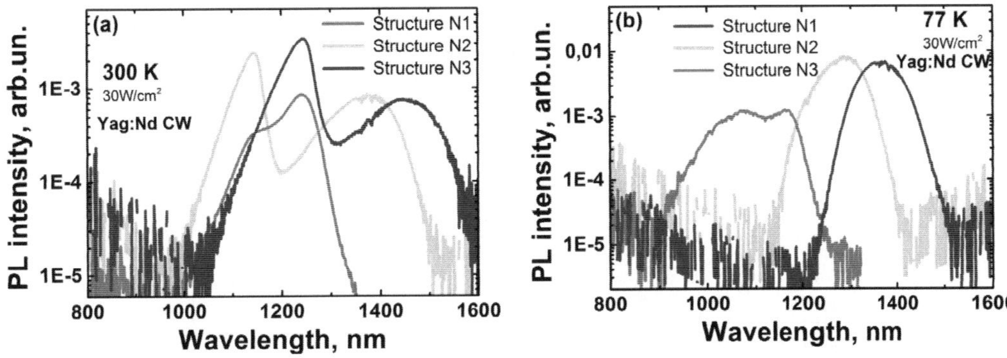

Figure 3. PL spectra for structure N1 (red), structure N2 (green) and structure N3 (blue) under low excitation power at: (a) – room temperature, (b) – 77 K.

Under high excitation density, the PL spectrum of structure N1 shows three additional peaks at 1040, 930 and 860 nm corresponding to the QDs excited state, QDs wetting layer and GaAs matrix, respectively (figure 4(a)). At the same time, the shape of the PL spectra of structures N2 and N3 do not significantly change but their integrated intensities become much higher in comparison with that of structure N1 (figure 4(a)). We suggest that this finding can be explained by saturation of the non-radiative recombination channels for metamorphic structures under high excitation density.

At 77 K under high excitation density, all the spectra show a matrix peak (figure 4(b)) attributed to saturation of the recombination via QDs. Under such conditions, the metamorphic structures also show a higher integrated intensity of the PL as compared to structure N1.

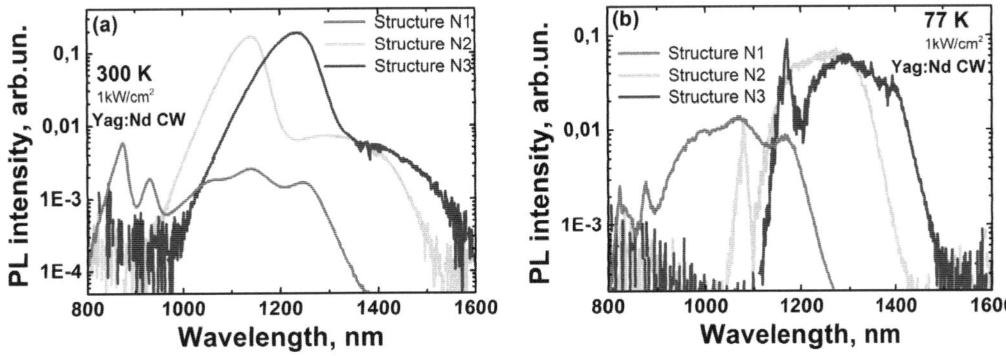

Figure 4. PL spectra for structure N1 (red), structure N2 (green) and structure N3 (blue) under high excitation power at: (a) – room temperature, (b) – 77 K.

4. Conclusion

The 2 ML-thick InAs insertions were epitaxially deposited by MOCVD into GaAs, metamorphic $In_{0.24}Ga_{0.76}As$ or metamorphic $In_{0.30}Ga_{0.70}As$ matrices. TEM studies have shown that in case of GaAs and $In_{0.24}Ga_{0.76}As$ matrices, the formation of SK QDs takes place, whereas for $In_{0.30}Ga_{0.70}As$, the InAs thickness is insufficient for QDs formation. The PL and TEM data has allowed to conclude that the density of QDs in $In_{0.24}Ga_{0.76}As$ is higher than that for GaAs. It has also been found that a smaller difference between the bandgap energy of InAs QDs or a QW and a matrix for metamorphic structures results in carrier thermal escape from the active region into the matrix with subsequent recombination there.

Acknowledgments
The study was supported by the Russian Foundation for Basic Research (grant #16-29-03127). Structural characterization by TEM was made on the equipment of the Federal Joint Research Centre «Material science and characterization in advanced technology» (Ioffe Institute, St. Petersburg, Russia).

References
[1] Bhattacharya P, Fornari R and Kamimura H 2011 Comprehensive Semiconductor Science and Technology, Six-Volume Set, 1st Edition *ELSEVIER S & T*
[2] Kirstaedter N, Ledentsov N N, Grundmann M, Bimberg D, Ustinov V M, Ruvimov S S, Maximov M V, Kop'ev P S, Alferov Zh I, Richter U et al. 1994 *Electron. Lett.* **30** 1416
[3] Lester L F, Stinz A, Li H, Newell T C, Pease E A, Fuchs B A and Malloy K J 1999 *IEEE Photon. Technol. Lett.* **11** 931
[4] Maximov M V, Zhukov A E Ed. S E Lyshevski 2014 Quantum Dot Lasers, In Dekker Encyclopedia of Nanoscience and Nanotechnology, Third Edition. *CRC Press: New York,* 3986–4000.
[5] Browne B, Lacey J, Tibbits T, Bacchin G, Wu T-C, Liu J Q, Chen X, Rees V, Tsai J and Werthen J-G 2013 *AIP Conference Proceedings***1556** 3-5
[6] Fujii H, Toprasertpong T, Wang Y, Watanabe K, Sugiyama M and Nakano Y 2014 *Prog. Photovolt: Res. Appl.* **22** 784–795
[7] Wheeldon J F, Valdivia C E, Masson D, Proulx F, Riel B, Puetz N, Desfonds E, Fafard S, Rioux B, Spring Thorpe A J, Arès R, Aimez V, Armstrong M, Swinton M, Cook J, Shepherd F, Hall T J and Hinzer K 2010 *Proc. SPIE* **7750** 77502Q
[8] Kalyuzhnyy N A, Mintairov S A, Salii R A, Nadtochiy A M, Payusov A S, Brunkov P N, Nevedomsky V N, Shvarts M Z, Martí A, Andreev V M and Luque A 2016 *Prog. Photovolt: Res. Appl.* **24** 1261–1271
[9] Wang Sh 2012 Lattice Engineering - Technology and Applications *CRC Press*
[10] Semenova E S, Zhukov A E, Mikhrin S S, Egorov A Yu, Odnoblyudov V A, Vasil'ev A P, Nikitina E V, Kovsh A R, Kryzhanovskaya N V, Gladyshev A G et al. 2004 *Nanotechnology* **15** S283 - S287
[11] Karachinsky L Ya, Kettler T, Novikov I I, Shernyakov Yu M, Gordeev N Yu, Maximov M V, Kryzhanovskaya N V, Zhukov A E, Semenova E S, Vasil'ev A P et al. 2006 *Semiconductor Science and Technology* **21**, 691-696
[12] Sasaki K, Agui T, Nakaido K, Takahashi N, Onitsuka R and Takamoto T 2013 *AIP Conf. Proc.* **1556** 22
[13] Mintairov S A, Emelyanov V M, Rybalchenko D V, Salii R A, Timoshina N K, Shvarts M Z and Kalyuzhnyy N A 2016 *Semiconductors* **50** 517
[14] Rybalchenko D V, Mintairov S A, Shvarts M Z and Kalyuzhnyy N A 2016 *Journal of Physics: Conference Series* **741** 012086
[15] Nadtochiy A M, Mintairov S A, Kalyuzhnyy N A, Rouvimov S S, Shernyakov Y M, Payusov A S, Maximov M V and Zhukov A E 2015 *Semiconductors* **49** 1090-1094

Emission properties of single-photon sources based on CdTe/ZnTe quantum dots

M V Rakhlin[1], K G Belyaev[1], S V Sorokin[1], S V Gronin[1], I S Mukhin[2,3], T V Shubina[1], S V Ivanov[1] and A A Toropov[1]

[1]Ioffe Institute, 26 Polytekhnicheskaya str., St. Petersburg, 194021, Russia
[2]St. Petersburg Academic University RAS, 8/3 Khlopina str., St. Petersburg, 194021, Russia
[3]ITMO University, 49 Kronverksky pr., St. Petersburg, 197101, Russia

E-mail: maximrakhlin@mail.ru

Abstract. We report on single photon emission from single CdTe/ZnTe quantum dots (QDs) grown by thermally-activated molecular beam epitaxy providing a reduced QD lateral density below 10^{10} cm^{-2}. For micro-photoluminescence spectroscopy studies and correlation measurements, the mesa structures of 200 and 500 nm diameters are fabricated by combination of electron-beam lithography and reactive plasma etching. Autocorrelation function of photons emitted from a single QD under cw excitation demonstrates antibunching with a value of $g^{(2)}(0)$ ~ 0.3 that is a signature of non-classical light.

1. Introduction

In the last decades, a lot of attention has been paid to sources of quantum light such as single photon emitters possessing non-classical photon statistics and sources of entangled pairs of single photons. These devices are key elements for the systems of quantum cryptography, information teleportation, and quantum computing. Self-organized single quantum dots (QDs) based on wide-gap II-VI compounds grown by epitaxial techniques (usually by molecular beam epitaxy – MBE), are considered as promising candidates for creation of the single photon sources, because they are less susceptible to blinking, which is the main drawback of colloidal QDs. So far, single-photon emission at elevated temperatures has been demonstrated for the structures with the epitaxial QDs made from various II-VI compounds including CdSe/Zn(S,Mg)Se [1-5] and CdTe/ZnTe [6,7]. Room temperature operation under optical [3] and electrical [4] pumping has been recently achieved. The serious obstacle hampering further progress in this field is a high lateral density inherent to the epitaxial QDs, which complicates spatial selection of a single excitonic line in a spectrum of a QD array, needed for the non-classical light observation. To distinguish the single-QD emission line, different approaches are exploited, including formation of etched mesa structures [1,2] or nanoapertures in a non-transparent mask [3,4,8], as well as insertion of a QD within a nanowire [5]. However, all these approaches require the relatively low QD density of at least 10^{10} cm^{-2} (the less the better), which can hardly be realized by conventional MBE.

As a promising approach to produce the arrays of II-VI QDs with a reduced density, the MBE technology exploiting thermal activation growth mode has been proposed [9,10]. This method implies the changing of surface energy with thermal annealing that facilitates the 2D-3D transition of a strained layer into a QD ensemble. It was first verified for the CdTe/ZnTe heterostructures [9]; then, some modification of this technique was applied to the structures based on CdSe/ZnSe [11]. The obtained lateral densities were about $4\cdot10^{10}$ cm^{-2} and (2 - 5) $\cdot10^{10}$ cm^{-2}, respectively, for a nominal insertion thickness in the 3 - 4.5 monolayers (MLs) range. The formed objects were not canonical Stranski-Krastanov QDs but rather flat nano-islands whose average sizes in vertical/lateral dimensions are 5 nm/50 nm for CdTe/ZnTe and 2 nm /20 nm for CdSe/ZnSe. Thus, the decrease of the lateral QD density was realized on account of increasing the sizes of the nano-islands. When the island size is markedly more than 10 nm, the exciton density of states (DOS) corresponds rather to that in a quantum well that can prevent undistinguished photon generation, since single photons emission assumes recombination of excitonic states possessing DOS of distinct QDs [12]. At low temperature, nevertheless, micro-photoluminescence (μ-PL) spectra exhibited narrow emission lines [9,10], probably related to fluctuating localization potential within the extended islands, and antibunching photon behaviour was demonstrated in the CdTe/ZnTe structures grown by this method [6,7].

In this paper, we investigate CdTe/ZnTe QDs grown by MBE using the thermal activation mode with and without exposure to Te flux. This procedure is considered as a method to influence basic QD emission properties. Data of μ-PL spectroscopy allow us to estimate the QD density, which turns out to be sufficient for spatial and spectral selection of single quantum emitters. For the confirmation of single photon statistics, the correlation function $g^{(2)}(\tau)$ was measured and analyzed.

2. Experiment

We have studied emission properties of two samples (labeled further as A and B) grown by MBE on InAs (001) substrates overgrown by a 0.2-μm-thick InAs buffer. The details of the growth procedure will be published elsewhere [13]. The sequence of deposited layers includes a 5-ML-thick ZnTe interface layer, a ~ 100 nm-thick $Zn_{0.9}Mg_{0.1}Te$ barrier, a 30 nm-thick ZnTe layer followed by a 3.1 - L-thick CdTe insertion, and a 20 nm-thick top ZnTe layer. The QDs were grown in a thermal activation mode, which implies growth of the strained CdTe layer on the ZnTe surface, coverage of the growth surface with a ~0.1 μm-thick amorphous Te layer at $T_S < 50°C$, desorbtion of the tellurium layer at ~ 220°C during the gradual T_S increase with a ramping rate of ~ 40°C/min, and the following gradual increase of T_S up to 300°C that was done slowly for 30-40 min either under exposure to Te flux (sample A) or without that (sample B).

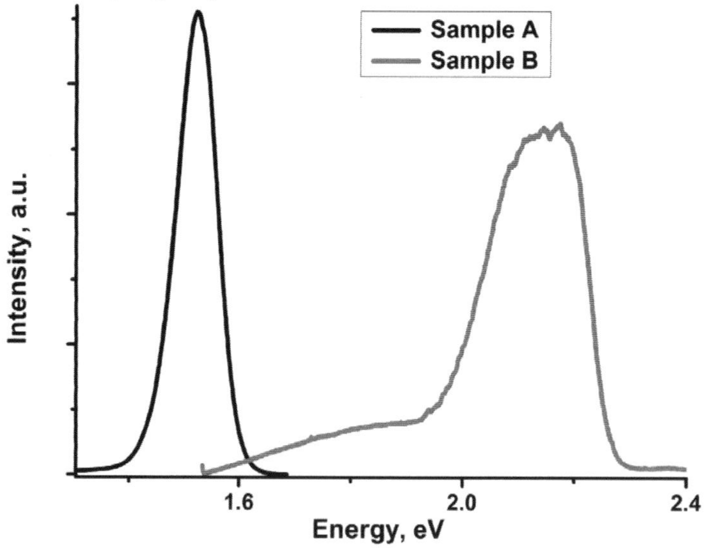

Figure 1. Normalized low temperature (77 K) spatially integrated PL spectra of structures grown either with exposure to Te flux (sample A) or without that (sample B).

The emission properties of QDs were investigated using cylindrical mesa-structures of 200 nm and 500 nm in diameter, which were fabricated by means of the combination of electron lithography and plasma etching. At first, the masks possessing the circle shape of different diameters were formed on the surface of the heterostructures with the QD layer using an e-beam-resist (MaN 2403). Then, the structures were etched by Ar plasma. Scanning electron microscopy studies have confirmed formation of cylindrical mesas with vertical sidewalls. The sample is mounted in a He-flow cryostat with an Attocube XYZ piezo driver inside, which allows us to optimize and precisely maintain the positioning of the chosen mesa with respect to a laser spot during a long time (few hours). The μ-PL measurements were carried out under optical excitation by a cw laser line (405 nm).The laser power density was 4 W/cm^2 before a cryostat window. Photon correlation measurements were performed in a Hanbury Brown – Twiss detection scheme exploiting two single-photon avalanche diodes (from Micro Photon Devices) possessing a full width at half maximum (FWHM) of photon timing resolution of about 40 ps.

3. Results and discussion

The cw PL spectra (integrated over the area of ~200 μ^2) of samples A and B, measured at 77 K are shown in figure 1. The emission line in sample A is markedly red-shifted relative to that in sample B, in spite of the same amount of deposited CdTe. This shift is a signature of larger QD sizes in sample A. In sample B grown without the Te flux exposure, the mobility of Cd adatoms is higher and their certain re-evaporation can take place. As a result, the QD has smaller sizes and blue-shifted emission. Moreover, FWHM of the emission line in sample B is more than twice larger than in sample A (200 meV versus 85 meV), that indicates a stronger QD size dispersion leading to the dispersion of the emission energies of individual QDs.

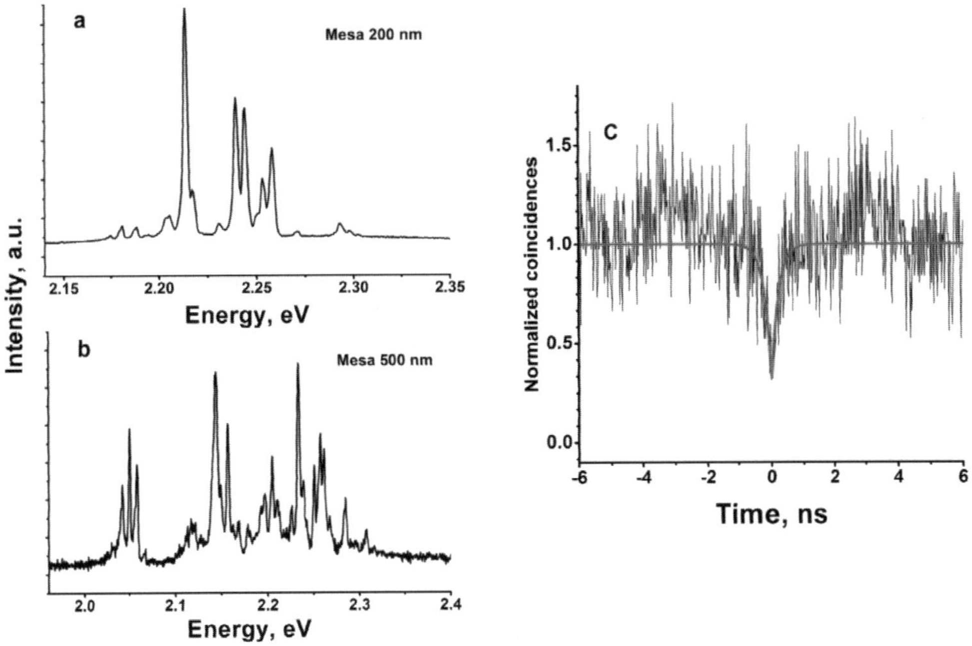

Figure 2. μ-PL spectra measured at 8 K inCdTe/ZnTe QD sample B using (a) 200-nm and (b) 500-nm mesas. (c) Normalized autocorrelation $g^{(2)}$ function of single photon emission measured in sample B at $T = 8$ K using the 200-nm mesa.

The µ-PL spectra of sample B reveal a number of relatively narrow lines, assigned to the emission of excitons and other electron-hole complexes – trions and biexcitons – from individual QDs (figure 2 (a, b)). The ballpark number of narrow excitonic lines in µ-PL spectra is estimated to be about several tens in the mesas with the 500-nm diameters. For the 200-nm mesas, the line number decreases to 2-7. We believe that the destroying of QDs adjusting to the sidewalls can contribute to such a decrease along with reducing the mesa diameter. The recorded number of lines in sample B corresponds to the QDs density as small as 10^{10} cm^{-2}. In contrast, the µ-PL spectra measured in sample A do not show pronounced narrow lines at all. This can indicate either the large density of QDs or the formation of flat extended nano-islands (quantum wells). The final conclusion needs detailed transmission microscopy studies (in progress).

Thus, sample B, grown without the Te flux exposure during the thermal annealing, demonstrates better optical characteristics from the standpoint of quantum light generation. Indeed, the distinct antibunching behavior has been recorded using the 200-nm mesas fabricated in this sample. Figure 2(c) shows the autocorrelation function $g^{(2)}(\tau)$ of a single QD, registered at $T = 8$ K. This function is obtained by fitting the experimental data with the equation

$$f(\tau) = a - b\exp\left(-\frac{|\tau|}{c}\right) \quad (1)$$

and normalizing by the obtained value of the parameter a [14]. The fitted value $c = 210$ ps gives estimation of the intrinsic width of the correlation function dip, which is essentially due to the lifetime of the involved QD exciton. For the value of the correlation function at zero time, the fitting gives the value of $g^{(2)}(0) = 0.32$ that evidences the dominance of single photon emission. We underline that, with high enough power density of excitation, the decay of different states, besides excitons, such as trions and biexcitons, can contribute to this value as discussed in [5-7]. Both proper subtraction of their contribution and optimization of design, aimed to increase the separation energies, can improve this characteristic.

4. Conclusions

We demonstrate single photon emission of CdTe/ZnTe QDs grown by thermally-activated MBE. It was achieved in a sample without Te flux exposure during the thermal annealing. We assume that the mobility of Cd adatoms is higher in this case as compared with QDs affected by the Te flux that results in the formation of QDs of smaller sizes but with more dispersed size statistics. This inhomogeneous QD ensemble exhibits a wide PL band (FWHM ~ 200 meV) with a maximum at 2.12 eV. Respective µ-PL spectra measured using 200-nm mesas demonstrate a countable number of narrow emission lines of single QDs corresponding to the dot density of about 10^{10} cm^{-2}. The value of the autocorrelation function $g^{(2)}(0) = 0.32$ measured in this sample is a signature of quantum light.

Acknowledgements

The authors gratefully acknowledge the financial support of the Russian Science Foundation (Project #14-22-00107).

References

[1] Sebald K, Michler P, Passow T, Hommel D, Bacher G and Forchel A 2002, *Appl. Phys. Lett.* **81** 2920

[2] Strauf S, Ulrich S M, Sebald K, Michler P, Passow T, Hommel D, Bacher G and Forchel A 2003, *Phys. Status Solidi* B **238** 321

[3] Fedorych O, Kruse C, Ruban A, Hommel D, Bacher G and Kümmell T 2012, *Appl. Phys. Lett.* **100** 061114

[4] Quitsch W, Kümmell T, T. Kummel, A. Kruse C, Hommel D and Bacher G 2014,*Phys. Status*

Solidi C **11** 1256
[5] Sallen G, Tribu T, Aichele T, Andre R, Besombes L, Bougerol C, Tatarenko S, Kheng K and Poizat J Ph 2009, *Phys. Rev. B* **80** 085310
[6] Couteau C, Moehl S, Tinjod F, Gerard J M, Kheng K, Mariette H, Gaj J A and Romestain R 2004, *Appl. Phys. Lett.* **85** 6251
[7] Nowak S, Goryca M, Kamzimierczhuk, Gaj J A, Golnik A, Kossacki P, Wojnar P and Karczewski G 2008, *Act. Polonica A* **114** 5
[8] Rakhlin M V, Belyaev K G, Sedova I V, Sorokin S V, Gronin S V, Usikova A A, Sitnikova A A, Brunkov P N, Ivanov S V and Toropov A A 2016, *Phys. Status Solidi C* **13** 514
[9] Tinjod F, Gilles B, Moehl S, Kheng K and Mariette H 2003, *Appl. Phys. Lett.* **82** 4340
[10] Tinjod F, Robin I C, Andre R, Kheng K and Mariette H 2004, *J. of Alloys and Comp.* **371** 63
[11] Mahapatra S, Kiessling T, Margapoti E, Astakhov G V, Renner J, Bass U, Bougerol C, Schmidt T, Bendounan A, Schmitt F, Schumacher C, Worschech L, Ossau W, Geurts J, Molenkamp L W, Reinert F, Forchel A and Brunner K 2007, *Phys. Status Solidi B* **4** 3129
[12] *Single Semiconductor Quantum Dots*. Ed. Michler P 2009, Series: Nanoscience and Nanotechnology, Springer-Verlag, Berlin, ch. 6,7
[13] Sorokin S V, Sedova I V, Gronin S V, Belyaev K G, Rakhlin M V, Toropov A A, Mukhin I S and Ivanov S V 2016, *Technical Physics Letters* (in print)
[14] Zwiller V, Blom H, Jonsson P, Panev N, Jeppesen S, Tsegaye T, Goobar E, Pistol M E, Samuelson L and Björk G 2001, *Appl. Phys. Lett.*.**78** 2476

Multi-particle effects in far-IR optical transmission spectra of Ge/Si quantum dots

D A Pashnev, R M Balagula, A N Sofronov, D A Firsov and L E Vorobjev

Peter the Great St.Petersburg Polytechnic University, Polytechnicheskaya 29, St.Petersburg 195251, Russia

e-mail: pashnevdaniil@gmail.com

Abstract. Photoinduced and equilibrium optical transmission was studied in undoped and δ-doped Ge/Si quantum dot structures in the far-infrared spectral range. Multi-particle excitations were observed in the absorption spectra of the structures with different doping levels. Photoinduced absorption spectra demonstrate the thermal distribution of the photoexcited holes over the single-particle states of the quantum dot ensemble in contrast with the equilibrium absorption spectra.

1. Introduction

The semiconductor epitaxial structures with quantum dots draw attention in the present time due to the specifics of the energy spectrum of charge carriers, which is discrete, and, therefore, the possibility of their use in optoelectronic devices. Structures with quantum Ge/Si dots are especially promising if we consider the possibility of their integration in advanced silicon technology.

Novel detectors of both near and mid-infrared radiation, operating via interband and intraband optical transitions of holes between the quantum-dot levels, can be created on the basis of systems with Ge/Si quantum dots. Detectors in the mid-IR range [1, 2] have advantages over similar devices based on quantum wells, such as the absence of strict selection rules for the radiation polarization, which leads to the possibility of using the simplest geometry of the normal incidence of radiation on the surface of the photosensitive device.

As most of the self-assembled quantum dot systems, the Ge/Si quantum dots (QDs) have significant size dispersion. It leads to the inhomogeneous broadening of the energy states of the quantum dot ensemble. In such conditions, the exact distribution of charge carriers over the dot ensemble is a significant issue. In the classic work [3], the nonthermal distribution is assumed in the framework of the random population approach for quantum dot systems based on the master equations. Under such conditions, at low temperatures dots with different sizes (and ground state energies) have the same population since there is a lack of interdot coupling. However, in structures with a dense array of quantum dots, it is not always the case, and the lateral correlation can exist. Indeed, the authors of paper [4] successfully applied the thermal (Fermi) distribution for the description of the temperature dependencies of the intraband absorption in the InAs/GaAs quantum dot system. However, the inhomogeneous broadening was not taken into account in the model used in reference [4].

In doped quantum dot structures, the dots with different charge states (different number of electrons or holes inside) can exist. In this case, one should consider the distribution of charge carriers over the *alternative multi-particle configurations* rather than over the single-particle states come from the quantum confinement.

In the present work, we study the optical transmission spectra of the dense array of Ge/Si quantum dots in the far infrared spectral range close to the very first interlevel resonance in the valence band. In this spectral range, the specific conditions are realized, when the optical transition energy is comparable with the inhomogeneous broadening of the interlevel resonance. We focused on both the equilibrium absorption spectroscopy and the steady-state non-equilibrium absorption spectroscopy under the interband optical excitation of the electron-hole pairs (so-called photoinduced absorption spectroscopy).

2. Experimental setup and samples

The MBE-grown structures with 10 periods of self-assembled GeSi/Si QDs were studied in this work. A surfactant (Sb) was used during QD growth process in order to increase the QDs density as described elsewhere [5]. We studied both the undoped structure and p-doped in a δ-layer structures with boron surface densities of $4\cdot10^{11}$, $8\cdot10^{11}$, and $1.2\cdot10^{12}$ cm^{-2}. The QDs have a pyramidal shape with an average height of h = 2.7 nm and the average base size of b = 14 nm. The surface density of QDs is $2\cdot10^{11}$ cm^{-2}.

We used the multipass sample geometry similar to the one used in our previous studies of the steady-state and transient photoinduced absorption in this structures in the mid-infrared spectral range [6]. The samples for optical measurements were mounted in a liquid nitrogen cooled cryostat with the controllable temperature T_L in the range between 80 and 320 K. The cryostat was equipped with a pair of polyethylene optical windows for the far-infrared probe radiation and a quartz optical window for the pump illumination.

The spectra of optical absorption in the far-infrared spectral range were recorded with a Bruker Vertex 80v Fourier transform spectrometer. A globar was used as a source of broadband infrared radiation. The mylar beamsplitter ensured the measurements in the spectral range of 100 - 500 cm^{-1} (approximately 10-60 meV). The intensity of the probe radiation, which has passed through the sample, was measured with a liquid He-cooled silicon bolometer.

A solid-state CW YAG:Nd laser with a wavelength of 532 nm and the maximal output optical power of 100 mW was used as a source of the pump illumination creating additional charge carriers in the samples. The photoinduced absorption spectra were recorded in a step-scan spectrometer mode with a SR830 Lock-In amplifier, which was modelocked to the chopping frequency of the pump radiation.

3. Experimental results and discussion

Let us consider first the equilibrium optical transmission spectra of the doped QDs that were measured at 80 K and 300 K. They are plotted in figure 1 for three structures with a nominal doping level (the ratio of the acceptor and QDs surface densities) of 2, 4, and 6 holes per dot and are surprisingly similar for all three structures.

Within the single-particle approximation, one should expect a completely different behaviour of the absorption peak related to the inhomogeneously broadened interlevel resonance of QDs with the increase of the total number of holes in the system. In an array of quantum dots with size dispersion, holes first populate the larger dots due to the lower ground state energies. The increase of the number of holes leads to population of smaller and smaller dots, where the interlevel resonance energy increases. As a result, for non-interacting holes one should expect the interlevel absorption spectrum to widen into a short-wave-length region with an increase of the nominal doping level, while the long-wave-length edge of the spectrum corresponds to the Gaussian distribution of the dots' sizes.

Consideration of the Coulomb interaction of the holes changes this picture dramatically. A charged dot with more than one hole inside should be treated as a dot with a quasi-particle that consists of several interacting holes (sometimes referred as 0D-plasmons for a large number of particles). Addition of one more hole changes (increases) the ground state energy of this quasi-particle (mainly due to the Coulomb repulsion), but the *energy differences between the ground and low excited states remain the same*. This fact was firstly mentioned by W. Kohn in the work [7] for cyclotron resonance of 3D electron gas with interaction, and later generalized for quantum dots (see, for example, the recent work [8] and references therein) with a parabolic confinement potential (the so-called "parabolic" quantum dot model, that can be applied to real QDs with finite potential barriers for description of the lowest excited states since their energies are usually significantly lower than the barrier height).

Thus, the interlevel resonance energy of charged dot does not depend on the number of holes inside, but is determined only by the confinement, i.e. by the geometry of the dot. However, the ground state energy of the confined few-hole quasi-particle does depend on the number of holes in this quasi-particle. In a broadened array of quantum dots with a low nominal doping, a certain hole can find an empty dot with ground state energy less than the one of the already charged dot. As a result, the distribution of the holes over the dot ensemble (which, in fact, is the thermal distribution over the alternative-multi-hole configurations) is more intricate than the thermal one (over the single-particle states). Increase of doping leads just to the increase of the hole number in certain quasi-particles (with the lowest energies), and the inter-(quasi-particle)-level absorption spectrum does not change. Experimental spectra of doped structures demonstrate exactly the same behaviour.

The experimentally observed temperature broadening of the far-IR transmission spectra (see fig 1, where the transmission data at 300 K are plotted with the dashed lines) proves again the lateral correlation in the distribution, because it clearly corresponds to redistribution of the holes between the dots with the heating of the structure.

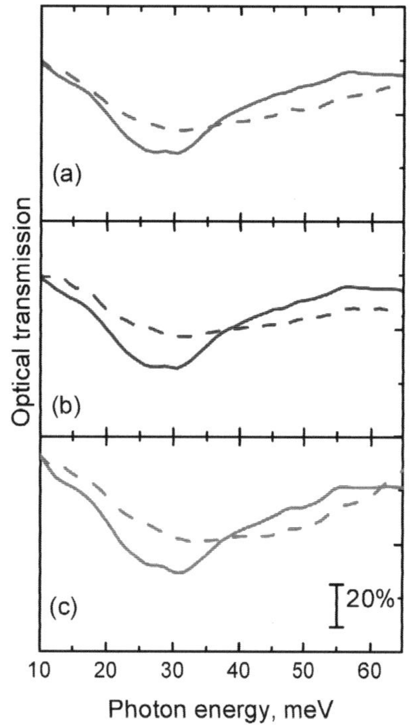

Figure 1. Transmission spectra of doped samples with a nominal doping of 2 (a), 4 (b) and 6 (c) holes per dot at T_L = 80 K (solid lines) and T_L = 300 K (dashed lines). Scale is the same for all panels.

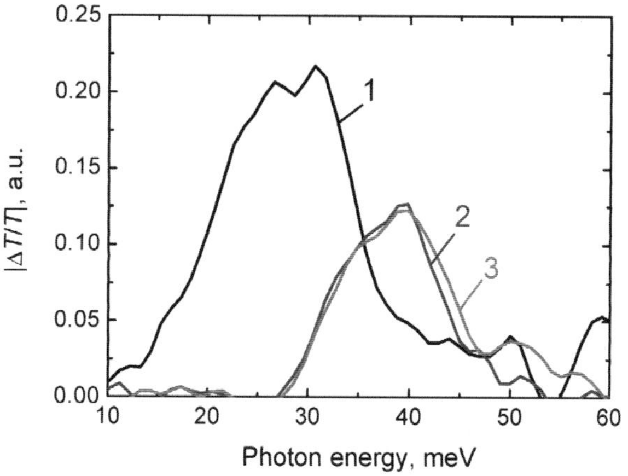

Figure 2. Photoinduced absorption spectra of the undoped structure (curve 1) and the structures with a nominal doping of 4 (curve 2) and 6 (curve 3) holes per dot at T_L = 80 K. Excitation intensity is the same for all curves and equal to 100 mW at a wavelength of 532 nm.

The photoinduced absorption spectra for the undoped structure and the two most heavily doped structures, measured at 80 K under the same excitation intensity, are shown in figure 2 as a ratio of the transmission change ΔT under interband photoexcitation and the equilibrium transmission T. In the case of photo-excitation, the capture of an additional hole into the QD does not change the charge state of the dot due to the simultaneous capture of an electron to the strain-induced 3D potential wells at the Si-side of the QD interface (see, for example, [9] for details on the Ge/Si QD interface band alignment and profiles). Thus, we effectively "turn off" the Coulomb interaction of the holes inside the dot. At the same time, the absorption spectra correspond exactly to the single-particle behavior discussed above. In the undoped sample, photoexcited holes occupy the dots with the lowest ground state energies (large dots). At the same excitation level, photoexcited holes in the doped sample occupy smaller dots with higher interlevel transition energies, so the photoinduced absorption peak widens to the short-wave-length region. Obviously, the extrinsic holes in doped samples do not contribute to the photoinduced change of the transmission, and we observe zero ΔT signal at $\hbar\omega < 30$ meV, where the equilibrium absorption exists.

4. Conclusion

In this work, we experimentally observe simultaneously a non-interacting holes' response and a 0D multi-hole excitations' response in far-IR equilibrium and photoinduced optical transmission spectra of Ge/Si quantum dot structures with size dispersion. The energy of the transition from the ground to one of the first excited states have been found to be independent on the number of holes in the multi-hole quasi-particle and equal to 30 meV. The qualitative differences of the hole distribution over the dot ensemble are indirectly registered in the absorption spectra. The temperature evolution of the few-particle excitation spectra proves the presence of the lateral correlation in the distribution of charge carriers over the dot ensemble.

Acknowledgements
This work was supported by grant of the President of the Russian Federation for young Candidates of sciences (MK-6064.2016.2) and Ministry of Education and Science of the Russian Federation (state assignment).

References

[1] Yakimov A I, Kirienko V V, Armbrister V A, Bloshkin A A and Dvurechenskii A V 2015 Appl. Phys. Lett. **107** 213502
[2] Yakimov A I, Kirienko V V, Bloshkin A A, Armbrister V A, Kuchinskaya P A and Dvurechenskii A. V. 2015 Appl. Phys. Lett. **106** 032104
[3] Grundmann M and Bimberg D 1997 Phys. Rev. B **55** 9740
[4] Bras F, Boucaud P, Sauvage S, Fishman G and Gérard J.-M. 2002 Appl. Phys. Lett. **80** 4620
[5] Tonkikh A, Zakharov N, Talalaev V and Werner P 2010 physica status solidi (RRL) Rapid Research Letters **4** 224
[6] Sofronov A, Vorobjev L, Firsov D, Panevin V, Balagula R, Werner P and Tonkikh A 2015 Superlattices and Microstructures **87** 53
[7] Kohn W 1961 Phys. Rev. **123** 1242
[8] Hayrapetyan D B, Kazaryan E M and Sarkisyan H A 2016 Physica E **75** 353
[9] Kurdi M El, Sauvage S, Fishman G and Boucaud P 2006 Phys. Rev. B **73** 195327

Intraband absorption and interband photoconductivity transients in Ge/Si quantum dots

O I Kirilenko, R M Balagula, A N Sofronov, D A Firsov and L E Vorobjev

Peter the Great St.Petersburg Polytechnic University, Polytechnicheskaya 29, St.Petersburg 195251, Russia

e-mail: kirilenko.oleg1993@yandex.ru

Abstract. In this work a series of relaxation curves of photoinduced intraband absorption and interband photoconductivity were measured in Ge/Si quantum dots in the temperature range from 80 K to 300 K and at different levels of the interband excitation. The low-temperature experimental curves show a two-step decay of a population of quantum dots. At high temperatures there is only one decay component. A simultaneous analysis of photoinduced intraband absorption and interband photoconductivity relaxation curves shows that quantum dots act as traps for non-equilibrium holes only at low temperatures.

1. Introduction

Nanostructures with Ge/Si quantum dots are considered as a promising material for the development of new optoelectronic devices. This is due to their optical properties in the near and mid-infrared range. This work is devoted to studying the relaxation of nonequilibrium charge carriers in structures with Ge/Si quantum dots after a short high-power optical interband excitation.

2. Experimental setup and samples

A structure with self-assembled Ge/Si quantum dots (QD) MBE-grown on a Si (001) substrate was studied. The structure contains 10 periods of quantum dot layers formed in the growth regime similar to the one described in [1]. Quantum dots have a pyramidal shape with an average height of 2.7 nm, and an average base size of 14 nm. No intentional doping was performed.

In our experiment, the non-equilibrium charge carriers are excited in the structure by a 10-nanosecond pulse laser with a wavelength of 532 nm. Dynamics of the population of the quantum dot states was studied by means of the mid-infrared intraband absorption decay measurements. The wavelength of the mid-IR radiation was tuned to resonance corresponding to the hole transitions from the QD ground state to the continuum of states. We measured a signal of the pump-induced change of optical mid-infrared transmission ΔT that is directly proportional to the absorption for small ΔT values.

At the same time, the interband photoconductivity signal decay was measured. The photoconductivity measurements allowed us to uncover the overall relaxation processes of free bulk carriers in the structure. The experimental setup provided the time resolution of 100 ns.

3. Experimental results

The series of relaxation curves of photoinduced intraband absorption and interband photoconductivity $\Delta\sigma$ after a short excitation pulse were measured in the temperature range from 80

K to 300 K at different interband excitation levels. The absorption decay curves for the maximum excitation power are plotted in figure 1 for low and high temperatures as time dependencies of the optical transmission change ΔT together with the interband photoconductivity signals $\Delta\sigma$ in the same semi-log scale.

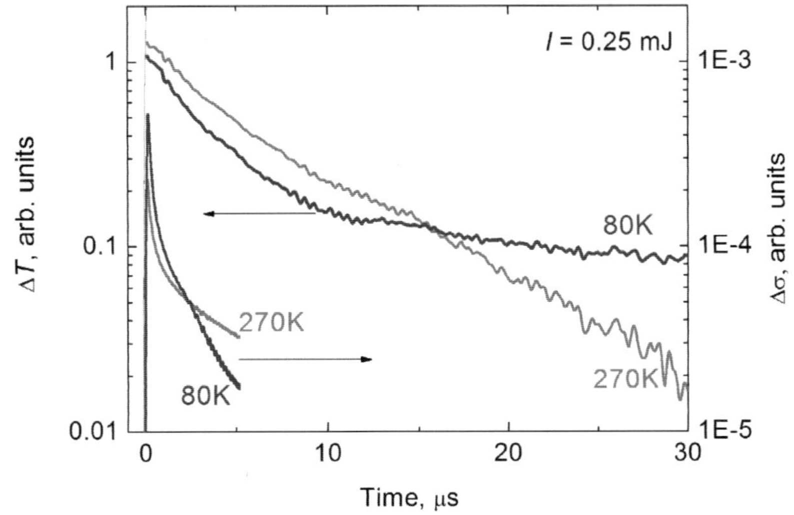

Figure 1. Time dependencies of the photoconductivity (right axis) and the photoinduced mid-IR absorption (left axis) signals after a pulsed excitation $I = 0.25$ mJ at two temperatures.

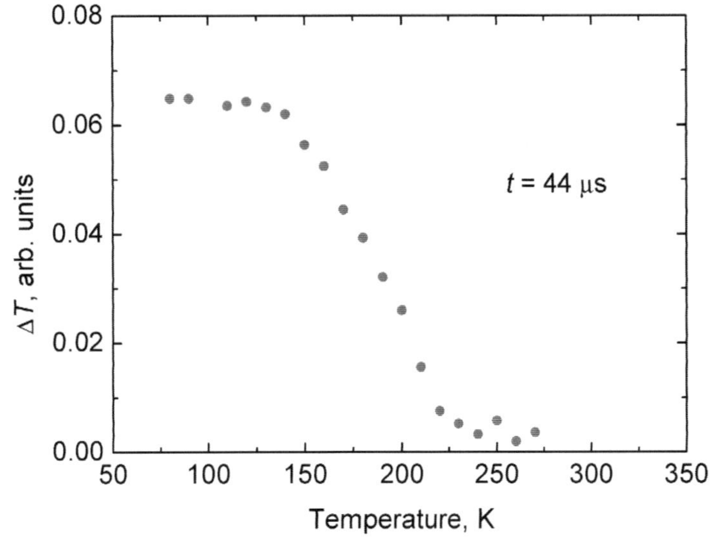

Figure 2. The dependence of the photoinduced absorption signal upon the temperature 44 microseconds after the excitation pulse. At low temperatures, the probability of ejection of holes from the quantum dot is small. This probability increases with temperature and holes begin to leave the dots. At room temperatures, there are no photoexcited carriers inside the dots.

The interband photoconductivity shows a non-linear recombination at the initial stage. Exponential parts of the photoconductivity decay curves have the characteristic decay times less than 10 μs for both temperatures.

The slope of the high-temperature absorption decay curve is comparable with the corresponding exponential part of the photoconductivity curve slope. Thus, the population of the QD ground states follows the bulk non-equilibrium charge carrier density which determines the photoconductivity.

In contrast, the low-temperature absorption relaxation curve approximately follows the exponential part of the photoconductivity decay only within the first 10 μs. Next, the absorption curve demonstrates a two-exponential decay behaviour with a slow relaxation time about one order of magnitude larger than the fast one. This experimental result clearly shows the presence of holes inside QDs when the bulk carrier density has already relaxed. Quantum dots act as traps for bulk charge carriers only at low temperatures, whereas at high temperatures this effect disappears. This behaviour can be clearly seen in figure 2, where the temperature dependence of the photoinduced absorption signal (proportional to the number of holes captured to QDs) 44 μs after the excitation pulse is plotted. We attribute this effect to the thermal escape of holes from the dots and their drag by the bipolar diffusion current of bulk charge carriers.

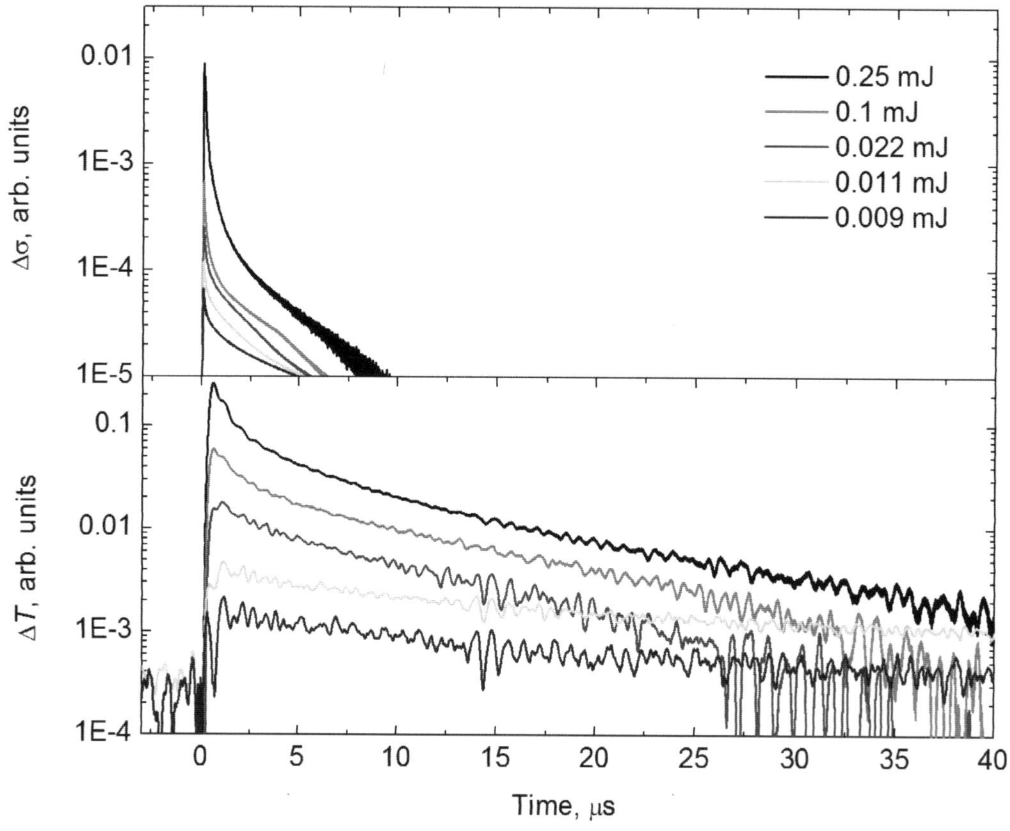

Figure 3. The time dependencies of photoconductivity (top panel) and photoinduced absorption (bottom panel) signals at different pump intensities at a temperature of 80 K.

The measured low-temperature time dependencies of the photoconductivity and photoinduced absorption signals are plotted in figure 3 for different excitation powers. The photoconductivity decay curves depend weakly on the excitation level, showing approximately the same characteristic decay times of the bulk carrier density below 10 μs for all excitation powers. However, the character of the absorption decay curves changes with the decrease of the excitation. Namely, there are no fast decay components for low-excitation QD population decay curves, but the characteristic decay time for the slow component becomes higher with the decrease of the excitation.

We attribute the disappearance of the fast QD population decay component at low excitation levels to the relatively low average energy of the bulk holes in the Si valence band, which becomes lower than the strain-induced energy barrier for hole capture to the QD (see, for example, [2] for details on the Ge/Si QDs heterointerface profile). The dependence of the slow relaxation time (which, in fact, is the lifetime of holes inside the QD) upon the excitation level can be related to the different QDs charge states since the total number of hole states in a QD is much higher than the total number of electronic states in strain-induced potential wells on a Si side of the QD interface [2].

4. Conclusion

In this work, we demonstrate experimentally that Ge QDs in a Si matrix effectively act as traps for non-equilibrium photo-excited holes only at low temperatures. The population of QDs follows directly the time dependence of the bulk photo-excited charge carrier density in silicon at high temperatures, so there are no holes in QDs when there are no holes in the bulk Si valence band. The characteristic lifetime of holes in QDs can be as long as 50 μs, and depends highly on temperature and initial excitation level.

Acknowledgements

This work was supported by grant of the President of the Russian Federation for young Candidates of sciences (MK-6064.2016.2) and Ministry of Education and Science of the Russian Federation (state assignment).

References

[1] Tonkikh A, Zakharov N, Talalaev V and Werner P 2010 *Phys. Status Solidi RRL* **4** 224
[2] Kurdi M El, Sauvage S, Fishman G and Boucaud P 2006 *Phys. Rev.* B **73** 195327

New plasma-assisted approach to the fabrication of $Cu(In,Ga)(S,Se)_2$ nanowires

D A Mokrov[1], S P Zimin[1], E S Gorlachev[2], I I Amirov[2], V V Naumov[2] and V F Gremenok[3]

[1] Yaroslavl State University, Sovetskaya Street 14, 150003 Yaroslavl, Russia
[2] Yaroslavl Branch of the Institute of Physics and Technology of the Russian Academy of Sciences, Universitetskaya Street 21, 150007 Yaroslavl, Russia
[3] Scientific-Practical Materials Research Center of the NAS of Belarus, P. Brovka Street 17, 220072 Minsk, Republic of Belarus

E-mail: dimonstr16@mail.ru

Abstract. In this work we have applied the low-energy high-density low-pressure radio-frequency inductively coupled argon plasma treatment to the $Cu(In,Ga)(S,Se)_2$ thin films grown on glass substrates using a selenization/sulfurization process. This approach allowed to obtain nanowires with an average height varying from 120 to 240 nm and with a 15-50 nm diameter. It is shown that the mechanism of the nanowire formation is plasma-assisted vapor-liquid-solid growth in combination with micromasking with In-Ga alloy nanoinclusions serving both as catalyst droplets and micromasks.

1. Introduction

Chalcopyrite compounds $Cu(In,Ga)Se_2$ (CIGS) and $Cu(In,Ga)(S,Se)_2$ (CIGSS) are promising materials for the fabrication of high-efficient thin film solar cell devices due to the unique combination of a high absorption coefficient, stability, radiation tolerance and a tunable direct band gap in the range from 1.04 eV for $CuInSe_2$ to 2.4 eV for $CuGaS_2$ [1, 2]. CIGSS-based thin film solar cells are reported to show an outstanding efficiency reaching 20.9% [3]. One of the most promising methods to further increase the solar cell efficiency is to apply a micro- or nanostructured absorbing CIGS/CIGSS layer. While it was reported to be fabricated using high-energy ion beam sputtering [4], we have proposed a novel, more versatile and technologically convenient approach of using low-energy inductively coupled plasma treatment which allowed us to carry out plasma-assisted formation of arrays of uniform nanotips with controlled geometrical parameters on the surface of CIGS layers [5, 6].

Here we apply this approach to CIGSS films and report the plasma-assisted fabrication of nanowires with varied height, diameter and density during inductively coupled argon plasma treatment of CIGSS films grown using a selenization/sulfurization growth method.

2. Experimental details

$Cu(In,Ga)(S,Se)_2$ thin films with a 2.5 μm thickness were grown on glass substrates using a selenization/sulfurization process. This method is based on simultaneous annealing of the metal layers in the combined Se and S vapor in an inert gas stream [2]. For the film growth, precursor Cu-In-Ga films were deposited onto glass substrates at 100°C by ion-beam plasma evaporation. The simultaneous selenization and sulfurization were conducted in a horizontal tube reactor under N_2 flow as a reaction with solid state sources of S and Se and had two consecutive annealing stages at 260°C and 540°C temperatures [2]. For the films, the Ga/(In+Ga) and S/(Se+S) ratios were 0.07 and 0.77, respectively.

Plasma treatment of the samples was carried out in a high-density low-pressure radio-frequency inductively coupled argon plasma reactor according to our experimental technique used for the nanostructuring of CIGS films [5, 6]. During the treatment, the argon gas flow was 10 sccm, the operating pressure was 0.07 Pa, and the RF bias power on the aluminum substrate holder was 300 W. These plasma reactor operating parameters allowed obtaining the low average energy of the bombarding argon ions of about 200 eV. The treatment duration was 60 s and 120 s.

The morphology of the as-grown and plasma-treated CIGSS film surface was studied with scanning electron microscopy (SEM) using a Supra 40 (Carl Zeiss) microscope.

3. Results and discussion

As-grown CIGSS films were polycrystalline, which is also typical for CIGS films [5, 6]. SEM images of the initial surface of the CIGSS thin film on a glass substrate are given in figure 1. On the CIGSS film surface, there were droplet-like spherical polycrystalline clusters of 1–3 μm size, as seen in figure 1(a), which were resulted from the used film growth technique. The grain size, as measured on flat surface areas such as shown in figure 1(b), was 65–220 nm. According to the high-resolution SEM data, the initial film surface also showed a presence of arrays of nanosize (5–15 nm) inclusions with a density of 1.5×10^{12} cm^{-2}, as seen in figure 1(b) as white dots. In Ref. [7] it was theoretically shown for the CIGS material that there should take place the segregation of Ga or In nanoinclusions, and it is expectable to have them also for the CIGSS films. Therefore, the nanoinclusions on the initial surface are assumed to be metallic In, Ga or In-Ga alloy nanoparticles.

(a)
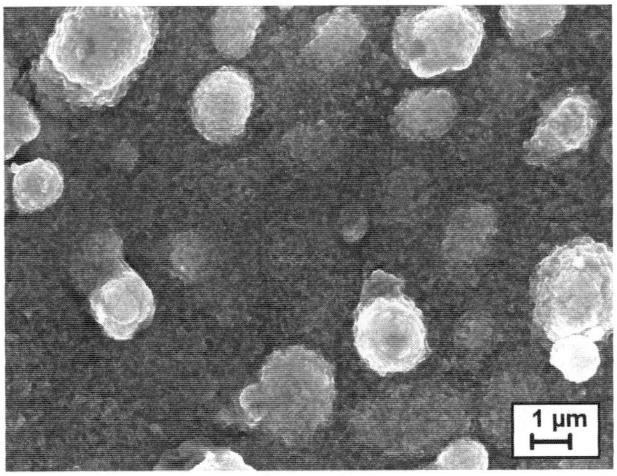

(b)

Figure 1. SEM images of the initial CIGSS film surface.

Plasma treatment of the CIGSS film surface resulted in its significant modification. While the spherical clusters remained practically intact, the grain-related micro-relief on flat areas was smoothed. Simultaneously, there also took place a formation of vertical nanostructures uniformly distributed over the entire plasma-treated area including spherical clusters (figure 2). After the plasma treatment of 60 s duration, the average height of the nanostructures was 120 nm, the diameter was 15–35 nm, and the density was 8×10^{10} cm^{-2}. After the plasma treatment of 120 s duration, the average height of the nanostructures was 240 nm, the diameter was 30–50 nm, and the density was 4×10^{10} cm^{-2}. Typical SEM images for the nanowires formed on the CIGSS surface after the 60 s and 120 s plasma treatments are given in figure 3 and figure 4, respectively. SEM data clearly show that the nanowires are characterized by round "caps" on their tops.

In order to explain the physical mechanism of the nanowire formation, we have to examine the main processes which take place on the surface of the CIGSS film during the plasma treatment. Firstly, a physical sputtering of film by low-energy Ar ions takes place. The surface is smoothed, and the sputtered atoms of selenium, sulfur, copper, indium, and gallium are distributed over it in a vapor phase. Secondly, plasma heats the film surface (up to ~ 500 K). The In-Ga nanoinclusions with eutectic temperature of 15.3 °C become liquid. As a result, the formation of In-Ga droplets, which are liquid during plasma treatment, occurs. These solidified droplets can be seen as "caps" on tops of the nanowires (figure 3, figure 4).

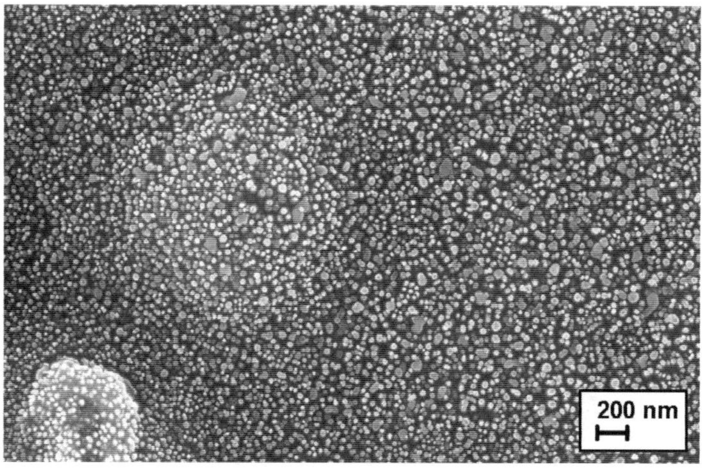

Figure 2. Top-down SEM image of the array of nanowires fabricated on the CIGSS film surface after a 60 s argon plasma treatment. A spherical cluster can be seen in the bottom-left corner of the image.

Figure 3. SEM image of the nanowires fabricated on the CIGSS film surface after a 60 s argon plasma treatment. Sample tilt during imaging is 70°.

During the plasma treatment the droplets serve as micromasks protecting the material under it from physical sputtering. On the other hand, when the high surface temperature and the high vapor density for the sputtered atoms provide sufficient conditions for the realization of vapor-liquid-solid growth [8], the droplets serve as liquid catalysts for the growth of solid nanowires under them. As a result, the nanowires can grow both in vertical direction and laterally with the ongoing treatment, which corresponds to the SEM data. The density decrease can be explained by eventual removal of the micromasks.

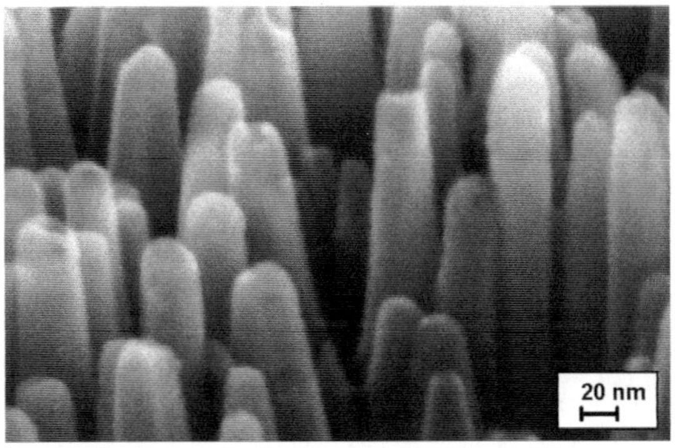

Figure 4. SEM image of the nanowires fabricated on the CIGSS film surface after a 120 s argon plasma treatment. Sample tilt during imaging is 70°.

4. Conclusions

We have experimentally shown a capability of fabricating large arrays of $Cu(In,Ga)(S,Se)_2$ nanowires with varied geometrical parameters using low-energy plasma treatment of thin films on glass substrates. The physical mechanism of CIGSS nanowire growth includes micromasking and the vapor-liquid-solid mechanism with In-Ga nanoinclusions serving both as micromasks and catalytic liquid droplets. Our approach is versatile, low-cost, applicable both to CIGSS and CIGS films, and is prospective for the development of thin film photovoltaic devices with high efficiency.

Acknowledgments

SEM investigations were carried out at the Facilities Sharing Center "Diagnostics of Micro- and Nanostructures", this work was supported by the Ministry of Education and Science of the Russian Federation.

References

[1] Kazmerski L L 1977 *Renewable and Sustainable Energy Reviews* **1** 71
[2] Tivanov M S, Zaretskaya E P, Gremenok V F, Ivanov V A, Mudriy A V, Ivaniukovich A V, Zalesski V B, Zykotinski S and Bente K 2007 *Moldavian J. Phys. Sci.* **6** 117
[3] Green M A, Emery K, Hishikawa Y, Warta W and Dunlop E D 2016 *Prog. Photovolt.: Res. Appl.* **24** 905
[4] Lee J-Y, Ahmed Sk F, Park M-H, Ha S-K, Park J-K, Yun J H, Ahn S J, Lee K-R, Choi W J, Moon M W and Yang C-W 2012 *Solar Energy Mater. Solar Cells* **105** 119
[5] Zimin S P, Gremenok V F, Gorlachev E S, Mokrov D A, Amirov I I, Naumov V V, Kim W Y, Moon C B, Chai S H and Jhun C G 2015 *Proc. 31st European Photovoltaic Solar Energy Conference and Exhibition* p 1365
[6] Zimin S P, Mokrov D A, Gorlachev E S, Amirov I I, Naumov V V, Gremenok V F and Bashkirov S A 2015 *Proc. Int. Conf. "Nanomeeting-2015"* p 388
[7] Ludwig C D R, Gruhn T, Felser C, Schilling T, Windeln J and Kratzer P 2010 *Phys. Rev. Lett.*

105 025702
[8] Zimin S P, Gorlachev E S, Amirov I I, Naumov V V and Bagiyeva G Z 2015 *Semicond. Sci. Technol.* **30** 035017

IOP Conf. Series: Journal of Physics: Conf. Series 816 (2017) 012029 doi:10.1088/1742-6596/816/1/01

Misfit stresses in a composite core-shell nanowire with an eccentric parallelepipedal core subjected to one-dimensional cross dilatation eigenstrain

S A Krasnitckii[1,2], D R Kolomoetc[1], A M Smirnov[2] and M Yu Gutkin[1-3]

[1]Peter the Great St. Petersburg Polytechnic University, Polytekhnicheskaya 29, St. Petersburg, 195251 Russia
[2]ITMO University, Kronverkskii pr. 49, St. Petersburg, 197101 Russia
[3]Institute of Problems in Mechanical Engineering, Russian Academy of Sciences, Bolshoi 61, Vasil. Ostrov, St. Petersburg, 199178 Russia

Email: smirnov.mech@gmail.com

Abstract. We present an analytical solution to the boundary-value problem in the classical theory of elasticity for a core-shell nanowire with an eccentric parallelepipedal core of an arbitrary rectangular cross section. The core is subjected to one-dimensional cross dilatation eigenstrain. The misfit stresses are found in a concise and transparent closed form which is convenient for practical use in theoretical modeling of misfit relaxation processes.

1. Introduction

Fabrication and studies of composite nanowires (NWs) is a necessary step in the development of modern nanotechnologies in optoelectronics, nanoscale field effect transistors, storage and data transmission devices, logic devices, sensors, etc. [1–10]. Physical properties of composite NWs depend on the shape, size, chemical composition, and types of crystalline lattices of the NW components as well as on the presence of various defects in their structure. In particular, the shape effects can be tightly related with peculiarities in misfit stress distribution over the NWs and with the mechanisms of misfit stress relaxation in them. However, in theoretical description of these mechanisms, they commonly use the model of cylindrically symmetric core-shell NWs [11–26] as it is much simpler for analytical modeling. The problem is that this simple approach strictly limits the variety of possible relaxation mechanisms available for theoretical examination. For example, it excludes the glide of straight misfit dislocations along the flat areas of the core-shell interface that is sometimes observed in experiments [27, 28]. It is worth noting that flat interface regions often form in radially inhomogeneous NWs [29–32].

It seems that the simplest case of a core with flat faces is the core in the form of a long parallelepiped. To our best knowledge, today there are two analytical solutions which describe the elastically strained state in a core-shell NW with a square core placed symmetrically with respect to the shell surface [33, 34]. The solution [33] was found in the model case of plane strain through the complex potentials method and illustrated by stress maps in Cartesian coordinates. The drawback of the work [33] is that the authors did not demonstrate the evidence of the boundary condition fulfillment. Moreover, the case of plane strain is obviously quite far from the case of three-dimensional mismatch of crystalline lattices in real core-shell NWs. In work [34], the stress field caused by dilatation eigenstrain (3D misfit strain), was found in the form of trigonometric series. However, the solution [34] does not allow to consider the case of dilatation anisotropy in composite NWs which consist of materials with different crystalline lattices.

The present work is aimed at the analytical calculation and numerical analysis of the misfit stresses in a core-shell NW with an eccentric parallelepipedal core of an arbitrary rectangular cross section, which is characterized by a one-dimensional cross dilatation eigenstrain with respect to the shell material. We show the analytical formulas for the misfit stress components applicable for practical use in theoretical modeling of anisotropic misfit stress and relaxation processes. We also demonstrate some examples of stress distribution in the NW cross section, from which the fulfillment of boundary conditions on the shell free surface is evident.

2. Model

Consider a long core-shell cylinder of radius R, which consists of an elastically homogeneous and isotropic shell and a core having the shape of a long parallelepiped (figure 1). The shear modulus G and the Poisson ratio ν are the same for the shell and the core. The core cross section is a rectangle given by the coordinates of his vertexes. The core is subjected to a dilatation eigenstrain ε^*_{yy}.

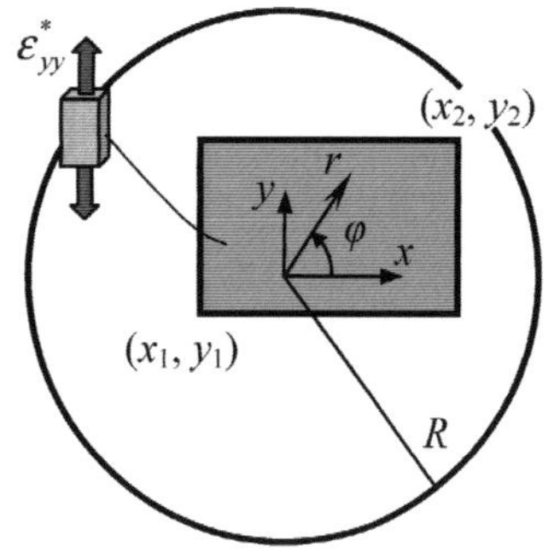

Figure 1. Cross section of a core-shell cylinder of radius R with the core in the form of a long parallelepiped of the rectangle cross section $|x_1 - x_2| \times |y_1 - y_2|$. The Cartesian (x, y, z) and cylindrical (r, φ, z) coordinate systems have the origin at the center of the cylinder cross section.

The desired stress field ${}^{\text{cyl}}\sigma^{(y)}_{ij}$ caused by the ε^*_{yy} eigenstrain can be represented by the sum of a similar stress field ${}^{\infty}\sigma^{(y)}_{ij}$ created by the core in an infinite medium, and an extra stress field $\sigma^{(y)}_{ij}$ which is needed to satisfy the boundary conditions on the shell free surface:

$$ {}^{\text{cyl}}\sigma^{(y)}_{ij} = {}^{\infty}\sigma^{(y)}_{ij} + \sigma^{(y)}_{ij}. \tag{1}$$

The boundary conditions are

$$ {}^{\text{cyl}}\sigma^{(y)}_{rr}(r = R) = 0 \text{ and } {}^{\text{cyl}}\sigma^{(y)}_{r\varphi}(r = R) = 0. \tag{2}$$

The non-vanishing stress components ${}^{\infty}\sigma_{ij}$ are given as follows [35]:

$$ {}^{\infty}\sigma^{(y)}_{ij} = C\ {}^{\infty}\Psi^{(y)}_{ij}\Big|^{x_0=x_2}_{x_0=x_1}\Big|^{y_0=y_2}_{y_0=y_1}, \tag{3}$$

where $C = \varepsilon^*_{yy}G/[2\pi(1-\nu)]$ and

$$ {}^{\infty}\Psi^{(y)}_{xx} = \frac{(x-x_0)(y-y_0)}{(x-x_0)^2+(y-y_0)^2}; \qquad {}^{\infty}\Psi^{(y)}_{yy} = -\frac{(x-x_0)(y-y_0)}{(x-x_0)^2+(y-y_0)^2} + 2\arctan\frac{y-y_0}{x-x_0}; $$

$$ {}^{\infty}\Psi^{(y)}_{xy} = -\frac{(x-x_0)^2}{(x-x_0)^2+(y-y_0)^2} - \ln\sqrt{(x-x_0)^2+(y-y_0)^2}; \qquad {}^{\infty}\Psi^{(y)}_{zz} = 2\nu\arctan\frac{y-y_0}{x-x_0}. \tag{4}$$

Any stress function Ψ_{ij} in the cylindrical and Cartesian coordinate system can be calculated through the complex potentials [36] as

$$\Psi_{rr} + \Psi_{\varphi\varphi} = \Psi_{xx} + \Psi_{yy} = 2[F'(\zeta) + \overline{F'(\zeta)}] = 4\,\mathrm{Re}\,F'(\zeta), \tag{5}$$

$$\Psi_{\varphi\varphi} - \Psi_{rr} + 2\mathrm{i}\Psi_{r\varphi} = [\Psi_{yy} - \Psi_{xx} + 2\mathrm{i}\Psi_{rr}]\mathrm{e}^{2\mathrm{i}\varphi} = 2[\overline{\zeta}F''(\zeta) + \chi''(\zeta)]\mathrm{e}^{2\mathrm{i}\varphi}, \tag{6}$$

where $F'(\zeta)$ and $\chi''(\zeta)$ are the complex potentials that are unknown analytical functions of the complex variable $\zeta = x + \mathrm{i}y = r\mathrm{e}^{\mathrm{i}\varphi}$, $\mathrm{i} = \sqrt{-1}$. $\overline{F'(\zeta)}$ and $\overline{\chi''(\zeta)}$ are the functions conjugate with $F'(\zeta)$ and $\chi''(\zeta)$, respectively.

Subtracting equations (5) from (6), we obtain the following formula which is convenient for the fulfillment of the boundary conditions:

$$\Psi_{rr} - \mathrm{i}\Psi_{r\varphi} = 1/2\left([\Psi_{xx} + \Psi_{yy}] - [\Psi_{yy} - \Psi_{xx} + 2\mathrm{i}\Psi_{xy}]\mathrm{e}^{2\mathrm{i}\varphi}\right) = F'(\zeta) + \overline{F'(\zeta)} - [\overline{\zeta}F''(\zeta) + \chi''(\zeta)]\mathrm{e}^{2\mathrm{i}\varphi}. \tag{7}$$

With taking into account the stress finiteness at $r \to 0$, we will search the functions $F'(\xi)$ and $\chi''(\xi)$ in the form of power series

$$F'(\zeta) = \sum_{n=0}^{+\infty} A_n\zeta^n, \qquad \chi''(\zeta) = \sum_{n=0}^{+\infty} B_n\zeta^n, \tag{8}$$

where A_n and B_n are complex constants in the general case. Now we could rewrite equation (7) as

$$\Psi_{rr} + \mathrm{i}\Psi_{rr} = A_0 + \overline{A_0} + \sum_{n=1}^{+\infty}\overline{A_n}\overline{\zeta}^n - \sum_{n=2}^{+\infty}\left[(n-1)A_n + \frac{1}{R^2}B_{n-2}\right]\zeta^n. \tag{9}$$

Let us find now the extra stress field $\sigma_{ij}^{(y)}$ in the same form as that given by equation (3):

$$\sigma_{ij}^{(y)} = C\,\Psi_{ij}^{(y)}\Big|_{x_0=x_1}^{x_0=x_2}\Big|_{y_0=y_1}^{y_0=y_2}. \tag{10}$$

The boundary conditions (2) with account for equation (7) are

$$[\Psi_{rr}^{(y)} - \mathrm{i}\Psi_{r\varphi}^{(y)}]_{r=R} = -1/2\left([{}^{\infty}\Psi_{xx}^{(y)} + {}^{\infty}\Psi_{yy}^{(y)}] - [{}^{\infty}\Psi_{yy}^{(y)} - {}^{\infty}\Psi_{xx}^{(y)} + 2\mathrm{i}\,{}^{\infty}\Psi_{xy}^{(y)}]\mathrm{e}^{2\mathrm{i}\varphi}\right)_{r=R} \tag{11}$$

Introduce new complex variables $\xi = R\,\mathrm{e}^{\mathrm{i}\varphi}$, $\alpha = x_0 + \mathrm{i}y_0$ and consider the terms in (11) separately:

$${}^{\infty}\Psi_{xx}^{(y)} + {}^{\infty}\Psi_{yy}^{(y)} = 2\arctan\frac{y - y_0}{x - x_0} = 2\arg(\xi - \alpha), \tag{12}$$

$$[{}^{\infty}\Psi_{yy}^{(y)} - {}^{\infty}\Psi_{xx}^{(y)} + 2\mathrm{i}\,{}^{\infty}\Psi_{xy}^{(y)}]\mathrm{e}^{2\mathrm{i}\varphi} = -\mathrm{i}\left[1 + \frac{\xi - \overline{\alpha}}{\xi - \alpha} + 2\ln(\xi - \alpha)\right]\frac{\xi^2}{R^2}. \tag{13}$$

To satisfy the boundary conditions (2), we should represent equations (12) and (13) in terms of the power series. Finally, equation (11) reads

$$[\Psi_{rr}^{(y)} - \mathrm{i}\Psi_{r\varphi}^{(y)}]_{r=R} = -\frac{\mathrm{i}}{2}\left\{\sum_{n=1}^{+\infty}\left[\left(1 - \frac{\alpha\overline{\alpha}}{R^2}\right) + \frac{1}{n} - \frac{2}{n+2}\frac{\alpha^2}{R^2}\right]\left(\frac{\alpha}{\xi}\right)^n - \frac{\alpha^2 + \alpha\overline{\alpha}}{R^2} - \sum_{n=2}^{+\infty}\frac{1}{n}\left(\frac{\overline{\alpha}\xi}{R^2}\right)^n\right\}. \tag{14}$$

Comparing the coefficients at ζ^n in both equations (9) and (14), we have

$$A_0 = \frac{\mathrm{i}}{2}\frac{\alpha^2 + \alpha\overline{\alpha}}{R^2}, \quad A_n = \frac{\mathrm{i}}{2}\left[\frac{R^2 - \alpha\overline{\alpha}}{R^2} + \frac{1}{n} - \frac{2}{n+2}\frac{\overline{\alpha}^2}{R^2}\right]\left(\frac{\overline{\alpha}}{R^2}\right)^n, \quad n = 1,2,3...$$

$$B_n = -\frac{\mathrm{i}\,R^2}{2}\left[\frac{\alpha\overline{\alpha} - 2\overline{\alpha}^2}{R^2} + (n+2)\frac{R^2 - \alpha\overline{\alpha}}{R^2} + \frac{2(n+1)}{(n+4)}\frac{\overline{\alpha}^2}{R^2}\right]\left(\frac{\overline{\alpha}}{R^2}\right)^{n+2}, \quad n = 1,2,3... \tag{15}$$

Substitution of equations (15) and (8) to (5), (6) gives the complex form for the stress function of the extra stress tensor:

$$\Psi_{rr}^{(y)} + \Psi_{\varphi\varphi}^{(y)} = 2\frac{x_0 y_0}{R^2} - 2\,\mathrm{Im}\left\{\frac{\overline{\alpha}\zeta}{R^2}\frac{R^2 - \alpha\overline{\alpha}}{R^2 - \overline{\alpha}\zeta} + \frac{2R^2 - \zeta^2}{\zeta^2}\ln\frac{R^2 - \overline{\alpha}\zeta}{R^2}\right\},$$

$$\Psi_{\varphi\varphi}^{(y)} - \Psi_{rr}^{(y)} + 2\mathrm{i}\Psi_{r\varphi}^{(y)} = \mathrm{i}\left\{\frac{\overline{\alpha}^2}{r^2} + \frac{R^2 - r^2}{r^2}\frac{2\overline{\alpha}^2}{R^2 - \overline{\alpha}\zeta} - \frac{\alpha\overline{\alpha} - r^2}{r^2}\frac{\overline{\alpha}\zeta}{R^2 - \overline{\alpha}\zeta} - \right.$$

$$\left. - \frac{R^2 - r^2}{r^2}\frac{(R^2 - \alpha\overline{\alpha})\overline{\alpha}\zeta}{(R^2 - \overline{\alpha}\zeta)^2} + 2\frac{3R^2 - 2r^2}{r^2}\frac{R^2}{\zeta^2}\ln\frac{R^2 - \overline{\alpha}\zeta}{R^2}\right\}. \tag{16}$$

Transforming equations (16), we come to the final components of the extra stress tensor:

$$\Psi_{rr}^{(y)} = -\frac{1}{2}\left\{q^2\frac{1-t^2}{t^2}\left[\sin 2\theta + 2\frac{\sin 2\theta - q^2t^2\sin(\varphi+\theta)}{\Lambda}\right] + \frac{q(1-q^2)(1+2t^2)\sin(\varphi-\theta)}{t\Lambda} + \right.$$

$$\left. + \frac{q(1-q^2)(1-t^2)(1-q^2t^2)\sin(\varphi-\theta)}{t\Lambda^2} + 2\left[\mathrm{P}\cos 2\varphi + 1\right]\arctan\Phi + \mathrm{P}\sin 2\varphi\ln\sqrt{\Lambda}\right\};$$

$$\Psi_{r\varphi}^{(y)} = -\frac{1}{2}\left\{\frac{q^2}{t^2}\cos 2\theta + 2q^2\frac{1-t^2}{t^2}\frac{\cos 2\theta - q^2t^2\cos(\varphi+\theta)}{\Lambda} + \frac{q(1-q^2)(1-t^2)(qt - \cos(\varphi-\theta))}{t\Lambda} + \right.$$

$$\left. + \frac{q(1-q^2)(1-t^2)(2q^2t^2 + (1+q^2t^2)\cos(\varphi-\theta))}{t\Lambda^2} - 2\mathrm{P}\left[\sin 2\varphi\arctan\Phi - \cos 2\varphi\ln\sqrt{\Lambda}\right]\right\};$$

$$\Psi_{\varphi\varphi}^{(y)} = \frac{1}{2}\left\{q^2\frac{1+t^2}{t^2}\sin 2\theta + 2q^2\frac{1-t^2}{t^2}\frac{\sin 2\theta - q^2t^2\sin(\varphi+\theta)}{\Lambda} + \frac{q(1-q^2)(6t^2+1)\sin(\varphi-\theta)}{t\Lambda} + \right.$$

$$\left. + \frac{q(1-q^2)(1-t^2)(1-q^2t^2)\sin(\varphi-\theta)}{t\Lambda^2} + 2\left[\mathrm{P}\cos 2\varphi - 1\right]\arctan\Phi + \mathrm{P}\sin 2\varphi\ln\sqrt{\Lambda}\right\};$$

$$\Psi_{r\varphi}^{(y)} = 2\nu\left\{\frac{q^2}{2}\sin 2\theta - \frac{q(1-q^2)t\sin(\varphi-\theta)}{\Lambda} + \frac{2\cos 2\varphi - t^2}{t^2}\arctan\Phi + \frac{\sin 2\varphi}{t^2}\ln\sqrt{\Lambda}\right\}; \tag{16}$$

where ρ and θ are polar coordinates of the core vertex, and

$$t = r/R; \quad q = \rho/R; \quad \Lambda = q^2t^2 - 2qt\cos(\varphi-\theta) + 1; \quad \Phi = \frac{qt\sin(\varphi-\theta)}{1 - qt\cos(\varphi-\theta)}; \quad \mathrm{P} = \frac{2t^2+3}{t^4}.$$

Thus, the problem under consideration (figure 1) is solved.

3. Results

To illustrate our solution, we show in figure 2(a)–(d) the distribution of stress components ${}^{\mathrm{cyl}}\sigma_{rr}^{(y)}$, ${}^{\mathrm{cyl}}\sigma_{r\varphi}^{(y)}$, ${}^{\mathrm{cyl}}\sigma_{\varphi\varphi}^{(y)}$ and ${}^{\mathrm{cyl}}\sigma_{zz}^{(y)}$ in a cross section of the core-shell NW. It is evident that the ${}^{\mathrm{cyl}}\sigma_{rr}^{(y)}$ and ${}^{\mathrm{cyl}}\sigma_{r\varphi}^{(y)}$ components satisfy the boundary conditions (2).

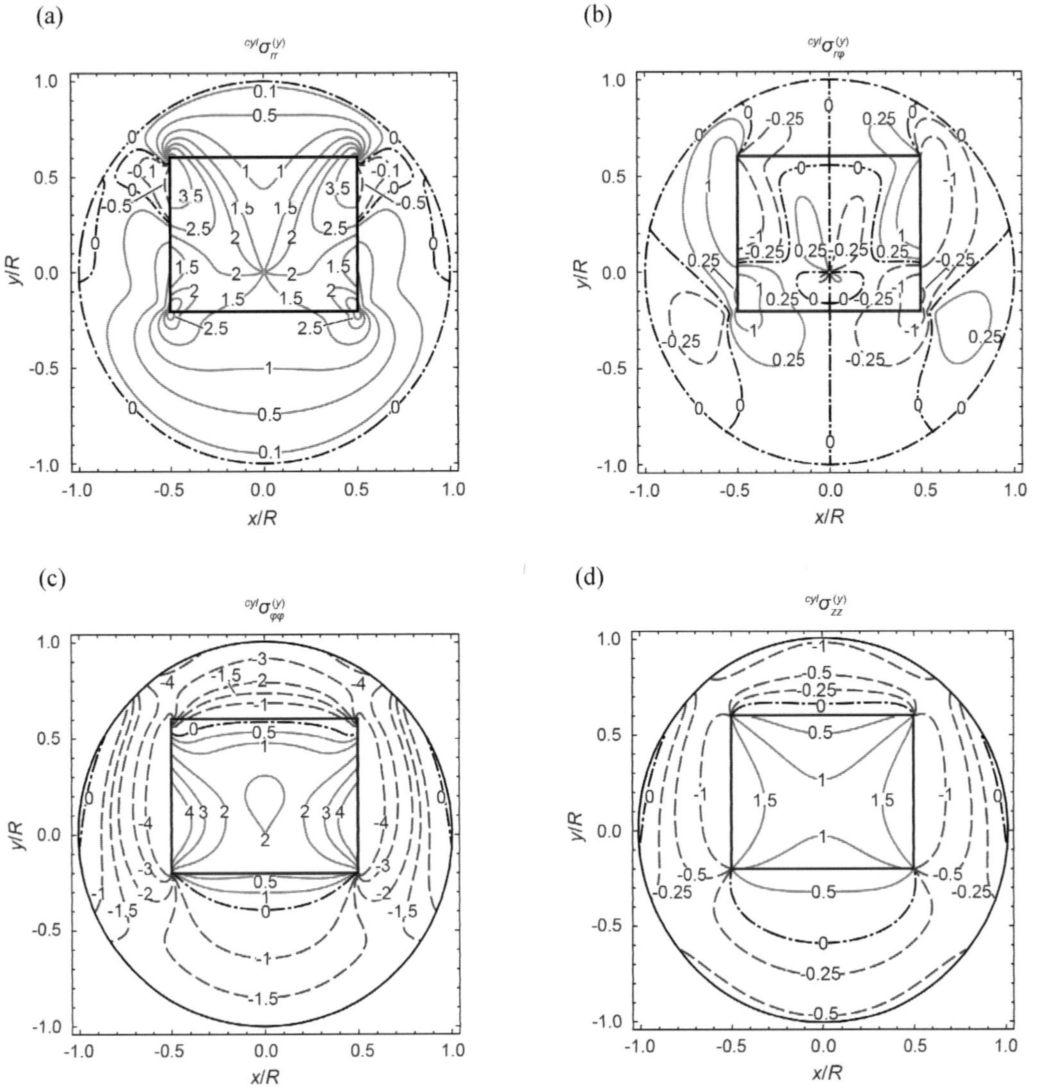

Figure 2. Distribution of the misfit stress components caused by the ε^*_{yy} eigenstrain in the cross section of the core-shell NW: (a) $^{cyl}\sigma^{(y)}_{rr}$, (b) $^{cyl}\sigma^{(y)}_{r\varphi}$, (c) $^{cyl}\sigma^{(y)}_{\varphi\varphi}$ and (d) $^{cyl}\sigma^{(y)}_{zz}$. The core position is defined by the coordinates of its vertices: $x_{1,2} = \pm\, 0.5R$, $y_1 = -\,0.2R$, $y_2 = 0.6R$. The stress values are given in units of $C = \varepsilon^*_{yy} G/[2\pi(1-\nu)]$.

4. Conclusions

We can conclude that our analytical solution, first, satisfies the boundary conditions of the problem, second, gives an opportunity to analyze the misfit stress distribution in detail, and, third, is represented in a concise and transparent closed form which is applicable for theoretical modeling of misfit stress relaxation processes.

Acknowledgments
S. A. Krasnitckii and A. M. Smirnov acknowledge the Russian Foundation of Basic Research (RFBR project No. 16-32-00521) for the support in providing the analytical and numerical calculations of the misfit stress distribution in the core-shell NW. M. Yu. Gutkin acknowledges the Russian Science Foundation (grant RSF No. 14-29-00086) for the support in the invention of the theoretical model of the core-shell NW and the formulation of the corresponding boundary-value problem to solve.

References

[1] Lauhon L J, Gudiksen M S and Lieber C M 2004 *Phil. Trans. R. Soc. Lond.* A **362** 1247
[2] Mieszawska A J, Jalilian R, Sumanasekera G U and Zamborini F P 2007 *Small* **3** 722
[3] Agarwal R 2008, *Small* **4** 1872
[4] Comini E, Baratto C, Faglia G, Ferroni M, Vomiero A and Sberveglieri G 2009 *Prog. Mater. Sci.* **54** 1
[5] Dubrovskii V G, Cirlin G E and Ustinov V M 2009 *Semiconductors* **43** 1539
[6] Barth S, Hernandez-Ramirez F, Holmes J D and Romano-Rodriguez A 2010 *Prog. Mater. Sci.* **55** 563
[7] Gao Q, Tan H H, Jackson H E, Smith L M, Yarrison-Rice J M, Zou J and Jagadish C 2011 *Semicond. Sci. Technol.* **26** 014035
[8] Garnett E C, Brongersma M L, Cui Y and McGehee M D 2011 *Annu. Rev. Mater. Res.* **41** 269
[9] Kenry C T 2013 *Prog. Mater. Sci.* **58** 705
[10] Amato M, Palummo M, Rurali R and Ossicini S 2014 *Chem. Rev.* **114** 1371
[11] Gutkin M Yu, Ovid'ko I A and Sheinerman A G 2000 *J. Phys. Condens. Matter* **12** 5391
[12] Sheinerman A G and Gutkin M Yu 2001 *Phys. Stat. Solidi* A **184** 485
[13] Ovid'ko I A and Sheinerman A G 2004 *Phil. Mag.* **84** 2103
[14] Liang Y, Nix W D, Griffin P B and Plummer J D 2005 *J. Appl. Phys.* **97** 043519
[15] Raychaudhuri S and Yu E T 2006 *J. Appl. Phys.* **99** 114308
[16] Aifantis K E, Kolesnikova A L and Romanov A E 2007 *Phil. Mag.* **87** 4731
[17] Colin J 2010 *Phys. Rev.* B **82** 054118
[18] Wang X, Pan E and Chung P W 2010 *Int. J. Plasticity* **26** 1415
[19] Chu H J, Wang J, Zhou C Z and Beyerlein I J 2011 *Acta Mater.* **59** 7114
[20] Gutkin M Yu, Kuzmin K V and Sheinerman A G 2011 *Phys. Stat. Solidi* B **248** 1651
[21] Haapamaki C M, Baugh J and LaPierre R R 2012 *J. Appl. Phys.* **112** 124305
[22] Chu H J, Zhou C Z, Wang J and Beyerlein I J 2013 *J. Mech. Phys. Solids* **61** 2147
[23] Dayeh S A, Tang W, Boioli F, Kavanagh K L, Zheng H, Wang J, Mack N H, Swadener G, Huang J Y, Miglio L, Tu K N and Picraux S T 2013 *Nano Lett.* **13** 1869
[24] Salehzadeh O, Kavanagh K L and Watkins S P 2013 *J. Appl. Phys.* **114** 054301
[25] Enzevaee C, Gutkin M Yu and Shodja H M 2014, *Phil. Mag.* **94** 492
[26] Gutkin M Yu and Smirnov A M 2015 *Acta Mater.* **88** 91
[27] Ding Y, Fan F, Tian Z and Wang Z L 2010, *J. Am. Chem. Soc.* **132** 12480
[28] Bhattarai N, Casillas G, Ponce A and Jose-Yacaman M 2013 *Surf. Sci.* **609** 161
[29] Wang Z L, Dai Z R, Gao R P, Bai Z G and Gole J L 2000 *Appl. Phys. Lett.* **77** 3349
[30] Hu J, Bando Y, Liu Z, Sekiguchi T, Golberg D and Zhan J 2003 *J. Am. Chem. Soc.* **125** 11306
[31] Teo B K, Li C P, Sun X H, Wong N B and Lee S T 2003 *Inorg. Chem.* **42** 6723
[32] Qian F, Li Y, Gradečak S, Park H G, Dong Y, Ding Y, Wang Z L and Lieber C M 2008 *Nature Materials* **7** 701
[33] Zou W-N, He Q-C and Zheng Q-S 2012 *Int. J. Solids Struct.* **49** 1627
[34] Krasnitckii S A, Smirnov A M and Gutkin M Y 2016 *J. Phys.: Conf. Ser.* **690** 012022
[35] Malyshev K L, Gutkin M Yu, Romanov A E, Sitnikova A A and Sorokin L M 1988 *Sov. Phys.-Solid State* **30** 1176
[36] Muskhelishvili N I 1977 *Some Basic Problems of the Mathematical Theory of Elasticity* (Leyden: Noordhoff Int. Publ.)

1/*f* magnetic noise in exotic ε-$In_{0.24}Fe_{1.76}O_3$ nanoparticles

S A Kostyuchenko[1] and A I Dmitriev[2]

[1] Lomonosov Moscow State University, Moscow, 119992, Russia
[2] Institute of Problems of Chemical Physics of Russian Academy of Sciences, Chernogolovka, 142432 Russia

E-mail: aid@icp.ac.ru

Abstract. The relation between the magnetic viscosity of exotic ε-$In_{0.24}Fe_{1.76}O_3$ nanoparticles and 1/*f* magnetic noise caused by the random thermally activated magnetization reversal of a single nanoparticle was found.

1. Introduction

The physics of magnetic noises is still an intriguing fundamental field of science. Magnetic noises manifest in the logarithmic dependence of slow magnetization reversal (magnetic viscosity) in different magnets [1–5]. The general concept of the logarithmic magnetization reversal as a consequence of a uniform energy barrier distribution $F(E)$ was developed in [6]. In the case of thermally activated processes, there exists a direct relation between an $F(E)$ distribution and a relaxation time distribution $F(\tau)$, which determines the spectrum of magnetic fluctuations [7]. The key characteristic of noise is spectral density $W(f)$, which characterizes the distribution of the random signal power over a frequency spectrum *f*. If $W(f)$ = const, the noise is called "white". In contrast to the noise having a spectrum $W(f)$ = const, the origin of flicker noise with a spectrum $W(f) \sim 1/f$ is poorly understood. Complex methods of detecting electric noise 1/*f* in thin films, heterostructures, and monoatomic graphene layers are being developed [8]. However, the physics of the magnetic 1/*f* noises is still retained beyond the scope of investigation. In this work, we study the slow relaxation of the magnetization of ε-$In_{0.24}Fe_{1.76}O_3$ nanoparticles ensembles. New nanomagnets based on the epsilon phase of iron oxide ε-Fe_2O_3 demonstrate a giant (up to 2.34 T) coercive force and magnetic anisotropy [9]. In addition, the ε-Fe_2O_3 phase exhibits magnetoelectric properties [10] and is known to be a good absorber of millimetre electromagnetic waves [11]. Magnetic properties of ε-$In_{0.24}Fe_{1.76}O_3$ nanoparticles ensembles were well studied in papers [12–14]. The aim of the study is to find the relation between the magnetic viscosity of ε-$In_{0.24}Fe_{1.76}O_3$ nanoparticles and 1/*f* magnetic noise caused by the random thermally activated magnetization reversal of a single nanoparticle.

2. Experiment

ε-$In_{0.24}Fe_{1.76}O_3$ nanoparticles with lengths of 80 nm and diameters of 35 nm were synthesized by two methods: synthesis by inverse micelles and the sol-gel method [15, 16]. Nanoparticles were grown in a solution of inverse micelles containing the nitrates of iron and indium. The ε-$In_{0.24}Fe_{1.76}O_3$ nanoparticles were studied by mass spectrometry, transmission electron microscopy, and x-ray diffraction [15, 16]. The nanoparticles had an orthorhombic crystal structure with four nonequivalent cation positions for the Fe ion. One of these has tetrahedral surroundings and the other three, octahedral ones. The indium ions are a substitution impurity and align themselves in octahedral iron ion sites. Time dependences of the magnetization $m(t)$ of samples were measured using a SQUID magnetometer MPMS 5XL Quantum Design. The sample was first magnetized in a magnetic field H = 50 kOe (exceeding the strength of the saturation field). The time of exposure of the sample to this magnetic field was 300 s. Following such treatment, the field directed opposite to the magnetic moment vector of the sample was switched on. The time of field reversal was approximately 90 s. After the reverse magnetic field was set, the time dependence of the magnetic moment was measured during 3400 s.

3. Experimental results

Figure 1 shows the magnetization increment dependences $\Delta m(t)$ at temperature T = 300 K in different magnetic fields directed opposite to the magnetic moment vector of the sample on a logarithmic time scale.

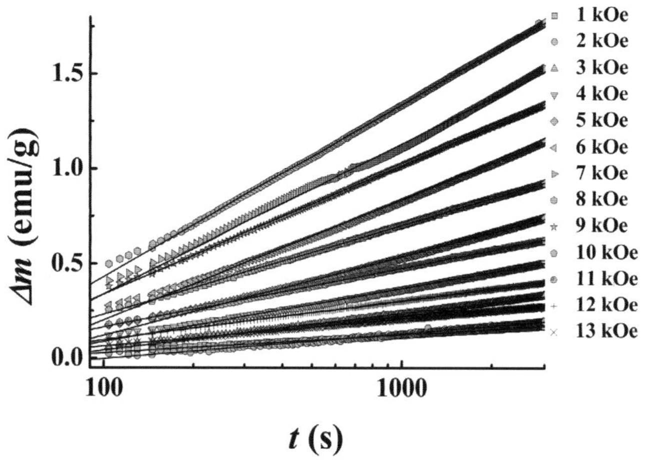

Figure 1. Time dependences of magnetization $\Delta m(t)$ at T = 300 K in different magnetic fields. Approximations are shown by solid lines.

It can be seen from figure 1 that $\Delta m(t)$ is straightened in semi-logarithmic coordinates. The tangent of the slope of $\Delta m(\ln t)$ dependences corresponds to magnetic viscosity $S = \mathrm{d}m/\mathrm{d}(\ln t)$. The solid lines represent the result of approximation with the following expression:

$$\Delta m(t) \sim S \ln t . \quad (1)$$

Slopes of $\Delta m(\ln t)$ curves and corresponding magnetic viscosities are non-monotonic functions of magnetic field (figure 1). Field dependences $S(H)$ at T = 300 K show maximums corresponding to 8 kOe (figure 2).

4. Discussion

4.1. Time dependences of magnetization. Magnetic viscosity

When the relaxation process caused by fluctuations is characterized by a single value of time τ, the time dependence of the magnetization $m(t)$ is described by the exponential function [6]:

$$m(t) = m_0 \exp\left(-\frac{t}{\tau}\right). \quad (2)$$

Actually, thermal activated fluctuations are characterized by a wide relaxation time distribution $F(\tau)$ density [6]:

$$F(\tau) = \frac{k_B T}{\tau(E_2 - E_1)}, \quad (3)$$

where E_2-E_1 – the range of activation energy of fluctuations. In this case, we have the expression:

$$m(t) = m_0 \frac{k_B T}{E_2 - E_1} \int_{\tau_1}^{\tau_2} \exp\left(-\frac{t}{\tau}\right) \frac{1}{\tau} d\tau . \quad (4)$$

From Equation (4) we obtain the expression for the magnetization increment:

$$\Delta m(t) \sim \frac{k_B T}{E_2 - E_1} \ln t . \quad (5)$$

When comparing Equation (5) with Equation (1) (which was used to describe the experimental $\Delta m(t)$ dependences), we can derive a formula to calculate magnetic viscosity S expressed in the units of initial magnetization in a given magnetic field:

$$S = \frac{k_B T}{E_2 - E_1} . \quad (6)$$

Thus, the logarithmic kinetics of the spontaneous magnetization reversal of the ε-$In_{0.24}Fe_{1.76}O_3$ nanoparticles is determined by the superposition of relaxation processes with a wide relaxation time distribution.

4.2. Spectral density of thermal fluctuations. Relation between Magnetic Viscosity and Magnetic Noise

When the relaxation process is characterized by a single value of time τ, the normalized spectral density of magnetic fluctuations is described by the Lorentz function:

$$W(f) = \frac{2\tau}{1 + (2\pi f \tau)^2} . \quad (7)$$

On the assumption of a wide relaxation time distribution $F(\tau)$ (see Equation (3)), the spectral density is found by integrating Equation (7):

$$W(f) = \frac{k_B T}{E_2 - E_1} \int_0^{\infty} \frac{\tau}{1 + (2\pi f \tau)^2} \frac{1}{\tau} d\tau = \frac{k_B T}{2(E_2 - E_1)} \frac{1}{f} . \quad (8)$$

Equation (8) determines the spectral density of flicker noise. Thus, $1/f$ noise is formed as a superposition of Lorentz spectra [17, 18]. When comparing Equations (8) and (6), we can find a relation between the magnetic viscosity and the spectral density of thermal fluctuations:

$$W(f) = \frac{S}{2f} . \quad (9)$$

Figure 2 shows the $W(f)$ dependences of the magnetic fluctuation spectrum that determines the kinetics of spontaneous magnetization reversal of the samples shown in figure 1 that were calculated by Equation (9). The envelopes of the spectral dependence curves correspond to the field dependence of the magnetic viscosity $S(H)$. Thus, when controlling the magnetic viscosity by changing the magnetic field, we change the “window” width of detecting a magnetic fluctuation spectrum.

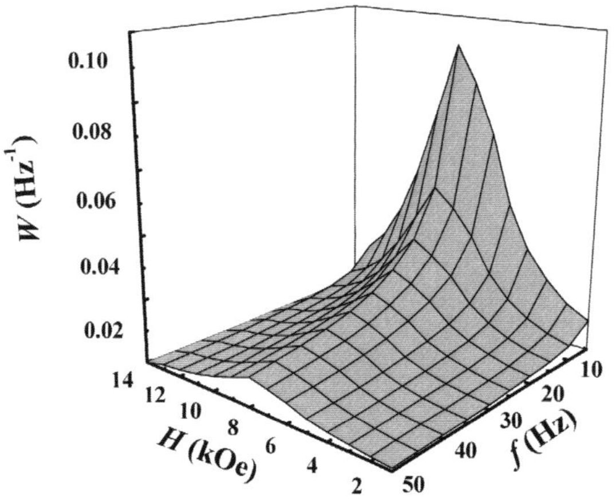

Figure 2. Calculated frequency dependences of the spectral density for magnetic viscosity in various magnetic fields at T = 300 K. The envelope of spectral density curves at a fixed frequency corresponds to the dependence of the magnetic viscosity on the magnetic field.

5. Summary

The logarithmic kinetics of the magnetic viscosity in exotic ε-$In_{0.24}Fe_{1.76}O_3$ nanoparticles is determined by the superposition of thermally activated magnetization reversal of a single nanoparticle with a wide relaxation time distribution. A relation between the magnetic viscosity and the spectral characteristics of magnetic fluctuations in nanoparticles ensembles was found.

Acknowledgments

Authors are grateful to R B Morgunov for discussion. The work was supported by the Russian Foundation for Basic Research, project no. 16-07-00863 a.

References

[1] Guy C N 1978 *J. Phys. F* **8** 1309
[2] Street R and Woolley J C 1949 *Proc. Phys. Soc. A* **62** 562
[3] Jinfang L, Shuming P, Helie L, Denglu H and Xiangfu N 1991 *J. Phys. D* **24** 384
[4] Collocott S J and Dunlop J B 2008 *J. Magn. Magn. Mater.* **320** 2089
[5] Gaunt P 1986 *J. Appl. Phys.* **59** 4129
[6] Dmitriev A I, Morgunov R B and Zaitsev S V 2011 *J. Exp. Theor. Phys.* **112 (2)** 317
[7] Dutta P and Horn P M 1981 *Rev. Mod. Phys.* **53** 497
[8] Balandin A A 2013 *Nature Nanotechnol.* **8** 549
[9] Ohkoshi S and Tokoro H 2013 *Bull. Chem. Soc. Jpn.* **86**, 897907
[10] Gich M, Frontera C, Roig A, Fontcuberta J, Molins E, Bellido N, Simon Ch and Fleta C 2006 *Nanotechnology* **17** 687
[11] Ohkoshi S, Kuroki S, Sakurai S, Matsumoto K, Sato K and Sasaki S 2007 *Angew. Chem.* **46** 8392
[12] Dmitriev A I and Morgunov R B 2015 *Low Temp. Phys.* **41** 917
[13] Dmitriev A I, Tokoro H, Ohkoshi S and Morgunov R B 2015 *Low Temp. Phys.* **41** 20
[14] Dmitriev A I, Koplak O V, Namai A, Tokoro H, Ohkoshi S, Morgunov R B 2014 *Phys. Solid State* **56** 1795
[15] Sakurai S, Kuroki S, Tokoro H, Hashimoto K and Ohkoshi S 2007 *Adv. Funct. Mater.* **17**, 2278
[16] Yamada K, Tokoro H, Yoshikiyo M, Yorinaga T, Namai A and Ohkoshi S 2012 *J. Appl. Phys.* **111** 07B506
[17] Du Pre F K 1950 *Phys. Rev.* **78** 615
[18] Van der Ziel A 1950 *Physica* **16** 359

Formation of one-dimensional ZnO structures on flexible substrates

A A Semenova, N A Lashkova, A I Maximov and V A Moshnikov

St Petersburg Electrotechnical University "LETI", 5 Prof. Popova str., St. Petersburg 197376, Russia

E-mail: anastassiyan@mail.ru

Abstract. Within the frames of the research described in the paper, *1D* ZnO nanostructures on flexible substrates were synthesized by the hydrothermal method. The paper describes technological features of obtaining oxide nanorods using low-temperature synthesis. Morphological properties of obtained structures were investigated.

1. Introduction

Zinc oxide ZnO is one of the most promising materials for formation of piezoelectric nanogenerators [1]. It is a wide band-gap (E_g = 3.36 eV) piezoelectric semiconductor, which has a *n*-type conductivity and is crystallized in a wurtzite type structure. The advantages of using this material are an environmental safety and ease of preparation as well as a possibility to use this material at high and ultra-high frequencies [2, 3].

The hydrothermal method [4, 5] is one of the simplest and most available methods suitable for synthesis of ZnO nanorods on polymeric substrates.

2. Experiment

The hydrothermal method is based on dissolving of substances, which are insoluble under normal conditions in water or aqueous solutions, at high temperatures and pressures. Main parameters of the hydrothermal synthesis, which determine the properties of the formed products, are the initial value of the solution pH, duration and temperature of synthesis.

In present work, zinc nitrate $Zn(NO_3)_2$ was used. Zinc nitrate is a salt of a strong acid and a weak alkali, so the hydrolysis reaction proceeds in two steps:

$$1)\ Zn(NO_3)_2 + H_2O \leftrightarrow Zn(OH)NO_3 + HNO_3$$
$$2)\ Zn(OH)NO_3 + H_2O \leftrightarrow Zn(OH)_2 + HNO_3$$

In this case, zinc ions are formed; its interaction with hydroxide ions results in formation of unstable $Zn(OH)_2$ converted into zinc oxide by the reaction $Zn(OH)_2 \rightarrow ZnO_{solid} + H_2O$.

Usually, NaOH, KOH, ammonium hydroxide NH_4OH or hexamethylenetetramine $C_6H_{12}N_4$ are used as an alkali. During the reaction, a zinc precursor is hydrolyzed, and, as a result, Zn^{2+} ions appear and react with the anions of alkali OH^-. During the reaction, various hydroxides can be formed: $ZnOH^+$ (aqueous), $Zn(OH)_2$ (aqueous), $Zn(OH)_2$, $Zn(OH)_3^-$ (aqueous) and $[Zn(OH)_4]^{2-}$.

$$Zn^{2+} + OH^- \leftrightarrow [Zn(OH)]^+, \quad (1)$$

$$Zn^{2+} + 2OH^{-} \leftrightarrow Zn(OH)_2, \quad (2)$$

$$Zn^{2+} + 3OH^{-} \leftrightarrow [Zn(OH)_3]^{-}, \quad 3)$$

$$Zn^{2+} + 4OH^{-} \leftrightarrow [Zn(OH)_4]^{2-}. \quad (4)$$

Interim hydroxides condense as oxides. One of the possible ways of *hydroxide – oxide* conversion is the process:

$$[Zn(OH)_4]^{2-} \leftrightarrow ZnO + H_2O + 2OH^{-}. \quad (5)$$

Overall, the chemical reaction can be written as:

$$Zn^{2+} + 2OH^{-} \leftrightarrow ZnO + H_2O. \quad (6)$$

It should be noted that in terms of the formation of one-dimensional structures based on ZnO, equation (2) with $Zn(OH)_2$ formation cannot be used because, as a result of the corresponding reaction, the uncharged zinc hydroxide is formed. The charged reaction products may be adsorbed on different charged crystal faces of growing structures or nucleation layer crystallites.

These chemical reactions are usually equilibrium reactions in aqueous solutions and proceed in the direction of minimizing the free energy of the entire system. The most high-energy surfaces of zinc oxide are the polar planes (0001), which end with Zn^{2+} or O^{2-} ions. Therefore, adsorption of new molecules, formed by the precursor hydrolysis, occurs on the polar planes in order to minimize the surface energy. In particular, the adsorption of molecules occurs due to formation of a monolayer which has an opposite polarity to the existing (0001) plane. Thus, on the surface of Zn^{2+} atoms, the precipitate hydroxides are formed in (3) and (4), then, according to reaction (5), ZnO is formed. At the end, the plane (0001) ends again with Zn^{2+} ions. Therefore, the growth of zinc oxide micro- and nanostructures occurs in the direction of *c*-axis, resulting in formation of an anisotropic structure.

In this work, hexamethylenetetramine (HMTA, $(CH_2)_6N_4$) was used for ZnO nanorods growth. In the work [5], it was discovered that HMTA molecules are a constant source of OH^- groups due to their slow release. At high concentrations of hydroxyl, the rate of nanorods growth increases and the solution is depleted of Zn^{2+} ions rapidly. This leads to the formation of tapered crystallites. Concentrations of reagents in the solution, used for growth of nanorods, were 25 mmol/l.

The samples were placed in the solution and incubated at 85°C for various times. After the nanorods synthesis, samples were washed with distilled water and dried in air.

Obtained structures were investigated using TESCAN MIRA (a scanning electron microscope (SEM) with an electron gun through a field-emission cathode (Schottky cathode)) and Seron Technology AIS 2300C.

Polyethylene terephthalate (PET) with ITO (60 Ohm/□) layer (Sigma Aldrich) was used to form ZnO nanostructures on flexible substrates.

3. Results and discussion

Experimental results are shown in figure 1. Figure 1 (a) shows formation of arrays and individual nanorods with a length of ≈ 500 nm and a diameter of ≈ 50 nm. In the synthesis of ZnO nanorods by the hydrothermal method on polymeric substrates, the cubic nanostructures grow in size from 50 nm to 150 nm uniformly over the entire surface of the substrate (figure 1 (b)).

In order to create a more uniform nanostructured ZnO coating on flexible substrates, 0.1 ml of $NH_3 \cdot H_2O$ was added to the initial solution (figure 2).

The change of pH during the synthesis of ZnO nanorods on polymeric substrates leads to a uniform coating of the individual zinc oxide nanorods (figure 2) with a length of up to 1 μm. In this case, the ammonia solution acts as an activator of the substrate surface formed by a conductive layer of ITO. Alongside this, the activation occurs due to formation of an adsorption layer consisting of hydroxyl groups (OH^-), which are the centers of zinc hydroxide adsorption (equation 1).

IOP Conf. Series: Journal of Physics: Conf. Series **816** (2017) 012031 doi:10.1088/1742-6596/816/1/012031

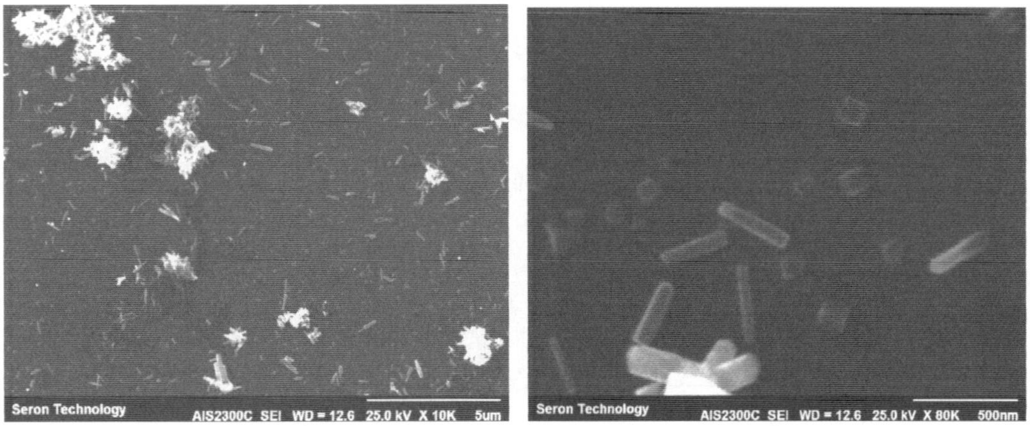

(a) **(b)**

Figure 1. ZnO nanorods on a PET substrate.

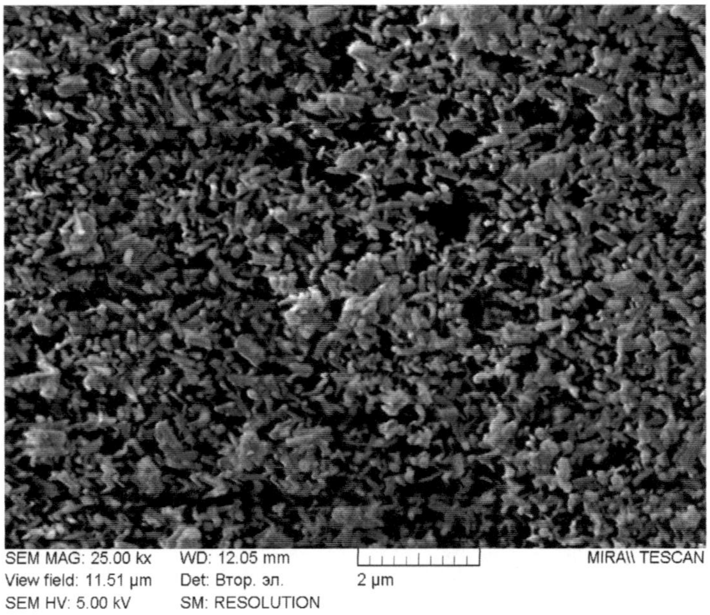

Figure 2. ZnO nanorods grown in alkaline conditions on a PET substrate.

To limit the growth of anisotropic crystalline properties in some dimensions, the surface-active substances (surfactants) are often used. The use of surfactants during hydrothermal synthesis of ZnO leads to significant changes in the crystals' morphology.

In the experiments, for synthesis in alkaline medium (5 ml of $NH_3{\cdot}H_2O$) on polymeric substrates, polyvinylpyrrolidone (PVP) was used. The experiments led to formation of needle-like crystals collected on the single growth center (figure 3).

As can be seen in figure 3, the average size of the needle structure is about 3 μm and individual nanoobjects have a hexagonal cut and taper in the direction of the growth center.

(a) **(b)**

Figure 3. ZnO nanostructures grown in alkaline conditions with added PVP.

4. Conclusions

Thereby, in this paper, *1D* nanostructures based on zinc oxide were formed on flexible conductive substrates via low temperature hydrothermal synthesis. Due to changes of pH, the individual nanorods and arrays of nanoobjects can be formed as well as the needle-like crystals collected on the single growth center.

Acknowledgments

The investigation was accomplished within the design part of the state job no. 16.2112.2014/K in the scientific sphere.

References

[1] Wang X 2012 *Nano Energy* **1** 13
[2] Klingshirn C F, Meyer B K, Waag A, Hoffmann A, Geurts J 2010 *Zinc Oxide. From Fundamental Properties Towards Novel Applications* (Berlin: Springer) p. 374
[3] Bobkov A A, Maximov A I, Moshnikov V A, Somov P A, Terukov E I 2015 *Semiconductors* **49** 1357
[4] Lashkova N A, Maximov A I, Ryabko A A, Bobkov A A, Moshnikov V A, Terukov E I 2016 *Semiconductors* **50** 1254
[5] McPeak K M, Le T P, Britton N G, Nickolov Zh S, Elabd Y A, Baxter J B 2011 *Langmuir* **27** 3672

Scanning tunneling microscopy of a graphene layer placed on a micro structured substrate

T Sh Ganbarova[1], **A V Babichev**[2, 3] **and S A Rykov**[1]

[1] Peter the Great Polytechnic University, 29 Politekhnicheskaya Str., St. Petersburg 195251, Russia
[2] Academic University RAS, 8/3 Khlopina Str., St.Petersburg, 194021, Russia
[3] ITMO University, 49 Kronverkskiy Prosp., St. Petersburg 197101, Russia

E-mail: rykov@rphf.spbstu.ru

Abstract. The features of the surface topography and the local energy spectrum of the graphene layer placed on a two-dimensional array of GaN pyramids are studied using scanning tunneling microscopy and local tunneling spectroscopy. This structure can be used as a test structure for the application of graphene as a flexible conductive transparent electrode in optoelectronics. Sagging of a graphene layer in the space between pyramids is found and the sagging value is estimated to be about 20% of the pyramid height of 1.1 μm. Local energy spectra show the change of the graphene electron energy spectrum due to the mechanical stress at sagging.

1. Introduction

Graphene can be used as an alternative inexpensive transparent contact in optoelectronic devices due to its high transparency, flexibility, low cost and availability of large area production by Chemical Vapor Deposition (CVD) [1–3]. The information about the morphology of a graphene layer used as an electrical contact is very important for the optimization of technology. Unfortunately, scanning electron microscopy (SEM) cannot be used due to high transparency of graphene for the electron beam. Scanning tunneling microscopy (STM) can help in this situation and provide the detailed information about the features of the graphene layer topography. At the same time, local tunneling spectroscopy (STS) can be used to study spatial distribution of the electron density in the graphene.

2. Experimental setup

We have studied the graphene layer placed on a two-dimensional array of GaN pyramids (figure 1) having the same morphology as the nanostructured LEDs [4]. The few-layered graphene has been grown by the CVD technique on a nickel layer as described in [5]. CH_4 was used as a carbon source for growth. The average thickness has been characterized by Raman spectroscopy and found to be equal to 4 monolayers. A GaN micro-pyramids array with hexagonal symmetry was fabricated by the metal organic vapor phase epitaxy technique using selective area growth [6]. The height of pyramids was 1.1 μm and the distance between pyramids was 1 μm. The pyramids were covered with a 300 nm thick SiO_2 layer to avoid the influence of the substrate on the electronic properties of graphene. To transfer the graphene on the SiO_2 covered micro-pyramid array, a wet transfer process has been used [2]. We have used a home built scanning tunneling microscope with an electromagnetic coarse approach system. All STM measurements were conducted at ambient conditions and at room temperature in the constant current mode with the bias voltage applied to the sample. Tungsten tips prepared by electrochemical etching have been used. Local tunneling spectra have been acquired in different points of the graphene surface.

Figure 1. SEM images of the graphene layer placed on the array of GaN micro-pyramids at 2 µm scale (a) and at 1 µm scale (b).

3. Results and discussion

The typical STM image of the graphene surface is presented in figure 2(a). The topography of graphene has the periodic structure replying the hexagonal symmetry of the GaN pyramids array. The graphene layer is suspended between tops of adjacent pyramids and has small sagging in the middle of this distance. The size of sagging is about 200 nm (about 20% of height of a pyramid). The profile of the cross-section of the graphene layer is presented in figure 2(b) (red line). The surface of graphene is almost flat around the top of a pyramid and then bends slowly with increasing of the distance from a pyramid. The graphene layer demonstrates thus rather high elastic stiffness in spite of its polycrystalline structure. We can see a sharper bending of this layer in the middle of the space between pyramids but even in this case the layer retains the surface integrity. We could not see any damages of the graphene layer in all STM images of our sample. This information can be useful for wider application of graphene as an electrode in optoelectronics.

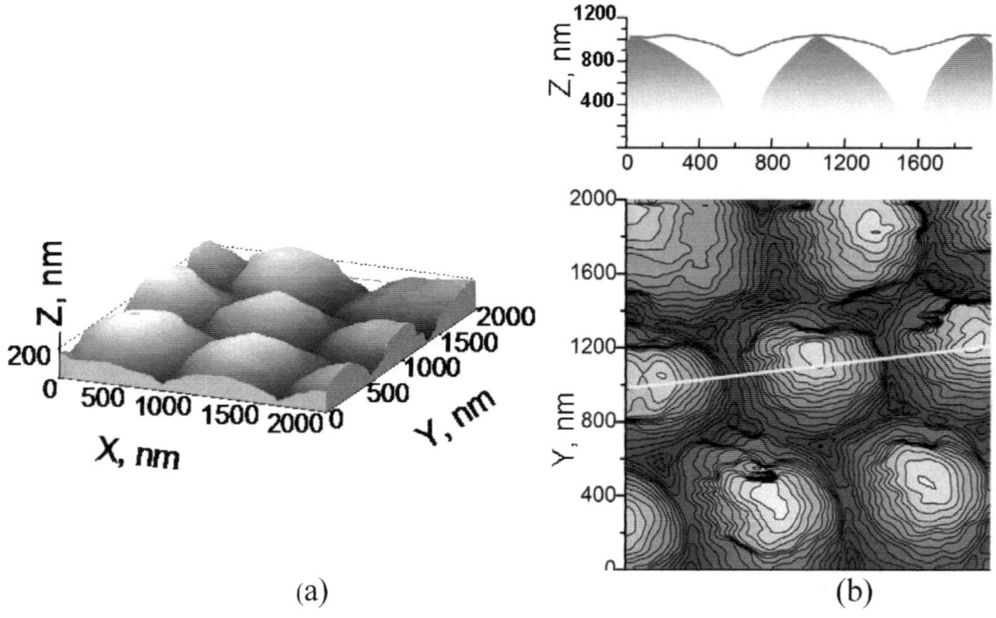

Figure 2. STM image of the graphene layer placed on the array of GaN micro-pyramids (a), cross-section of the layer and profile of this section with sketchy images of GaN pyramids. The image was acquired at a bias voltage of 0.2 V and a tunneling current of 0.1 nA.

Local tunneling spectra of the electron density were acquired for different areas of the graphene surface. The experimental data for the graphene layer on the top of a GaN pyramid and in the area between pyramids are presented in figure 3. The dependence of the tunneling current on the bias voltage for the flat surface of graphene on the top of a pyramid (curve 1 in figure 3(a)) is almost linear. The same dependence for the area between pyramids (curve 2 in figure 3(a)) is significantly nonlinear. A linear curve indicates the metallic conductance in this area of the graphene layer. The well-known tunneling differential conductance reflects the energy dependence of the electron density of states (DOS), and the zero bias voltage corresponds in this case to the Fermi energy in the sample. Curve 1 in figure 3(b) indicates the constant DOS in graphene above the Fermi energy and the increase of DOS below it. A nonlinear current-voltage dependence indicates the presence of some features in DOS. Curve 2 in figure 3b shows the appearance of a wide minimum of DOS at the Fermi energy. It is unlikely to be an energy gap since the differential conductance is not zero at all bias voltages. This minimum is not symmetric relative to the zero bias voltage: DOS is higher at positive bias voltages. The similar minimum in local tunneling spectra was presented in [7]. This transformation of the electron density of states can be connected with the local mechanical stress in the graphene layer due to its sagging between pyramids [8].

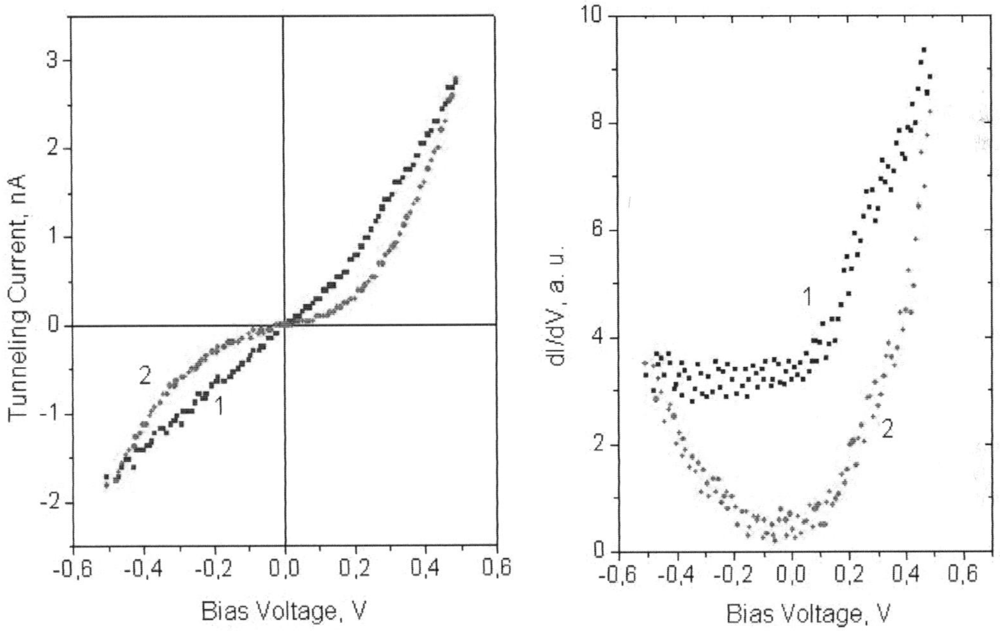

(a) (b)

Figure 3. Local tunneling current-voltage characteristics (a) and tunneling differential conductance dependencies on bias voltage (b). 1 – for the graphene layer on the top of a pyramid, 2 – for the graphene layer in the area between pyramids.

4. Conclusion

This work proves that the scanning tunneling microscope can be a powerful tool to study the features of the surface topography and the local energy spectrum of graphene placed on any micro structured substrates. In the case of GaN pyramids as a substrate, we did not find any damages of the graphene layer. The layer remains conductive in spite of its sagging between pyramids. It confirms good prospects of using graphene as a flexible conducting transparent electrode in optoelectronics.

Acknowledgements

The research was partly supported by the Russian Foundation of Basic Research (grant 15-02-08282A).

References

[1] Choe M, Cho C Y, Shim J P, Park W, Lim S K, Hong W K, Lee B H, Lee D S, Park S J and Lee T 2012 *Appl. Phys. Lett.* **101** 031115
[2] Babichev A V, Zhang H, Lavenus P, Julien F H, Egorov A Y, Lin Y, Tu L W and Tchernycheva M 2013 *Appl. Phys. Lett.* **103** 201103
[3] Xu Q, Yang M, Cheng Q, Zhong J, Wu S, Zhang Z, Cai W, Zhang F and Wu Z 2014 *Appl. Phys. Express* **7** 082103
[4] Wang L, Ma J, Liu Z, Yi X, Zhu H and Wang G 2014 *ACS Photonics* **1** 421
[5] Yu Q, Lian J, Siriponglert S, Li H, Chen Y P and Pei S S 2008 *Applied Physics Letters* **93** 113103
[6] Stankevič T *et al* 2015 *J. Appl. Crystallogr.* **48** 344
[7] Yeh C, Teague M L, Boyd DA, Bockrath M W, Velasco J and Lau C N 2010 *ECS Transactions* **28** 115
[8] Majid M J and Savinskii S S 2011 *Technical Physics Letters* **37** 519

IOP Conf. Series: Journal of Physics: Conf. Series **816** (2017) 012033 doi:10.1088/1742-6596/816/1/012033

Quasi-streamer mode of delayed avalanche breakdown initiated by technological imperfections

M S Ivanov[1,2], N I Podolska[2,3] and P B Rodin[2]

[1] Peter the Great St. Petersburg Polytechnic University, 29 Polytechnicheskaya str., St. Petersburg, 195251, Russia
[2] Ioffe Institute, 26 Politechnicheskaya str., St. Petersburg, 194021, Russia
[3] Saint-Petersburg Branch of the Joint Supercomputer Centre, 26 Politechnicheskaya str., St. Petersburg, 194021, Russia

E-mail: lygeon@gmail.com

Abstract. Numerical simulations of spatially non-uniform picosecond-range switching of high-voltage silicon diodes with technological imperfections have been performed. It is shown that spatially inhomogeneous distribution of process-induced deep-level centers that trigger ionization process easily provoke current localization. A fluctuation of concentration with a relative amplitude of 10% leads to almost complete localization of current. The switching time sharply decreases (from ~100 ps to ~10 ps) with the increase of the amplitude of the inhomogeneity and the decrease of its size.

1. Introduction

Propagation of the superfast impact ionization TRAPATT-like front represents one of the possible mechanisms of the delayed avalanche breakdown in high-voltage p^+–n–n^+ structures [1]. This phenomenon underlies the operation of picosecond range pulse power electronics semiconductor devices [2] and manifests itself in superfast switching of a diode structure from a blocking to a conducting state on condition of a rapid growth of the reverse voltage applied to the structure. A plane TRAPATT-like front is unstable with respect to small perturbations [1], which makes the spatio-temporal dynamics of the switching very sensitive to small technological inhomogeneities of the diode structure. The subsequent current localization can play a constructive role reducing the switching time [3]. The aim of this study is to investigate the influence of the technological inhomogeneity responsible for the generation of initial carriers that trigger avalanche ionization process in overvoltaged diodes on the switching dynamics.

2. The model

We consider a reversely biased p^+–n–n^+ structure connected in series with a load R = 50 Ω. The circuit and the structure are schematically presented in figure 1. A steep voltage ramp must be applied to the diode to achieve the delayed avalanche breakdown mode [1]. In our simulations the external voltage is approximated as a linear voltage ramp $V(t) = V_0 + At$, where V_0 = 1 kV is the initial bias and A = 2 kV/ns. The parameters of the structure are taken as follows: the diameter of the structure d = 1 mm, n-base length W = 100 μm, base doping level N_d = 10^{14} cm^{-3}. These structure and circuit parameters roughly correspond to those in recent experiments [4].

IOP Conf. Series: Journal of Physics: Conf. Series **816** (2017) 012033 doi:10.1088/1742-6596/816/1/012033

Numerical simulations are performed under an assumption that initial free carriers are generated by field-enhanced thermionic electron emission from deep electron traps (thermal defects) as it has been suggested in [5]. Amongst all possible technological inhomogeneities, the inhomogeneity of concentration N_{PI} of these deep-level centers over the cross section of the device is expected to be the most important because these centers influence the very beginning of the front propagation process.

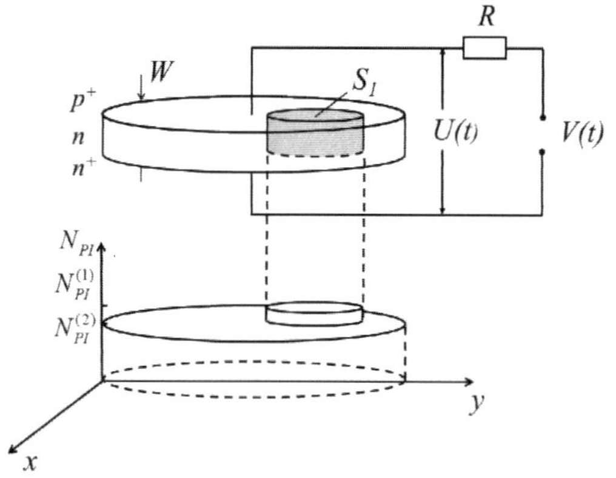

Figure 1. Sketches of p^+–n–n^+ structure with the external circuit and thermal defects distribution over the cross section of the structure.

Taking into account that the structure width W is much smaller than the diameter, we neglect possible inhomogeneity of the thermal defects concentration N_{PI} along the cathode-anode direction and model the spatial inhomogeneity of N_{PI} by dividing the structure cross section into two parts. These are a part with higher concentration $N_{PI}^{(1)} = N_{PI}^{(0)}(1+\delta)$ and area S_1, and a part with lower concentration $N_{PI}^{(2)} = N_{PI}^{(0)}(1-\delta)$ and area S_2. In the following we refer to these parts as to the "active" part and the "passive" part, respectively. In the framework of the long-wavelength approximation, it is equivalent to connecting two diodes with these parameters in parallel. The long-wavelength approximation is valid as far as $\sqrt{S_1} > W$, so S_1 cannot be chosen arbitrary small. The average value of the deep centers concentration is chosen as $N_{PI}^{(0)} = 10^{12}\,\mathrm{cm}^{-3}$. Parameter δ is the

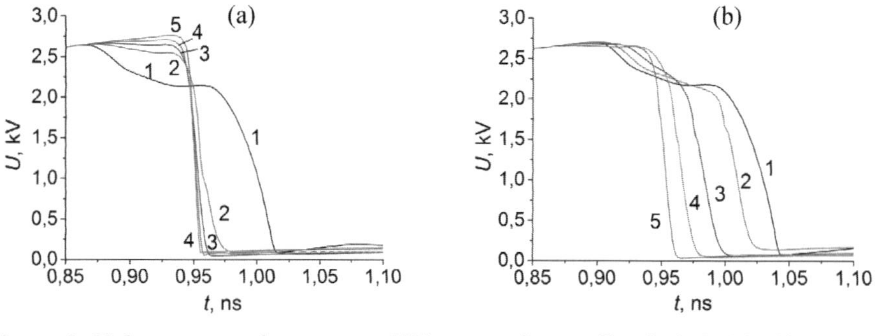

Figure 2. Voltage across the structure $U(t)$ versus time at K = 1, 4, 8, 16, 64 (curves 1–5, respectively) and fixed δ = 0.5 (a); and at δ = 0, 0.1, 0.2, 0.3, 0.5 (curves 1–5, respectively) and fixed K = 8 (b).

relative amplitude of the inhomogeneity. Another important parameter is the inverse relative size $K = (S_1 + S_2)/S_1$ of the inhomogeneity.

3. Results

The dependences of the voltage across the device $U(t)$ on parameters K and δ are shown in figure 2. At $t < 0.85$ ns the voltage increases linearly, following the external voltage applied to the structure and the load. It is readily seen that the switching process becomes much faster with the increase of either K or δ.

Numerical simulations do not agree with experiment as far as we assume that switching is homogeneous. Namely, the non-monotonic voltage waveforms $U(t)$ predicted by numerical simulations at low values of K and δ (e.g. curve 1 for the uniform switching in figure 2) correspond to slow switching process and are not observed in experiments. In contrast, for higher values of K and δ the results of numerical simulations and experiments [4] are in good agreement (e.g. curves 3,4,5 in figure 1). It is an argument in favour of the inhomogeneity of the switching. The switching time rapidly decreases with the increase of K and δ, which is shown in figure 3(a). This dependence saturates when the relative amplitude of the fluctuation δ reaches ~0.3 and K reaches ~6. Localization of current in the "active" part of area S_1 is even more sensitive to the value of δ: inhomogeneities with $\delta > 0.1$ lead to almost complete current localization irrespective of their size K (figure 3(b)). In figure 3(b) the current through the "active" part S_1 at the moment when the external voltage reaches 4 kV is shown. Note that the total current through the structure is mostly determined by the load $R = 50\ \Omega$ and equals ~80 A. It implies the winner-takes-it-all character of the competition between the "active" part of the structure, where the ionizing front has better starting conditions, and the lagging ionizing front in the "passive" part.

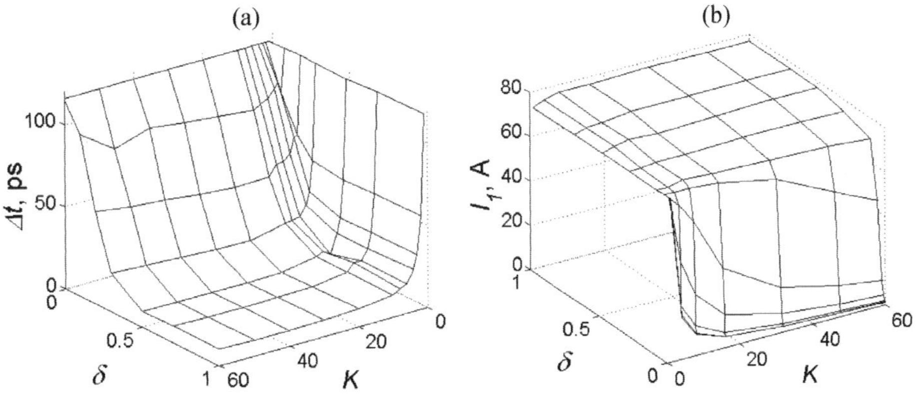

Figure 3. Dependence of the switching time (a) and the current in the conducting channel (b) on K and δ.

Our simulations reveal that current localization induced by technological inhomogeneity leads also to a significant change in the conductivity modulation mechanism. Namely, the switching process becomes divided into two stages. The first stage is the propagation of the impact ionization front in the "active" part of the structure (figure 4(a)). This stage is similar to the homogeneous switching scenario, but the voltage does not drop during the front passage. As a result, the electric field behind the front increases (figure 4(a), curves 3, 4). The front starts in the "passive" part as well, but it starts later, fails to spread through the whole base length and stops in the middle of the structure. Thus, the "passive" part of the device effectively plays a role of a parallel capacity. At the second stage of switching, the discharge of this capacity, associated with the "passive" part, through the conducting channel formed by the front passage in the "active" part, occurs. At this stage, the secondary avalanche breakdown of the "active" part takes place (figure 4(b)). It is accompanied by the rapid voltage drop on the device. Obviously, the second stage is not observed for the homogeneous switching process that is finished when the ionizing front passed across the diode base.

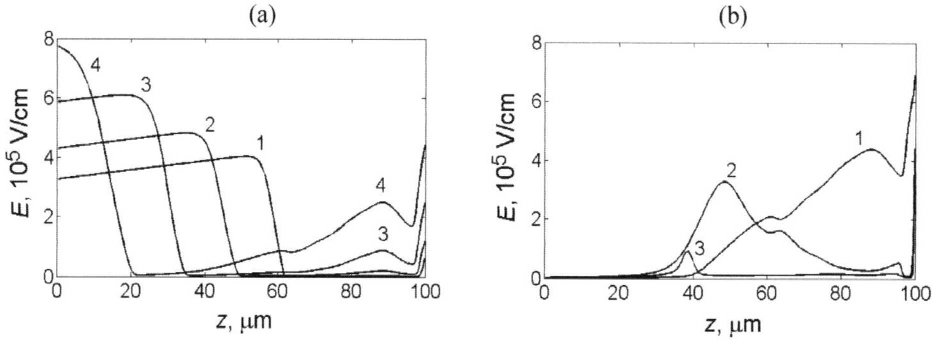

Figure 4. Electric field $E(z,t)$ at the front propagation stage (a) and at the secondary breakdown stage (b). Curves 1–4 in section (a) correspond to time moments t = 0.97, 0.985, 0.993 and 0.997 ns (a), curves 1–3 in section (b) correspond to t = 1.0, 1.01 and 1.02 ns. Parameter values are $K = 8$ and $\delta = 0.1$, which corresponds to curve 2 in figure 2(b).

4. Conclusions

The presence of technological imperfections qualitatively changes the switching scenario and reduces the switching time of picosecond-range avalanche diodes operating in the delayed avalanche breakdown mode. Instead of homogeneous switching due to propagation of the TRAPATT-like ionizing front over the whole cross-section of the device, our numerical simulations predict localization of current in the area with a higher concentration of deep electron traps responsible for triggering impact ionization. Although the ionizing front is triggered over the whole area of the device, its lagging part fails to compete with the leading part. This winner-takes-it-all dynamics eventually leads to formation of a conducting channel only in the location of the technological inhomogeneity. Nearly total localization of current in this area occurs if the relative amplitude of the inhomogeneity exceeds 10% (figure 3(b)), and it is not sensitive to the relative size of the inhomogeneity. Switching in such a spatially inhomogeneous quasi-streamer mode is faster than uniform switching, but a significant decrease of the switching time requires higher relative amplitude of the inhomogeneity and a smaller relative size than those required for current localization (figure 3(a)). The decrease of the switching time occurs not because the ionizing front propagates faster, but because switching occurs after the front passage during the secondary avalanche breakdown. At this stage, the "passive" part of the device, where the front passage has been terminated, recharges through the conductive channel formed by the front passage in the "active" part of the device. It implies that technological imperfections may play a constructive role leading to a faster switching process.

References

[1] Levinshtein M, Kostamovaara J and Vainshtein S 2005 *Breakdown Phenomena in Semiconductors and Semiconductor Devices* (London: World Scientific)
[2] Grekhov I V 2010 *IEEE Trans. Plasma Sci.* **38** 5 1118–23
[3] Rodin P B, Minarsky A M and Grekhov I V 2012 *Tech. Phys. Lett.* **38** 6 535–9
[4] Brylevskii V I, Smirnova I A, Rodin P B and Grekhov I V 2014 *Tech. Phys. Lett.* **40** 4 357–60
[5] Rodin P, Rodina A and Grekhov I 2005 *J. Appl. Phys.* **98** 094506

Temperature stabilization of spin-LEDs with a CoPt injector

A V Rykov, M V Dorokhin, P B Demina, A V Zdoroveyshchev and M V Ved'

Research Institute of Physics and Technology, Lobachevsky State University, Nizhni Novgorod, 603950, Russia

rikov@nifti.unn.ru

Abstract. Spin light-emitting diodes based on GaAs/InGaAs/AlGaAs heterostructures with a ferromagnetic CoPt injector have been fabricated and studied. To obtain stability of diodes at higher temperatures, heterostructures with three different combinations of a quantum well and barriers have been studied. A stable 2% circular polarization degree of electroluminescence was obtained at temperatures up to 300 K.

1. Introduction

Spin light-emitting diodes (spin-LEDs or SLEDs) based on A3B5 semiconductors are one of the key elements of spintronics [1]. Operation of SLEDs is based on electrical injection of spin-polarized carriers from a magnetized ferromagnetic injector layer followed by radiative recombination of injected carriers with unpolarized ones. The recombination process is accompanied by the emission of circularly polarized light. The degree of circular polarization is set by magnetization of the injector and can be controllably changed, for example, by an external magnetic field [1, 2].

Currently, the main task to be solved within the technology of spin light-emitting diodes is preservation of both high intensity and high polarization degree for a wide range of temperatures (including 300 K). The specific type of ferromagnetic injector layers including a thin interfacial MgO dielectric layer was developed in [3, 4] to obtain a high polarization degree. The question of high electroluminescence intensity at room temperature is very often ignored although most of SLEDs active regions tend to reveal temperature quenching of luminescence due to the thermal escape of carriers from the active region at room temperature.

In the present paper, we compare three types of SLEDs active regions based on AlGaAs/GaAs and/or GaAs/InGaAs heterostructures. We demonstrate that a relatively high confinement potential allows us to obtain room temperature luminescence. Spin light-emitting diodes based on structures under consideration emit circularly polarized light at 300 K thus the developed structures can be used for fabrication of SLEDs which would meet the mentioned requirements for their operation.

2. Structures fabrication and characterization

The series of heterostructures with different combinations of a quantum well and barriers was produced to study temperature quenching of luminescence in a quantum well (QW). The structures were grown by metal-organic chemical vapour deposition (MOCVD) process at a low pressure (100 mbar) and at 650°C on semi-insulating GaAs. The first batch of structures (No.1-3) includes the following layers: a 500 nm p-$Al_xGa_{1-x}As$ buffer layer, a 11 nm $In_{0,22}GaAs$ QW and a 100 nm $Al_xGa_{1-x}As$ cap layer. The parameter x has been varied for $Al_xGa_{1-x}As$ barriers which acted as a confinement potential for carriers in a QW. The second batch of structures differs from the first one by its active region. Instead of a usual InGaAs QW, the two-step GaAs/InGaAs/GaAs QW was grown (structures 4 and 5). The test structure of a InGaAs QW in a GaAs matrix was grown as a reference (*T*).

Results of photoluminescence (PL) measurements at room temperature are shown in figure 1. The line at ~ 995 nm dominates in all spectra, this line corresponds to radiative transitions in the quantum well. Structures with higher confinement barriers (a higher Al content in the barrier) demonstrate a higher luminescence intensity of the QW peak relatively to the reference structure. The only exception is structure 1, in which a slight increase of the barrier height (10% Al in AlGaAs) is prevailed by the luminescence quenching effect caused by a large concentration of irradiative centers on the InGaAs/AlGaAs heterointerface. This effect is also responsible for the fact that structures 1 and 2 have significantly lower PL intensities than structures 4 and 5, respectively, despite the same AlGaAs barrier height.

PL spectra of the studied structures at 77 K are shown in figure 2. The same QW peak dominates in the spectrum; the emission wavelength is ~ 950 nm due to the bandgap increase with decreasing temperature. It is worth mentioning that thermal ejection of carriers is insignificant at liquid nitrogen temperature. Therefore, the quality of the structure and, in particular, the quality of the QW/barrier heterojunction take a top priority. Thus, the reference structure has the highest PL intensity and is followed by structures 4 and 5 with two-step QWs, as expected.

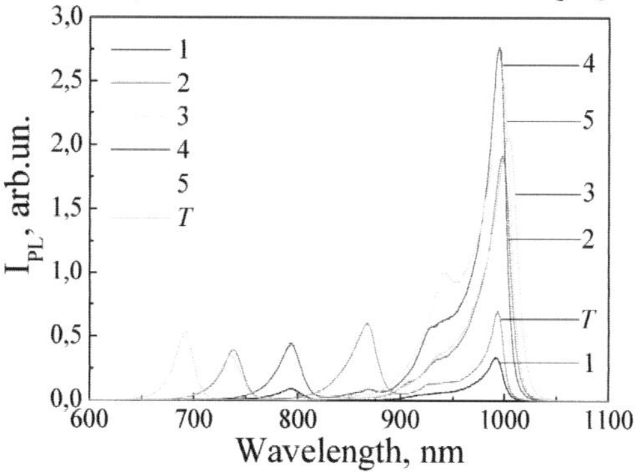

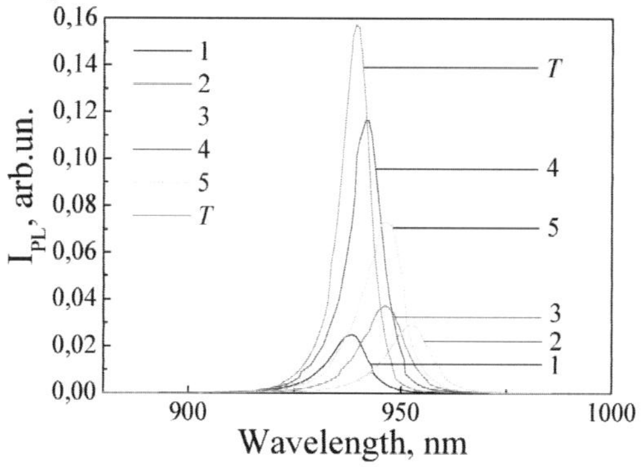

Figure 1. PL spectra measured at 300 K for the following structures: 1 - InGaAs/$Al_{0.1}$GaAs, 2 - InGaAs/$Al_{0.2}$GaAs, 3 - InGaAs/$Al_{0.3}$GaAs, 4 - InGaAs/GaAs/$Al_{0.1}$GaAs, 5 - InGaAs/GaAs/$Al_{0.2}$GaAs, *T* – InGaAs/GaAs (the reference sample).

Figure 2. PL spectra measured at 77 K. PL intensity of the reference sample (*T*) divided by 2 for convenience.

3. Electroluminescence and circular polarization study

Spin light-emitting diodes (spin-LEDs) with a ferromagnetic CoPt injector were produced basing on the results shown in the previous section (figure 3). The diode structures were grown on a p^+-GaAs substrate and included a two-step $In_{0.23}$GaAs/GaAs QW and barriers with Al content $x = 22\%$. The top barrier for a QW (a spacer) included a 10 nm $Al_{0.22}$GaAs layer and a 10 nm GaAs cap. The cap layer prevents redundant oxidation of AlGaAs due to the inter-operational air exposure during SLED fabrication. A ferromagnetic Schottky CoPt contact was deposited on the structure surface by layer-by-layer deposition of Co and Pt at 200°C using an electron beam [5]. To improve the efficiency of spin injection and the crystalline quality of the ferromagnetic layer (a semiconductor interface), a tunnel-thin Al_2O_3 layer was deposited (as a diffusion barrier [6]) on the GaAs surface before depositing the ferromagnetic CoPt layer.

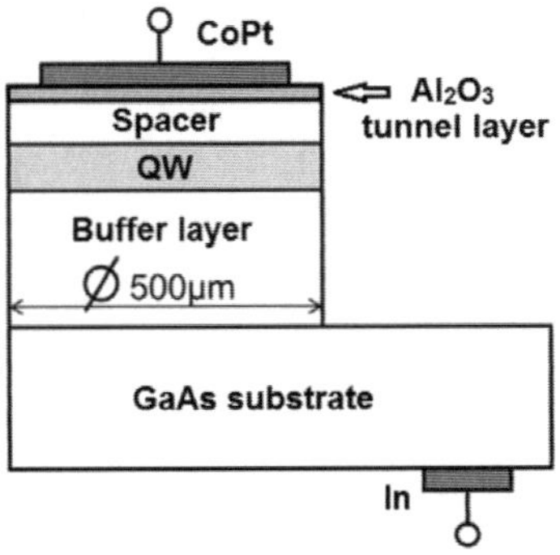

Figure 3. The scheme of a spin-LED device.

For electroluminescence (EL) measurements, a forward bias was applied to a sample CoPt/*p*-GaAs Schottky contact (a negative potential to the CoPt with respect to the In base). The EL circular polarization degree dependence on magnetic field was studied in the temperature range of 10 ÷ 300 K. Magnetic field was applied perpendicularly to structures' plane and has been varied in the range from -200 to 200 mT. The circular polarization degree of EL was defined as:

$$P = (I_+ - I_-)/(I_+ + I_-). \tag{1}$$

Here I_+ and I_- are the intensities of σ^+- and σ^--circularly-polarized components of EL, measured at a spectra maximum corresponding to radiative transitions in a QW. Emission of circularly polarized radiation with a wavelength corresponding to the QW is the evidence of spin-polarized carriers in the QW. In the situation when we apply a negative voltage to the structure, electrons of CoPt are injected into the conduction band of GaAs through the tunnel-thin Al_2O_3 film [7]. When we put the structure into magnetic field, the effective density of states of ferromagnetic material increases for carriers with «majority» spin and decreases for carriers with «minority» spin. That is why the carriers with mainly «majority» spin are injected into a semiconductor. In the end, spin-polarized electrons diffuse into the QW, then recombine with non-polarized holes supplied from the substrate.

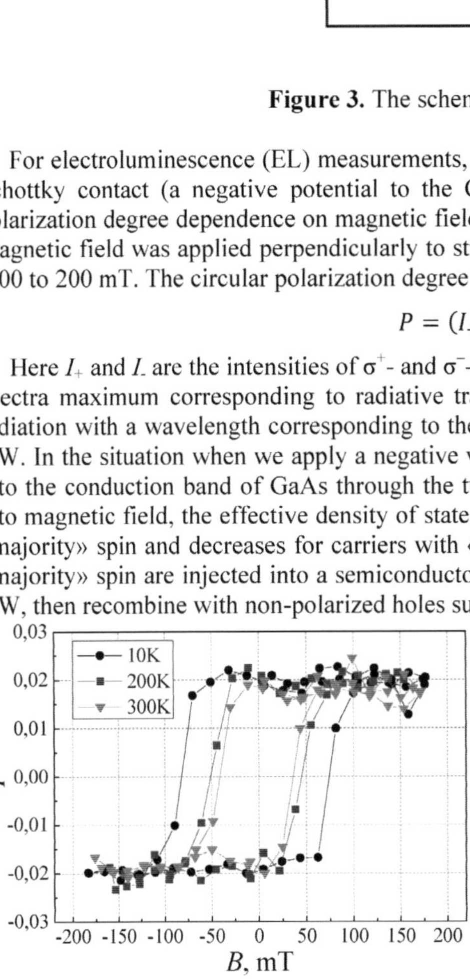

Figure 4. EL circular polarization degree of diodes with InGaAs/GaAs QWs in an $Al_{0,22}GaAs$ matrix.

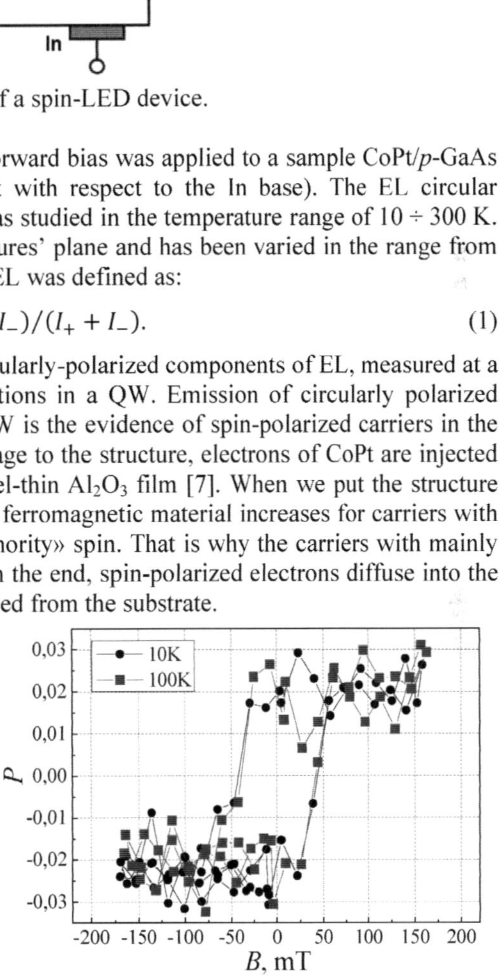

Figure 5. EL circular polarization degree of diodes with InGaAs QWs in a GaAs matrix (reference sample).

Electroluminescence studies of diodes with a CoPt injector have shown a 2% circular polarization degree at temperatures from 10 K up to 300 K. We believe that the observed EL circular polarization is caused by spin injection process described above. Other mechanisms that could be responsible for the circularly polarized emission (mainly magnetic circular dichroism) are insignificant in our measurement scheme. Indeed, the EL emission was collected from the back of the sample, i.e. the detected part of the EL signal transferred not through the CoPt layer but through the transparent GaAs substrate.

We note that there was no sign of temperature decrease of the polarization degree in the temperature range from 10 to 300 K according to the error levels (figure 4). A hysteresis loop characteristic for the ferromagnetic CoPt layer was observed [5]. The coercive field is temperature dependent, the width of the loop decreases with temperature increase. Diodes with a conventional InGaAs QW in a GaAs matrix with a 20 nm GaAs spacer were used as a reference for the studied SLEDs. Diodes with this kind of configuration of the active region and with the CoPt Schottky contact have a relatively weak electroluminescence, which affects the polarization degree measurements and causes a high dispersion of results (figure 5). In addition, a significant temperature quenching of EL prevents us from observation of electroluminescence and a circular polarization at room temperature. Spin-LEDs with the two-step InGaAs/GaAs QW in an AlGaAs matrix demonstrated a stable electroluminescence with polarization at room temperature. The studied SLEDs have shown a higher EL intensity and a significantly lower dispersion of a polarization degree magnitude in comparison to conventional SLEDs.

Thus, we have manufactured spin-light emitting diodes based on different active regions and investigated the circularly polarized emission of these diodes. We have demonstrated that both temperature quenching and non-radiative recombination influence the EL intensity and prevent EL observation at room temperature. Fabrication of the high-quality InGaAs/GaAs/AlGaAs quantum well allows us to improve the quality of interfaces and to increase the QW confinement potential. This results in stable observation of EL circular polarization at room temperature.

Acknowledgments
This study was supported by the Russian Foundation for Basic Research (project no. 16-07-01102_a).

References

[1] M. Holub, P. Bhattacharya 2007 *J. Phys. D: Appl. Phys.* **40** R179–R203
[2] A.T. Hanbicki, B.T. Jonker, G. Itskos, G. Kioseoglou, and A. Petrou 2002 *Appl. Phys. Lett.* **80** 1240
[3] G. Salis, R. Wang, X. Jiang, R. M. Shelby, S. S. P. Parkin, S. R. Bank, and J. S. Harris 2005 *Appl. Phys. Lett.* **87** 262503
[4] S. H. Liang, T. T. Zhang, P. Barate, J. Frougier, M. Vidal, P. Renucci, B. Xu, H. Jaffrès, J.-M. George, X. Devaux, M. Hehn, X. Marie, S. Mangin, H. X. Yang, A. Hallal, M. Chshiev, T. Amand, H. F. Liu, D. P. Liu, X. F. Han, Z. G. Wang, and Y. Lu 2014 *Phys. Rev. B.* **90** 085310
[5] A. V. Zdoroveyshchev, M. V. Dorokhin, P. B. Demina, A. V. Kudrin , O. V. Vikhrova, M. V. Ved' , Yu. A. Danilov , I. V. Erofeeva, R. N. Krjukov , and D. E. Nikolichev 2015 *Semiconductors* **49** 1601
[6] A. I. Bobrov, Yu. A. Danilov, M. V. Dorokhin, A. V. Zdoroveyshchev, N. V. Malekhonova, E. I. Malysheva, D. A. Pavlov, and S. Saeid 2015 *J. Surf. Invest.: X-ray, Synchrotron Neutron Tech.* **9**, 706
[7] N. V. Baidus', P. B. Demina, M. V. Dorokhin, B. N. Zvonkov, E. I. Malysheva, and E. A. Uskova 2005 *Semiconductors* **39**(1) 17

Room temperature spin injection in a light-emitting diode based on a GaMnSb/*n*-GaAs/InGaAs tunnel junction

M V Dorokhin[2], M V Ved'[1,2], E I Malysheva[2], P B Demina[2], A V Zdoroveyshchev[2], A V Kudrin[1,2] and Yu A Danilov[1,2]

[1]Lobachevsky State University of Nizhny Novgorod, Nizhny Novgorod, 603950 Russia
[2] Physical-Technical Research Institute of Nizhny Novgorod, Lobachevsky State University, Nizhny Novgorod, 603950 Russia

E-mail: mikhail28ved@gmail.com

Abstract. The circularly polarized electroluminescence from an InGaAs/GaAs light-emitting diode with a (Ga,Mn)Sb/n^+GaAs tunnel junction was investigated in the temperature range of 10-300 K. The obtained polarization degree is believed to be driven by the spin injection of electrons from the valence band of ferromagnetic (Ga,Mn)Sb into the conduction band of n^+GaAs. The temperature dependencies of the polarization characteristics are determined by the ferromagnetic properties of the (Ga,Mn)Sb layer. The room temperature circularly polarized emission is prospective for the fabrication of a spin light-emitting diode with an injector based on a diluted magnetic semiconductor.

1. Introduction

Diluted magnetic semiconductors (DMS), such as (III,Mn)V, are considered to be prospective materials for implementation in devices using the spin degree of freedom [1-4]. The advantages of DMS are the good embeddability into most semiconductor spintronic structures and easy control of ferromagnetic properties by the variation of the growth parameters (Mn doping level, fabrication temperature, additional acceptor doping) [1-7]. However, the application of (III,Mn)V materials is limited by two major obstacles. First one is a hole type of conductivity in (III,Mn)V, since short spin relaxation times for holes (as compared to electrons) significantly limit the efficiency of spin transport [6,7]. The second obstacle is a low Curie temperature of DMS not exceeding 250 K (for (Ga,Mn)As) [3,4].

In order to implement spin injection of electrons from p-type (III,Mn)V layers, the Zener diode structure was suggested in [8,9]. In the cited papers, the efficient injection of spin polarized valence band electrons from a reverse biased (III,Mn)V/n-AlGaAs tunnel junction into a GaAs quantum well (QW) was obtained.

The second obstacle has not been overcome yet, to our knowledge. The only (III,Mn)V layers possessing ferromagnetic properties at room temperature are the two-phase systems containing Mn-V or Mn-III clusters in a (III,Mn)V matrix [10-12]. These clusters are proven to be responsible for the observation of a ferromagnetic signal from both SQUID magnetometer [11] and anomalous Hall effect [12] measurements. We note that the two-phase DMS are not often considered as the element of spintronic devices, which is probably due to the low spin injection efficiency from DMS layers caused by spin scattering at the MnV/(III,Mn)V, MnV/III-V heterointerfaces.

In the present paper, we report on the fabrication of (Ga,Mn)Sb/GaAs layers that demonstrate ferromagnetic properties up to room temperature and on the implementation of these layers in a spin light-emitting diode. The fabricated diodes emit circularly polarized light with a circular polarization degree of ~ 0.54% at 10 K, which was found to be slowly decreasing with the temperature increase reaching the value of 0.32% at 300 K.

2. Experimental technique

The structures were fabricated by the two-stage epitaxial growth method [13,14]. First, the layers of a light-emitting diode were grown by metal-organic vapor phase epitaxy (MOVPE) on a *p*-GaAs substrate at 600°C. The following sequence of the layers was formed: a *p*-GaAs buffer, an $In_{0,16}Ga_{0,84}As$ quantum well (QW) (dopped with C, $p \sim 8\times10^{17}cm^{-3}$, width of $d_{QW} = 16$ nm), a 6 nm undoped GaAs, *n*–GaAs ($n \sim 10^{17}cm^{-3}$, width of 50 nm), n^{+}GaAs ($n \sim 10^{18}$ cm^{-3}, width of 20 nm), n^{++}GaAs ($n \sim 10^{19}$, width of 20 nm). At the next stage, a 30 nm (Ga,Mn)Sb layer was grown in the same reactor at 340°C by laser sputtering of GaSb and Mn targets. The Mn content was estimated by the formula:

$$Y_{Mn} = \frac{t_{Mn}}{(t_{Mn}+t_{GaSb})}, \quad (1)$$

where $t_{Mn}(t_{GaSb})$ is the sputtering time of Mn(GaSb). In the present study $Y_{Mn} = 0.5$. The decrease of the growth temperature is required to prevent Mn diffusion into structure [13].

The final treatment of light-emitting diodes included the deposition of an Au contact on a (Ga,Mn)Sb surface and 500 μm mesa formation by means of photolithography and chemical etching. The back Ohmic contact was fabricated by means of sparkling of In foil. The schematic diagram of our structures is shown in figure 1 (a).

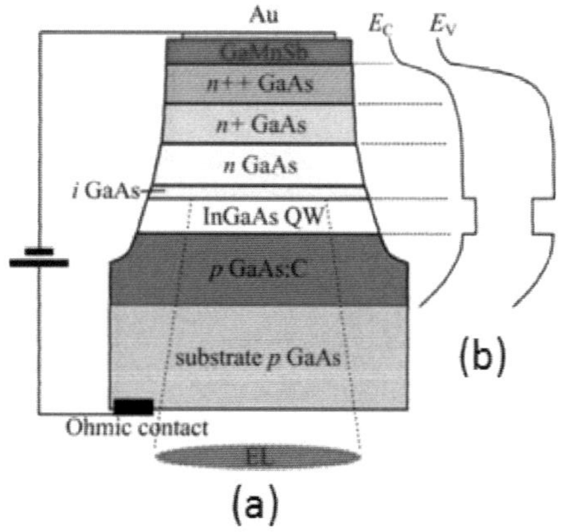

Figure 1. (a) The schematic diagram of the spin-LED structure; (b) the band diagram of a *p*-(Ga,Mn)Sb/*n*-GaAs/InGaAs junction with the applied bias.

The measurements of each sample started from the test investigations of *I-V* characteristics at 300 K. The electroluminescence (EL) spectra of the samples were measured in the temperature range of 10 - 300 K. The magnetic field $B = -0.3 \div +0.3$ T was normal to the QW plane (Faraday geometry). The EL emission was collected from the back of a sufficiently transparent substrate (figure 1 (a)). For the EL measurements, the negative (with respect to the substrate) potential was applied to the top Au contact thus the reverse bias of a (Ga,Mn)Sb/n^{++}GaAs tunnel barrier was achieved causing electron injection from the valence band of (Ga,Mn)Sb into the conduction band of n^{++}GaAs, as in [15].

The degree of circular polarization of the EL emission $P_C(B)$ was calculated by:

$$P_C = \frac{(I_+ - I_-)}{(I_+ + I_-)}, \quad (2)$$

where $I_+(I_-)$ are the emission intensities of the optical transition in the QW at σ^+ (σ^-) polarization.. In order to certify the spin injection process, the control measurements of circularly polarized photoluminescence (PL) were also carried out. Spin injection is not involved in the PL process, thus the PL polarization degree (if any) cannot be attributed to the spin injection. The magnetization of the samples was investigated at 300 K by an alternating gradient force magnetometer with a sensitivity of about 10^{-7} emu in a magnetic field of up to ± 180 mT. The magnetic field was applied perpendicular to the sample surface. The details of the gradient force magnetometry technique were described in [16].

3. Experimental results and discussion

I-V characteristics of the diode are shown in figure 2. In the applied measurement geometry the diode represents two *p-n* junctions in the opposite bias conditions. The kink followed by the rapid increase of the diode current is observed at the *I-V* curves when the negative bias $U \approx -0.6 \div -1.2$ V is applied to the top contact. Such a kink probably corresponds to the situation when the top of the valence band of (Ga,Mn)Sb coincides with the bottom of the conduction band of *n*+GaAs (figure 1 (b)). In this mode, tunneling of electrons from (Ga,Mn)Sb into GaAs takes place, and the holes are injected into the InGaAs QW from the forward-biased *p*-substrate/*n*-GaAs junction [9]. When the temperature is increased, the kink voltage decreases evidencing the role of the thermally activated tunneling mechanism at higher temperatures.

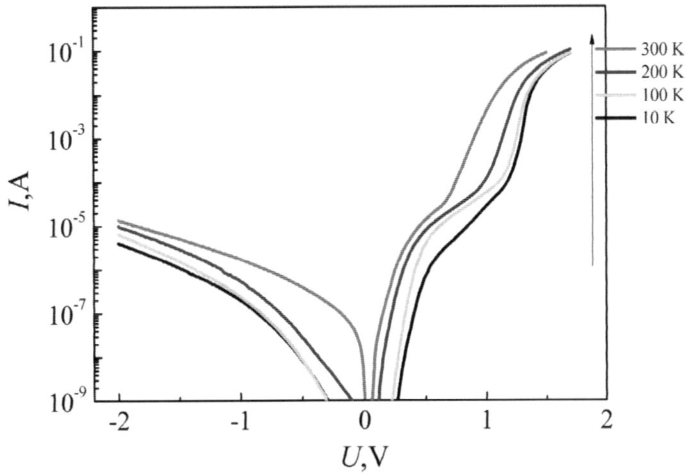

Figure 2. The *I-V* characteristics of the investigated diode structure measured at different temperatures.

The EL spectrum of the investigated sample is shown in figure 3. One can see a peak at 1.375 eV dominating in the spectrum. This line corresponds to the main radiative transition in the quantum well. When the magnetic field is applied, the intensities of the polarized EL components σ^+ (σ^-) differ from one another, thus one can determine the circular polarization of the EL emission with eq. (2). The dependencies of the circular polarization degree on magnetic field, measured at 10 and 300 K, are shown in figure 4 (plots 1 and 2, respectively).

IOP Conf. Series: Journal of Physics: Conf. Series 816(2017) 012035 doi:10.1088/1742-6596/816/1/01203

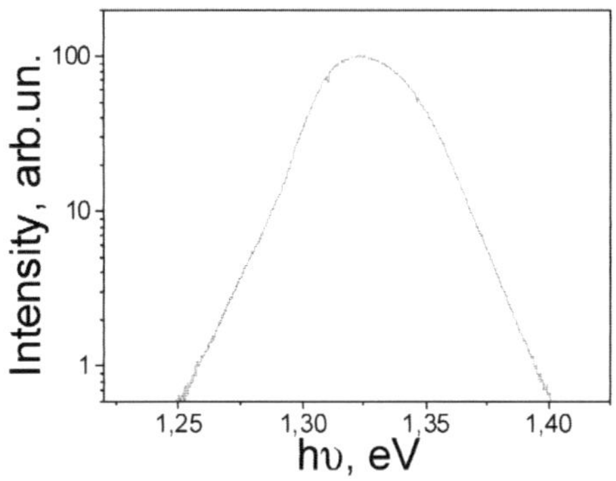

Figure 3. The electroluminescence spectrum of the sample measured at 77 K at a diode current of 10 mA.

The graphs in figure 4 demonstrate non-monotonic dependencies with a hysteresis loop. Note that the $P_C(B)$ dependence measured at 300 K is similar to the magnetization vs B dependence measured at the same temperature (figure 4: circles, curve 2 and a line, curve 3, respectively). We also note that the PL polarization degree of the investigated sample does not exceed 0.1 % (figure 4, curve 4). Both results are believed to be clear evidence of the spin injection [2] being the major reason for the observed EL circular polarization. Taking into account the polarity of the applied bias, we suppose the spin injection of electrons from (Ga,Mn)Sb into the InGaAs/GaAs heterostructure to be similar to that into the (Ga,Mn)As/GaAs one [10,18]. The contribution of magnetic circular dichroism to the polarization (as in [19]) is believed to be negligible since the EL emission is collected from the back of the sample and thus does not pass through the ferromagnetic layer. Also, it is worth noticing that the change in n^{++}-GaAs doping parameters (increase from 10^{19} to ~1.5×10^{19} cm^{-3}) leads to a nearly twofold decrease in P_{EL} (not shown to prevent data overload in figure 4). This cannot be explained by the magnetic circular dichroism effect but is in good agreement with the Zener diode tunneling mechanism [19].

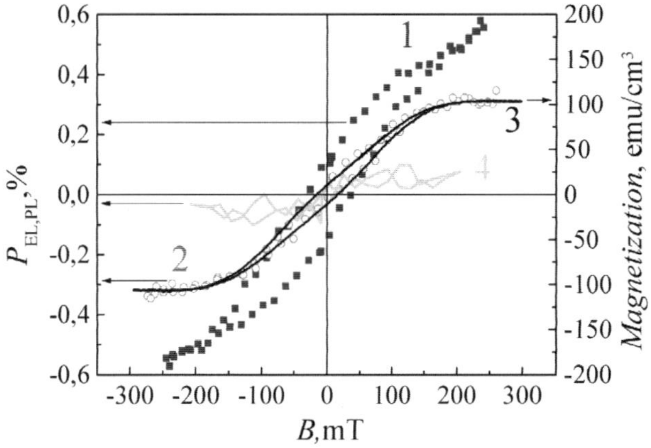

Figure 4. The magnetic field dependencies of the EL polarization degree (symbols, curves 1, 2) at a diode current of 20 mA) and of the magnetization (line, curve 3) of the investigated sample measured at temperatures of 10 K (1) and 300 K (2, 3). Curve 4 corresponds to the magnetic field dependence of the photoluminescence polarization degree of the investigated sample measured at 300 K.

The temperature dependence of the polarization degree is shown in figure 5. In the temperature range of 10 ÷ 50 K the polarization degree is temperature independent, at $T > 50$ K a monotonic linear decrease of P_C is observed. At $T = 300$ K, the polarization degree $P_C = 0.3$ %. The parameters of the hysteresis loop are also temperature-dependent (insert in figure 5). One can see that both the polarization degree in remanence (at $B = 0$) and the coercive field (at which $P_C = 0$) decrease with increasing temperature.

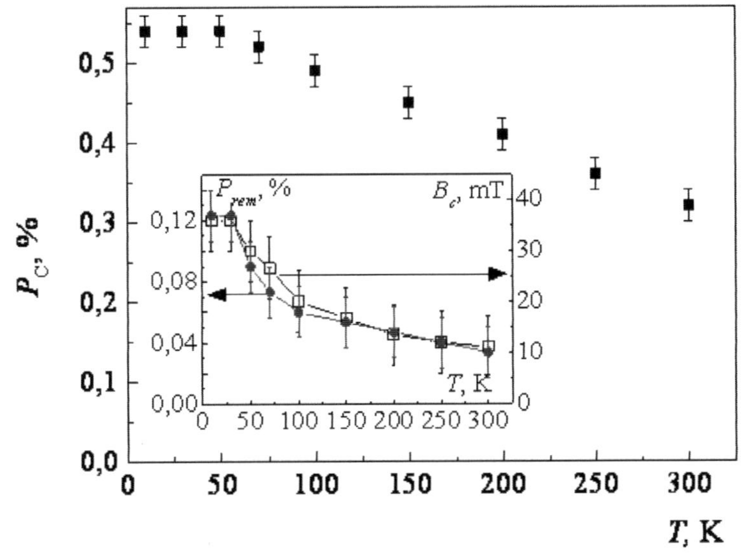

Figure 5. The temperature dependence of the EL polarization degree of the investigated sample measured in a magnetic field of 250 mT and a current of 20 mA. The insert shows the temperature dependence of the polarization degree in remanence (closed circles) and the coercive field (open squares) measured in the same conditions.

The temperature dependencies are driven by the magnetic properties of a (Ga,Mn)Sb layer. The monotonic decrease of the P_C with temperature is relatively extended and thus cannot be attributed to reaching the Curie temperature (for which an abrupt decrease of P_C is more typical [6]). One can suggest that the $P_C(T)$ dependence is associated with the increasing role of spin scattering [2] for higher temperatures. Also, the decrease of both the polarization degree in remanence and coercive field with the increase of T (insert in figure 5) can probably be attributed to the variation of magnetic properties of (Ga,Mn)Sb. A similar effect has been reported in [17] for the samples with MnAs clusters embedded into a GaAs matrix. In the cited paper, it was reported on a fivefold decrease of the coercive field and a nearly 60 percent decrease of magnetization with the temperature increasing from 5 to 300 K which is in good correlation with our result. However, no explanation on such magnetic properties variation was suggested. We believe that, in our case, MnGa or MnSb clusters are responsible for the preservation of ferromagnetic properties up to 300 K. The open question is whether there is an interaction between the clusters causing ferromagnetic ordering. To clarify the issue some additional investigation of ferromagnetic properties of (Ga,Mn)Sb films is to be carried out.

Two important features of our results are the spin injection and ferromagnetism of a (Ga,Mn)Sb semiconductor both obtained at room temperature. Previously, it has been shown that the room temperature ferromagnetism of (Ga,Mn)Sb layers, grown by laser sputtering in the identical conditions, is due to the MnGa clusters, which are formed in a (Ga,Mn)Sb matrix [18]. Such systems were rarely used for the spin injection, however, we demonstrated that the spin-polarized electrons can be injected from (Ga,Mn)Sb into GaAs even at room temperature, which, to our knowledge, is the highest spin injection temperature for the structures based on the diluted magnetic semiconductors [2,6,9,20]. A low polarization degree is probably due to the poor (Ga,Mn)Sb/GaAs interface quality [18], which, however, can be improved by the proper selection of the growth parameters.

4. Conclusion

In conclusion, we have investigated the circularly polarized electroluminescence from an InGaAs/GaAs light-emitting diode with a (Ga,Mn)Sb/n^+GaAs tunnel junction. The obtained polarization degree is believed to be due to the spin injection of electrons from the valence band of ferromagnetic (Ga,Mn)Sb into the conduction band of n^+GaAs. The effect persists up to 300 K, which is prospective for the fabrication of a room-temperature operating spin light-emitting diode on the basis of a diluted magnetic semiconductor.

Acknowledgments

Authors would like to acknowledge Dr. B.N. Zvonkov for the fabrication of investigated samples.
This study was supported by the Russian Foundation for Basic Research (project 15-02-07824), Ministry of Education and Science of Russian Federation (project 8.175.2017/PP) and Russian Federation President's grant (MK-8221.2016.2).

References

[1] Dietl T and Ohno H 2014 *Rev. Modern Phys.* **86** 187
[2] Holub M and Bhattacharya P 2007 *Phys. D: Appl. Phys.* **40** R179
[3] Jungwirth T, Wang K Y, Mašek J, Edmonds K W, Jürgen König, Jairo Sinova, Polini M, Goncharuk N A, MacDonald A H, Sawicki M, Rushforth A W, Campion R P, Zhao L X, Foxon C T and Gallagher B L 2005 *Phys. Rev. B.* **72** 165204
[4] Nazmul A M, Amemiya T, Shuto Y, Sugahara S and Tanaka M 2005*Phys. Rev. Lett.* **95** 017201
[5] King C S, Zemen J, Olejnık K, Horak L, Haigh J A, Novak V, Irvine A, Kucera J, Holy V, Campion R P, Gallagher B L and Jungwirth T 2011 *Phys. Rev. B.***83** 115312
[6] Young D K, Gupta J A, Johnston-Halperin E, Epstein R, Kato Y and Awschalom D D 2002 *Semicond. Sci. Tech.* **17** 275
[7] Hilton D J and Lang C L 2002 *Phys. Rev. Lett.* **89** 146601
[8] Andersen S E, Sorensen B S, Rasmussen F B, Lindelof P E, Sadowski J, Guertler C M and Bland J A C 2003 *J. Appl.Phys.* **94** 3990
[9] Kohda M, Ohno Y, Matsukura F and Ohno H 2006 *Physica E.* **32** 438
[10] Chakraborty A, Bouzerar R, Kettemann S and Bouzerar G 2012 *Phys. Rev. B.***85** 014201
[11] Lawniczak-Jablonska K, Wolska A, Klepka M T, Kret S, Gosk J, Twardowski A, Wasik D, Kwiatkowski A, Kurowska B, Kowalski B J and Sadowski J 2011 *J. Appl.Phys.* **109** 074308
[12] Kudrin A V, Shvetsov A V, Danilov Yu A, Timopheev A A, Pavlov D A, Bobrov A I, Malekhonova N V and Sobolev N A 2014 *Phys.Rev.B.* **90** 024415
[13] Dorokhin M V, Malysheva E I, Zdoroveishev A V, Danilov Yu A and Kudrin A V 2012 *Semiconductors* **46** 1518
[14] Zvonkov B N, Vikhrova O V, Danilov Yu A, Demina P B, Dorokhin M V, Podolskii V V, Demidov E S, Drozdov Yu N and Sapozhnikov M V 2008 *J. of Optical Technology* **75** 389
[15] Dorokhin M V, Malysheva E I, Zvonkov B N, Zdoroveishchev A V, Danilov Yu A, Nikolichev D E, Boryakov A V and Zubkov S Yu 2014 *Technical Physics* **59** 1839
[16] Flanders P J 1988 *J. Appl. Phys.* **63** 3940
[17] M. Moreno, A. Trampert, B. Jenichen, L. Däweritz, and K. H. Ploog 2002 *J. of Applied Physics* **92** 4672
[18] Bobrov A I, Pavlova E D, Kudrin A V and Malekhonova N V 2013 *Semiconductors* **47** 1587
[19] Van Dorpe P, Liu Z, Van Roy W, Mostnt V F, Sawicki M, Borghs G, De Boeck J 2004 *Applied Physics Letters* **84** 3495
[20] Holub M, Shin J, Chakrabarti S and Bhattacharya P 2005 *Appl. Phys. Lett.* **87** 091108

Light absorption by an atomic force microscope probe

V A Sharov[1], **M S Dunaevskiy**[1,2], **N V Kryzhanovskaya**[3], **Yu S Polubavkina**[3] **and P A Alekseev**[1]

[1] Ioffe Institute, 26 Politekhnicheskaya str., St. Petersburg 194021, Russia
[2] ITMO University, 49 Kronverksky pr., St. Petersburg 197101, Russia
[3] Saint Petersburg National Research Academic University of RAS, 8 Khlopina str., St. Petersburg 194021, Russia

E-mail: npoxep@gmail.com

Abstract. In this paper a probe of an atomic force microscope (AFM) served as a detector of optical radiation. We investigated absorption of optical radiation in the visible range by commercially available Si and Si_3N_4 probes. It has been shown that the radiation in the far field is mainly absorbed in the probe tip. By scanning a laser beam with a diameter of ~ 0.7 microns by the AFM probe, a lateral resolution of ~ 1.5 microns was demonstrated. Numerical modeling of evanescent wave absorption by AFM probes showed enhancement of the absorption in the probes as compared with a flat surface. A Si_3N_4 probe more effectively absorbs the evanescent radiation.

1. Introduction

Development of semiconductor lasers, optical waveguides and other structures for photonics and optoelectronics requires information about optical radiation distribution in them. Conventionally, scanning near-field optical microscopy (SNOM) is used for experimental studies of optical radiation distribution in the near field at the interface of two media, such as laser/air or optical fiber/air [1]. In this method, the sharpened optical fiber probe collects near-field radiation. An AFM probe with a subwavelength-diameter aperture is also used for collecting of near-field radiation [2]. These well-known techniques are limited in spectral range by transparence of the SNOM optical system.

Recently, we demonstrated the detection of optical radiation in the near [3, 4] and the far field [5] by conventional apertureless AFM probes. In our method, an oscillating probe absorbs optical radiation. Absorbed radiation heats the probe and shifts its oscillation frequency and phase. The frequency shift is proportional to the absorbed optical radiation. The method allows the detection of optical radiation in the near field in the visible and infrared ranges with subwavelength resolution.

The resolution of the method in the far field has not been studied. In addition, commercially available probes can be made from Si or Si_3N_4, which are transparent in the visible range and can be coated by various metal layers. At small distances between the probe and the surface, the probe is capable to capture evanescent waves emerging at the surface/air interface [6]. Effects of the probe shape, material and metal coating on the absorption of the optical radiation in the far field and evanescent waves are unstudied. The aim of this work was to experimentally and theoretically examine the characteristics of the visible light absorption by Si and Si_3N_4 probes in the near and the far fields.

2. Methods

In this study, we investigated a Si probe PPP-FMR (Nanosensors) and a Si_3N_4 probe with a metal coating (60 nm Au / 15 nm Cr) NPG10 (Bruker). The probes consist of a cantilever beam and a sharp pyramid mounted on it. Optical absorption measurements in the far field were carried out for Si_3N_4 probes using a Ntegra Spectra (NT-MDT) microscope. This microscope combines the functionality of confocal optical and atomic force microscopes. Figure 1 (a) shows a scheme of the experiment. A pyramid of the oscillating probe was illuminated by a laser beam with a wavelength λ = 527 nm. Investigations were carried out by varying the focused laser beam diameter on the probe tip. Additionally, the scanning system moved the probe with respect to the laser beam in the XY plane, i.e. the laser beam scanned the probe. During the scanning, the phase signal of the probe oscillations was recorded. Thus, a series of images of the laser radiation absorption in the probe was obtained at different diameters of the laser beam.

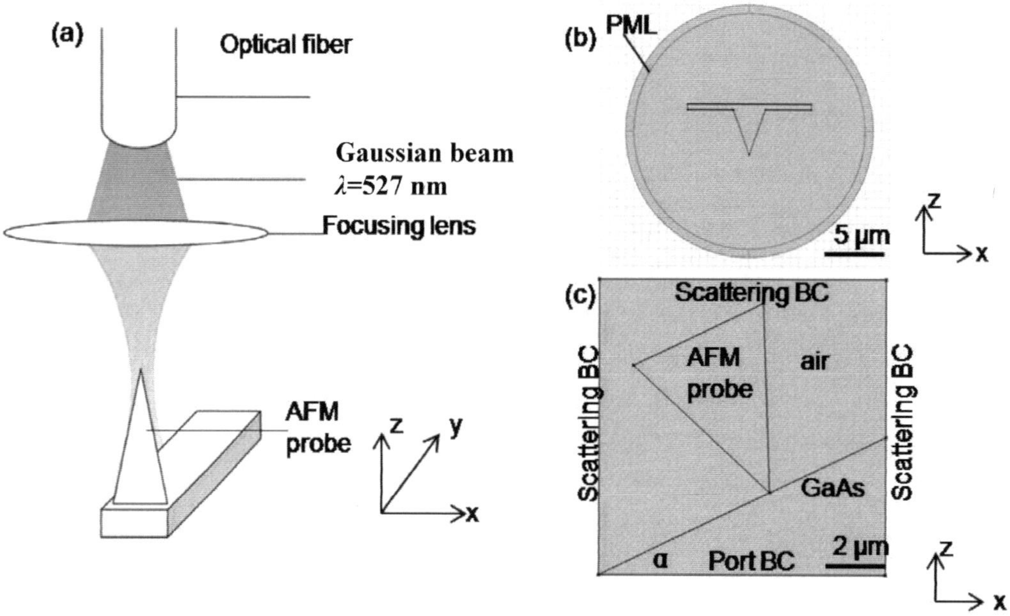

Figure 1. (a) Scheme of the experiment (the AFM probe scanning the laser beam of different diameters). Geometrical models for numerical simulations of light absorption by the AFM probe (b) – in the far field, (c) absorption of an evanescent wave.

For numerical simulation of the electromagnetic waves absorption by the AFM probe, the *Electromagnetic waves, frequency domain* module in the Comsol Multiphysics software was used. To calculate the absorption of radiation in the far field, the model was set as a two-dimensional region bounded by a circle with a perfectly matched layer (PML) on the boundary (figure 2(b)). In the center of the circle there was a probe that can be approximated by a triangle and was parameterized by the real and imaginary parts of the refractive index of the material of which it was made. It is assumed that all the fields have a harmonic dependence on time and that all material properties are linear with respect to the field strength. The solution of the Maxwell's equations is searched in the form:

$$E = E_{relative} + E_{background} ,$$

where the background field is known, and the relative field is the field, after the addition of which to the background one, it turns out that the field satisfies Maxwell's equations. In the simulation, the field of the Gaussian beam is used as the far field ($E_{background}$) (built-in function in the used module). The width of the beam ω depends on the distance z from the beam waist ω_0 as follows:

$$\omega(z) = \omega_0 \sqrt{1 + \left(\frac{\lambda z}{\pi \omega_0^2}\right)^2}.$$

By using this dependence, one can simulate different diameters of the laser beam by changing the coordinate of its waist.

To simulate the evanescent wave field, the model was set as a two-dimensional square area bounded at the top and sides by the scattering boundary condition (BC), and for the lower boundary condition the Port BC was used, which is a planar electromagnetic wave (figure 1(c)). For the evanescent field, the border GaAs /air was introduced in the model, inclined at an angle α to the lower boundary of the model. At angles α greater than the angle of total internal reflection at the GaAs/air interface, the evanescent wave appeared and its absorption in the probe was calculated. GaAs Refractive index at λ = 527 nm was equal to n = 4.058, GaAs was chosen as a classical material for semiconductor lasers.

The standard function *ewfd.Qe* (hereinafter referred to simply as *Q*) of the Comsol Multiphysics package was used as an indicator of the absorbed energy. This function corresponds to the electromagnetic power transformed to the heat (averaged over the time period of the wave). From the Poynting's theorem, this function is equal to:

$$Q = \int J \cdot E dV.$$

3. Results and discussion

3.1. Far field light absorption

Figures 2 (a), (b), and (c) show a series of phase images of the Si_3N_4 probe by a scanning laser beam with different degrees of focusing (with a beam diameter of 0.7 μm, 3 μm and 5 μm, respectively). Since the phase shift is proportional to absorbed radiation, it follows from figure 2(a) that the light is absorbed mainly in the pyramid of the probe (a bright spot on the image). Moreover, the pyramid edges are also visible in the figure. Defocusing of the laser beam blurs the image of the pyramid and the cantilever of the probe. Figure 2(d) shows cross sectional profiles corresponding to figures 2 (a), (b) and (c), built along the dotted lines in figure 2 (b). Interestingly, in a particular probe, the pyramid was shifted by 2 μm to the right from the center of the cantilever, it was taken into account in the numerical simulation. Figure 2 (e) shows the experimental and calculated profiles of absorbed optical radiation and they are in good agreement with each other.

Since Si_3N_4 does not absorb optical radiation with λ = 527 nm, it should be absorbed by a Au / Cr coating which is covering the pyramid and the cantilever. Since the thickness of the coating is the same everywhere, the absorption should also be uniform. However, in figure 2 the absorption is non-uniform. A possible explanation for this inhomogeneity is the enhancement of the optical radiation absorption in the probe tip, connected with its geometry [7] and a bilayer metal coating. Moreover, one should consider various optical radiation reflection coefficients from the surface of the probe cantilever and pyramid edges. Increased absorption of optical radiation in the probe tip opens up the possibility of its use for mapping of the optical radiation distribution in the far field. Indeed, according to the Rayleigh criterion [8] the Si_3N_4 probe can resolve two laser beams with diameters of 0.7 μm at a distance of 1.5 μm from each other (figure 2 (f)).

3.2. Absorption of the evanescent wave

Figure 3(a) shows the dependence of the light absorption by Si and Si_3N_4 probes from the distance between the probe tip and the surface (z- dependence). The angle of light incidence α was 25° (figure 1). Note that the angle of total reflection for our model was ~14°. For comparison, the same dependence was calculated for the Si plane of rectangular shape. The figure shows that both the nitride and silicon probes increase the absorption of evanescent waves in comparison with the flat Si plane of rectangular shape. The same behaviour was observed in the absorption of the optical radiation in the far field. It is interesting to note that the z-dependence for the Si_3N_4 probe decays more slowly than for the Si probe, indicating that the nitride probe is more effective in detecting evanescent radiation.

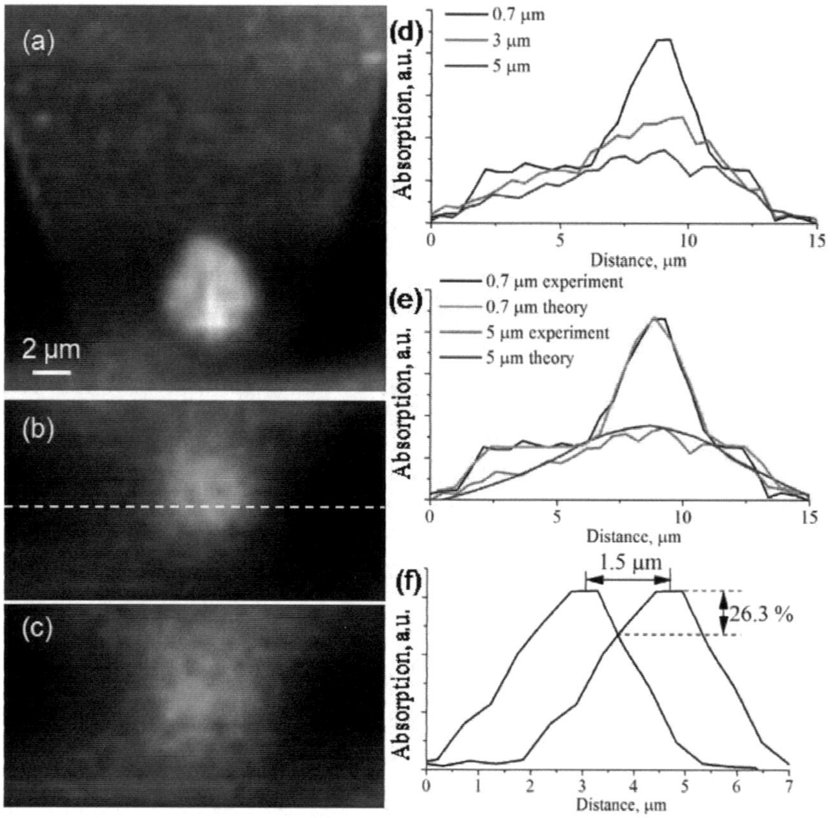

Figure 2. Scanning a laser beam by the Si_3N_4 probe. Phase shift images obtained during scanning the laser beam with a diameter of (a) 0.7 μm; (b) 3 μm, (c) 5 μm; (d) corresponding cross-sectional profiles; (e) experimental and numerically simulated cross-sectional profiles; (f) illustration to the Rayleigh criterion of optical resolution.

Figure 3 (b) shows the absorption of evanescent radiation by Si and Si_3N_4 probes depending on the angle of incidence α. The distance between the probe and the surface was 50 nm. The figure shows that the absorption ratio $Q_{Si3N4}/Q_{Si} > 1$ at $\alpha = 15°$; 20°; 30° and $Q_{Si3N4}/Q_{Si} < 1$ for $\alpha = 25°$ and 35°.

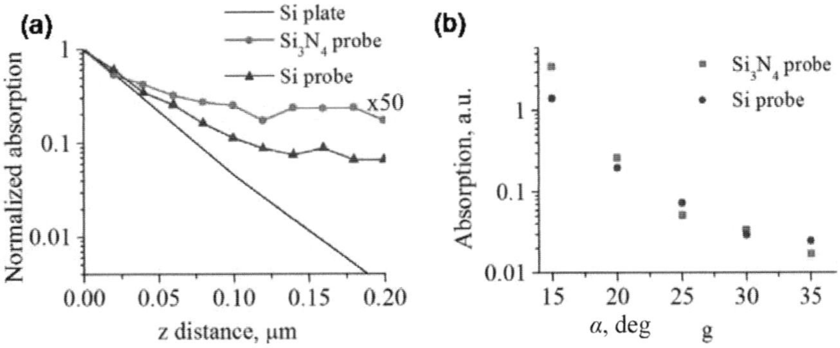

Figure 3. (a) Normalized dependence of evanescent wave absorption on the z-distance between the probe and the surface for Si_3N_4 and Si probes and for a Si plane of rectangular shape (black curve). (b) Dependence of evanescent wave absorption on light incidence angle α for Si_3N_4 and Si

This result shows that under certain α, a Si_3N_4 probe is more efficient than a Si probe and vice versa. Interestingly, this ratio Q_{Si3N4}/Q_{Si} varies non-monotonically with α.

4. Conclusions

To conclude, we investigated the absorption of optical radiation (λ = 527 nm) by Si and Si_3N_4 AFM probes. The absorption of the laser beam with varying diameters by a Si_3N_4 probe was experimentally studied. It was shown that in the far field for the laser beam with a diameter of 0.7 µm, the probe provides a resolution of ~ 1.5 microns. Such a high resolution is due to increased absorption of light in the tip of the probe.

Numerical simulation of evanescent wave absorption by Si and Si_3N_4 probes showed the enhancement of absorption, as compared with a flat absorber. The Si_3N_4 probe absorbs evanescent radiation more efficiently. The study of evanescent radiation absorption depending on the incidence angle α revealed that the dependence $Q_{Si3N4}(\alpha)/Q_{Si}(\alpha)$ is non-monotonic because of the influence of the probe material.

Acknowledgments

Authors acknowledge A.A. Bogdanov for fruitful discussion. This study was supported by Presidential Grant MK-7001.2016.2. M.S.D. acknowledge for financial support the Government of Russian Federation (Grant 074-U01).

References

[1] Betzig E and Trautman J K 1992 *Science* **257** 189
[2] Jiang Y, Wang A, Ren B, and Tian Z-Q 2008 *Langmuir* **24** 12054
[3] Dunaevskiy M S, Alekseev P A, Baranov A N, Monakhov A M, Teissier R, Arinero R, Girard P and Titkov A N 2013 *Appl. Phys. Lett.* **103** 053120
[4] Alekseev P A, Dunaevskii M S, Slipchenko S O, Podoskin A A and Tarasov I S 2015 *Tech. Phys. Lett.* **41** 870
[5] Dunaevskiy M, Dontsov A, Alekseev P, Monakhov A, Baranov A, Girard P, Arinero R, Teissier R and Titkov A 2015 *Appl. Phys. Lett.* **106** 171105
[6] Pohl D W, Denk W, and Lanz M 1984 *Appl. Phys. Lett.* **44** 651
[7] Cao L, White J S, Park J-S, Schuller J A, Clemens B M and BrongersmaM L 2009 *Nature Mater.* **8** 643
[8] Michalet X 2006 *PNAS* **103** 4797

IOP Conf. Series: Journal of Physics: Conf. Series **816** (2017) 012037 doi:10.1088/1742-6596/816/1/012037

Development and approbation of nanoscalpels based probes for atomic force microscopy in the field of plasmonics

M V Zhukov[1,2], I S Mukhin[1,3] and A O Golubok[1,2]

[1] Department of Material Science and Nanotechnology, ITMO University, 197101 Kronverkskiy pr., 49, Saint-Petersburg, Russian Federation
[2] Department of Scanning Probe Microscope and Spectroscopy, Institute of Analytical Instrumentation RAS, 198095 Ivan Chernykh, 31-33, Saint-Petersburg, Russian Federation
[3] Laboratory of Renewable Energy Source, St Petersburg Academic University RAS, 194021 Khlopina, 8/3, Saint-Petersburg, Russian Federation

E-mail: cloudjyk@yandex.ru

Abstract. The formation and approbation of the specialized nanoscalpel (NS) probes were carried out. The possibility of producing diffraction gratings on gold substrates for plasmonics applications was shown. The optimal lithography parameters were revealed by using NS probes. It was shown that NS probes allow improving the lithography results such as the depth and width of the incisions as compared to standard probes.

1. Introduction

In recent years, the probes based on single crystalline or amorphous whiskers are very promising to become one of the main instruments to study samples with developed topological structures by atomic force microscopy (AFM). Unlike nanotubes, whisker structures can be formed directly in the place of the electron beam exposure, they are relatively cheap and easy to manufacture, and besides, such probes have different structures due to usage of various precursor gases and have different forms [1, 2].

In the literature, there are a lot of works devoted to producing functionalized probe sensors based on whisker structures, which are designed to perform specific tasks, such as carrying out the lithography with high resolution and cell nanosurgery [3], as well as the manipulation of objects at the nanoscale level by atomic force microscopy [4]. One of the most interesting fields of such research is plasmonics.

The investigations devoted to excitation, propagation and subsequent detection of surface waves (plasmon polaritons) on the modified metal substrates are of great interest nowadays. This phenomenon is widely used in nanosensorics for studying and characterization of particles of nanometer size. Development of plasmon sources and simulations of effects with surface waves by interacting with the objects is actively investigated [5].

Herewith, scanning near-field optical microscopy, in particular, the photon scanning tunneling microscopy, is basically used for the detection of plasmon waves, while nanomodified periodic structures on metal substrates, such as periodically structured channels [6] and the diffraction gratings [7], for the formation of plasmonic waves.

In this article, we consider the possibility of creating sources of plasmon waves (nanoscale gratings) by a high-precision surface modification using the atomic-force lithography with specialized probes based on nanoscalpel-formed whisker structures.

2. Experimental setup

Creation of the NS structures on the tops of probes and control of their parameters were performed by means of a scanning electron microscope (SEM) Inspect (FEI, USA). Studies of the features of NS probes were carried out by means of a scanning probe microscope (SPM) Ntegra Aura (NT-MDT, Russia) in the contact and semi-contact modes. Samples of polycarbonate and gold substrates were investigated and modified to create diffraction gratings with a fixed period. The SPM measurements were performed for both specialized NS and standard probes.

3. Results and discussion

NS is a 2D carbon nanostructure grown by vacuum deposition with a focused electron beam imitating the shape of a blade or a scalpel. This geometry defines the high stability of this type of probes in force impacts directed along the long axes of the nanostructures. NS probes can be used for visualization of samples of relief, as well as for their local modification.

Parameters of a NS grown on the pyramid tops of the probes can be varied within wide limits by changing the accelerating voltage of the electron beam, its diameter, the path of the beam and duration of deposition. It was revealed that the optimum NS width is 300–400 nm, the thickness is 60–80 nm, and the length does not exceed 400 nm while a NS operates in a force lithography mode. The single NS was formed in the vacuum chamber of a microscope at a pressure of about 10^{-3} Pa, an acceleration voltage of 30 kV, and an electron beam diameter of (1–3) nm.

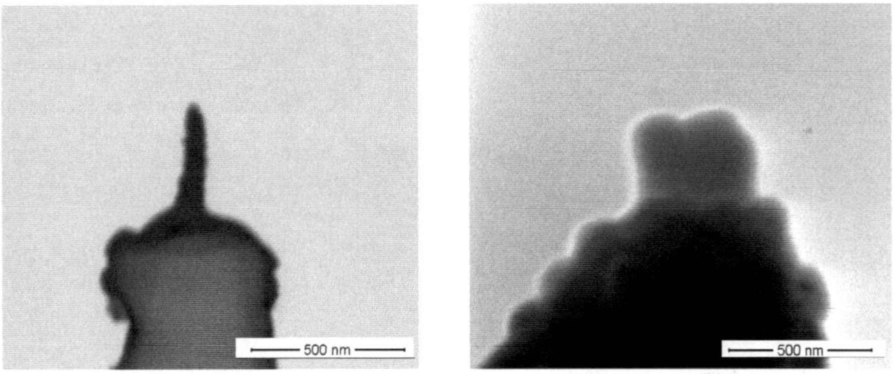

(a) (b)

Figure 1. SEM image of the NS probe after 3 scans and lithography of a 10 nm gold film: a front view (a) and a side view (b).

For calculating the critical force (F_{cr}) that impacts the NS during operation, it is possible to use the approximation of longitudinal critical force for a rectangular thin plate [8] presented by the following equation:

$$F_{cr} \approx \frac{4Eba^3}{3l^2\sqrt{8(1+\sigma)}}, \qquad (1)$$

where E is the Young's modulus, l is the NS length, a is the NS thickness, b is the NS width, σ is the Poisson's ratio.

Inserting typical values $E \approx 0.6 \cdot 10^{11}$ N/m^2, $l \approx 4 \cdot 10^{-7}$ m, $a \approx 6 \cdot 10^{-8}$ m, $b \approx 3 \cdot 10^{-7}$ m, $\sigma \approx 0.5$ [9], we obtain a value of F_{cr} of about 9 μN. Since the typical values of the "probe–sample" interaction forces in atomic force lithography are in the range 0.1–8 μN, NS probes are applicable to high-precision lithography on samples with different stiffness. According to the experimental data, the NS probe maintains about 2–3 cycles (20–30 patterns) of lithography (figure 1) retaining its structure and stiffness, which is comparable with the stability of the standard probes.

Semi-contact probes with a hardness of the order of 4.5–5.5 N/m were used during the atomic force lithography. A thin film of 10 nm gold sputtered on the surface of polycarbonate was used as a working substrate. The film of gold was chosen since this material is applied in the fabrication of nano-sized local sources of plasma waves such as diffraction gratings in plasmonics.

Below we consider the variation in the lithography process by altering the clamping (impact) forces (figure 2). Changing the time of probe standing at the same point had no effect on the structure, and it was chosen to be about $t = 300$ ms, the recommended lithography speed was about 2 μm/s. At a lower speed, there is a significant broadening of the grooves without improvement in depth, whereas at a high speed, reducing the penetration depth occurs without substantially reducing the width of the grooves. When the lithography strength is more than 3 μN, a noticeable shift of the material at the starting point of each line was observed. A possible effect of sticking a probe and contamination of the surface by shifting of the material can occur while increasing the impact forces.

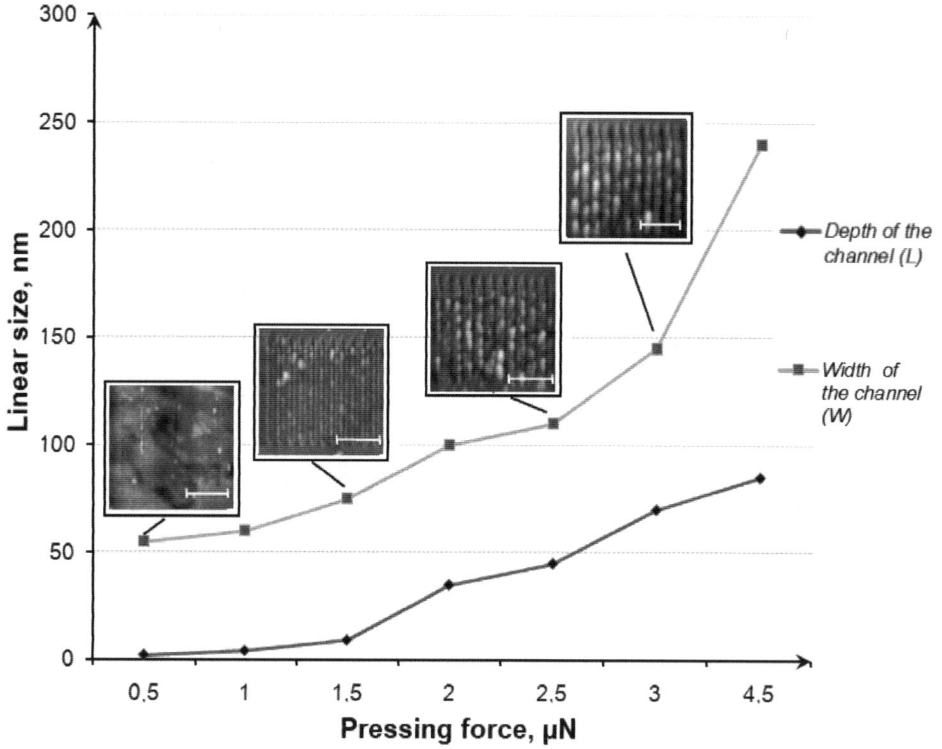

Figure 2. The graph of average features of diffraction gratings (L, W) with AFM insertions fabricated by a NS probe at different impact forces (the scale bar is about 1 μm).

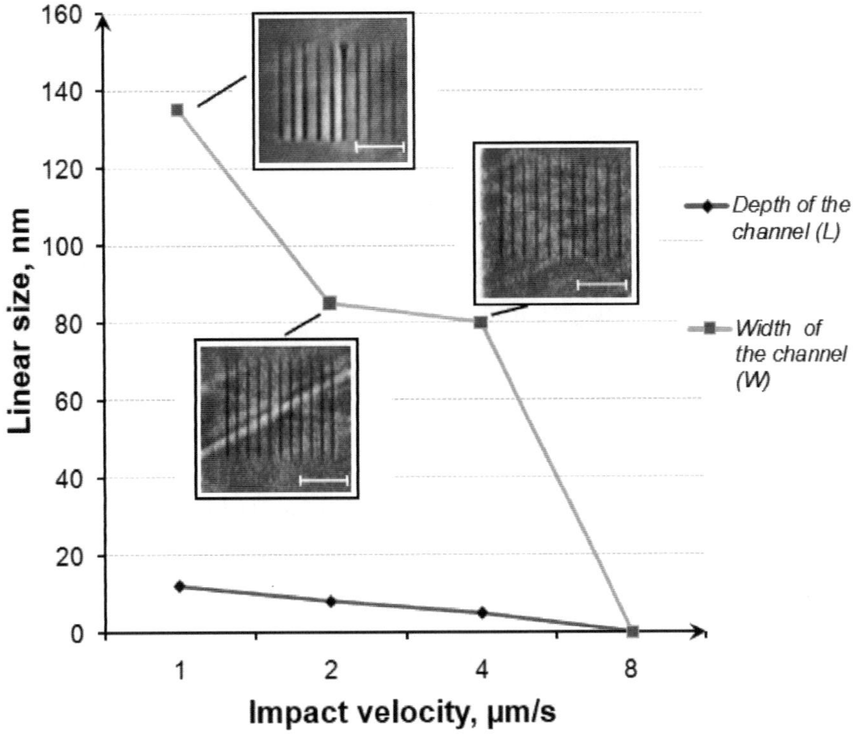

Figure 3. The graph of average features of diffraction gratings (*L, W*) with AFM insertions fabricated by a NS probe at different lithography speeds (the scale bar is about 1 μm).

After finding the optimal parameters of lithography impact forces (1–1.5 μN), altering the lithography speed was carried out (figure 3). At a lithography speed of about 1 μm/s, a strong change of the gratings' widths along the lines was found. At a speed of 2 μm/s, the perfect ratio of the depth and width of gratings was found. While increasing the lithography speed up to 4 μm/s, the decreasing of penetration depth without reduction of the line width was revealed. The further increasing of the lithography speed up to 8 μm/s reveals no grating structure which can be explained by the effect of probe slipping over the surface due to a high speed of lithography.

After finding the optimal parameters of the atomic force lithography, the comparison of lithography with standard and NS probes was carried out (figure 4). The standard probes resolution limit was found in the area of 5 μm with periods of grating of about 500 nm, a line length of about 4 μm, a line depth of about 4 nm, and a line width of about 160-180 nm. The optimal area of NS probes was about 3 μm with a grating's period of about 250 nm, a line length of about 2 μm, a line depth of about 11–12 nm, and a line width of about 70-80 nm.

It was found that the depth of grating grooves obtained by the standard probe is 3 times smaller than the one obtained using the NS probe. Furthermore, a noticeable outgrowth of material (17 nm) around grooves was found using the standard probe, which can be associated with the geometry of the standard probe, expanding from the top to the bottom. In other words, the standard probes push the material outward from the groove with increasing the depth of the line while the NS probes have a constant cross-section and don't have this effect.

At the same time, the limit of resolution of the NS probes was in the range of 25–50 nm, which was identified by lithography of pair lines with a gradual reduction of the distance between them. It should be noted that fabricated gratings can potentially be used in tasks of plasmonics, in particular, for determining the parameters of multipath particle plasmons at the interface of the diffraction grating. The possibility of using NS probes for manipulating particles was discussed previously [10].

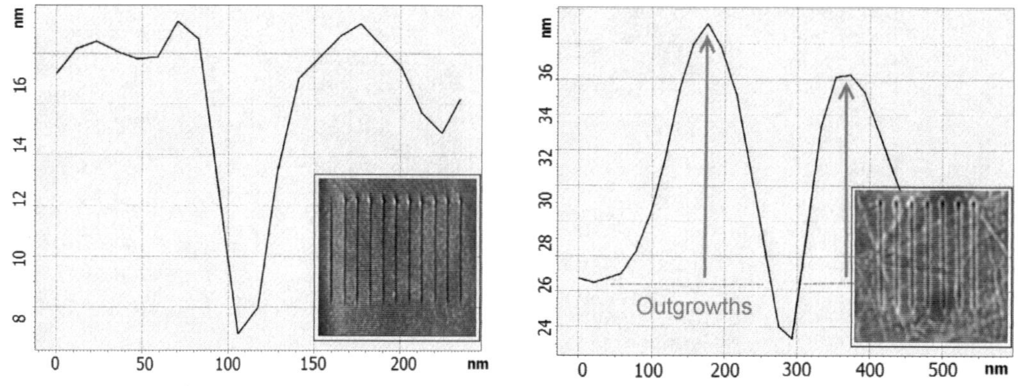

(a) (b)

Figure 4. AFM images of the surface of the gold substrate and typical cross-sections of one line of diffraction gratings obtained by lithography with NS probes (a) and standard probes (b). A red mark shows the build-up material around grooves for the standard probe lithography.

It should be noted that the results of lithography are strongly affected by the stiffness and the parameters of the samples. Probes with a low coefficient of stiffness require a large deflection of the probe to achieve the lithography effect, while at a higher coefficient of stiffness, lower deflection values are required. Therefore, it is recommended to use probes with high stiffness for a strong impact and probes with low stiffness for removal of thin layered material.

4. Conclusions

The formation and approbation of the specialized nanoscalpel based probes were carried out. Improvement of lithography results by using the nanoscalpel probes in comparison with the standard probes was shown. The optimal parameters of lithography (a force impact, a speed) were revealed, and the diffraction gratings on gold substrates for plasmonics research with an accuracy of about 25–50 nm were produced.

The studies demonstrated that nanoscalpel probes can significantly improve the results of lithography process that is interesting for the current nanoelectronics and plasmonics research.

Acknowledgments
The work was carried out with the support of the Russian Foundation for Basic Research (16-32-00806, 16-32-60094), the leading universities of the Russian Federation (grant 074-U01).

References

[1] Ashiri I, Gartsman K, Cohen S R and Tenne R 2003 *AIP Conf. Proc.* **685** 545
[2] Levichev V V, Zhukov M V, Mukhin I S, Denisyuk A I and Golubok A O 2013 *Technical Physics* **58** 1043
[3] Beard J D and Gordeev S N 2011 *Nanotechnology* **22** 175303
[4] Custance O, Perez R and Morita S 2009 *Nature Nanotechnology* **4** 803
[5] Cortes R and Coello V 2009 *NANO: Brief Reports and Reviews* **4** 201
[6] Wu J J 2010 *Progress in Electromagnetics Research* **104** 113
[7] Bonch-Bruevich A M, Libenson M N, Makin V and Trubaev V V 1992 *Optical Engineering* **31** 716

[8] Landau L D and Lifshitz E M 1970 *Theory of Elasticity, A Course of Theoretical Physics* (UK: Pergamon Press)
[9] Zussman E, Chen X, Ding W, Calabri L, Dikin D A, Quintana J P and Ruoff R S 2005 *Carbon* **43** 2175
[10] Mukhin I S, Fadeev I V, Zhukov M V, Dubrovskii V G and Golubok A O 2015 *Ultramicroscopy* **148** 151

Luminescence peculiarities of InGaN/GaN dichromatic LEDs

D S Arteev, A V Sakharov, A E Nikolaev, S O Usov, W V Lundin and A F Tsatsulnikov

Ioffe Institute, Politekhnicheskaya 26, 194021 St-Petersburg Russia

E-mail: dima0724@gmail.com

Abstract. The series of InAlGaN LED structures containing two different quantum wells emitting at wavelengths of ~ 430 nm and ~ 490 nm was grown. The influence of the quantum wells order, thickness of the barrier between the wells and its doping level on the optical properties was studied by photo- and electro-luminescence measurements. It was found that the quantum well with longer-wavelength emission is preferable to be located farther from a p-layer than the shorter-wavelength quantum well to obtain emission from both QWs. Variation of the thickness of the barrier between the QWs and its doping level allows controlling the intensity ratio of two emission peaks.

1. Introduction

The first GaN-based light-emitting diodes were developed in 1971 [1]. Since 1990s, when I. Akasaki, H. Amano, and S. Nakamura fabricated the first high-brightness blue LED, advancement of LED technology has allowed to create high efficiency white solid state light sources. Luminous efficacy of contemporary commercially available white LEDs is above 150 lm/W, when the best ones show efficacy as high as 300 lm/W [2]. A conventional method of white LEDs fabrication is converting blue (~ 440 nm) LED's radiation into yellow-red one, using a phosphor. Mixing blue and yellow lights results in white light with desirable parameters. This method allows achieving high luminous efficiency at low cost. Some disadvantage of these LEDs is a relatively low color rendering index (CRI). The situation can be improved by a proper phosphor mixture, but the usual phosphor-based white LED spectrum has a "gap" in the range of 480 nm to 500 nm, as shown in figure 1. One of the ways to improve CRI over 95 is to use UV LED and several phosphors to get more filled spectrum, but in this case the chip fabrication becomes more complicated and the Stokes' losses increase lowering the total efficiency.

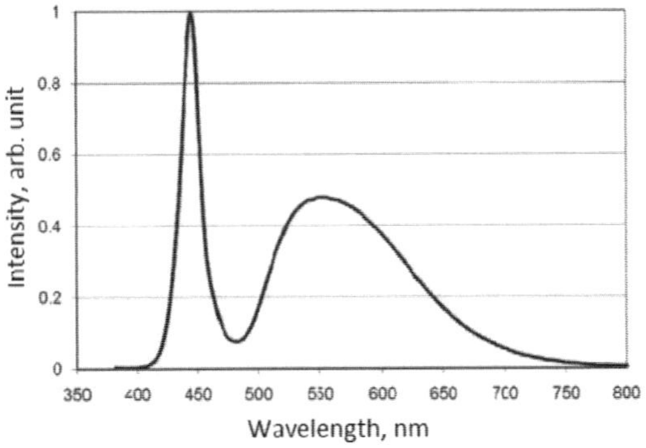

Figure 1. The typical spectrum of a phosphor-converted GaN-based white LED

In this paper, we propose an approach for developing a InAlGaN-based LED with two quantum wells emitting simultaneously at wavelengths of ~ 430 nm and ~ 490 nm to be used as a pumping source for a white phosphor-converted high-CRI LED.

2. Experiments

Samples were grown by metal-organic chemical vapour deposition (MOCVD) on sapphire (0001) substrates. Trimethilgallium, triethilgallium, trimethilaluminum, and trimethilindium were used as the group-III source materials, ammonia was used as the group-V source material. Monosilane and bis(cyclopentadienyl)magnesium were used as doping sources for n- and p-type layers, respectively. Hydrogen and nitrogen were used as carrier gases.

The initial sample was grown as follows. A thick Si-doped n-GaN buffer layer was firstly grown, then a 12 period InGaN/GaN (1 nm/1 nm) short-period superlattice (SPSL) was grown by the InGaN-conversion technique [3]. Afterward, a 20 nm-thick GaN layer was deposited, followed by an active region consisting of two 3 nm-thick InGaN quantum wells (QWs) separated by an 8 nm-thick GaN barrier with an emission wavelength of ~ 430 nm and ~ 490 nm for the first and second QWs (QW1 and QW2), respectively. Thereafter, an unintentionally doped 5 nm-thick GaN layer was grown, a Mg-doped 18 nm-thick $Al_{0.15}Ga_{0.85}N$ electron blocking layer (EBL) was deposited next. At last, the structure was capped with a p-type GaN layer with a thickness of 180 nm. According to the XRD-analysis and simulation using Synopsys Sentaurus TCAD, the indium composition in QW1 and QW2 is about 13% and 19%, respectively. The design of other samples was optimized by changing the relative position of the QWs and varying the thickness of the barrier between the QWs and its doping.

Photoluminescence (PL) studies were performed using excitation by a CW He-Cd laser (325 nm), electroluminescence (EL) studies were performed both on-wafer using In contacts and on processed chips.

3. Results and discussions

In figure 2, PL and EL spectra of the initial sample are shown. As one can see, the PL peak of QW2 has a much higher intensity compared to the QW1 peak. The EL peak of QW1 is almost unnoticeable. Obviously, carriers are mostly injected into QW2 located closer to the p-layers and having more indium in the composition. This situation is fundamentally different from the one observed for blue/green dichromatic LEDs grown in the same geometry [4], where emission from both QWs was observed and the short-wavelength emission was dominant.

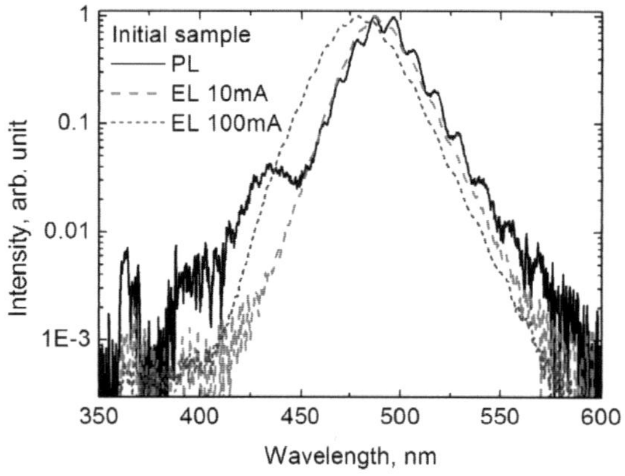

Figure 2. The photoluminescence and electroluminescence (at 10 mA and 100 mA) spectra of the initial sample.

To create the identical conditions for both quantum wells, decreasing the thickness of the barrier layer between wells to improve holes injection was suggested. Two samples with a 1.5 nm-thick barrier were grown. The first sample has the same quantum wells order as the initial sample. In the second sample, the quantum well with a longer emission wavelength (QW2) is farther from the p-layers than QW1. As figures 3 and 4 show, there is practically no emission from QW1 even for PL for both samples. By comparison of the spectra in figures 2, 3 and 4, it can be concluded that there is a redistribution of carriers between quantum wells by direct or thermal-assisted tunneling, when almost all carries are accumulated in the deeper QW.

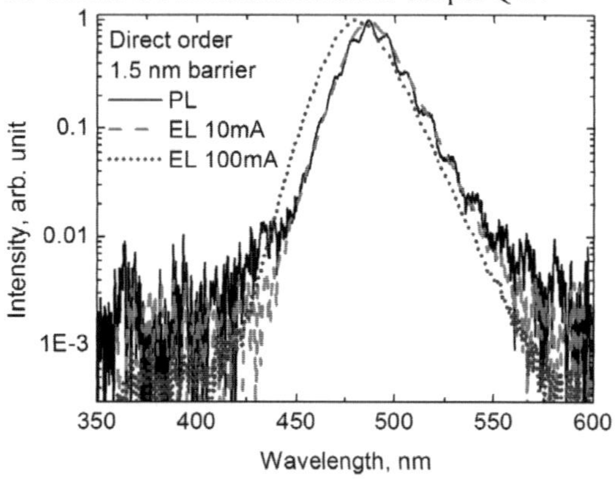

Figure 3. The PL and EL spectra of the sample with a 1.5 nm-thick barrier and a direct wells order.

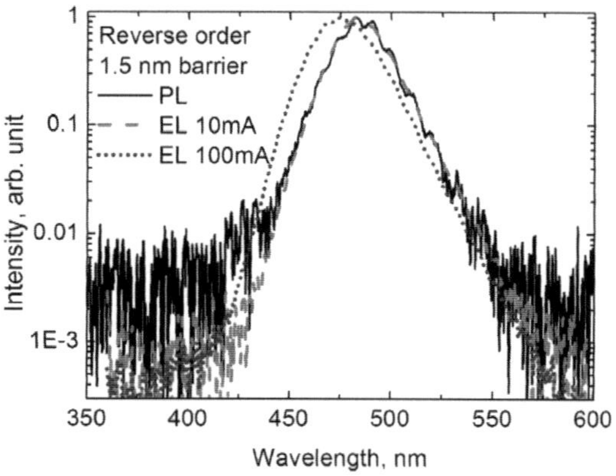

Figure 4. The PL and EL spectra of the sample with a 1.5 nm-thick barrier and a reverse wells order.

The next step was growing two samples with different quantum wells order similar to the previous ones, but with a thicker barrier between the quantum wells. The thickness of the barrier was increased to 44 nm. As one can see in figures 5 and 6, the PL spectra of both samples have two emission peaks, but only the EL spectrum of the sample with a reverse wells order has a short-wavelength peak, while the direct order sample's spectrum does not. The difference between the EL and PL spectrum can be attributed to a strong modification in the band diagram for biased (EL) and unbiased (PL) LED. It is also should be noted that the PL peak intensity ratio slightly differs for these samples, but this effect can be attributed to the nonuniform absorption of UV laser light.

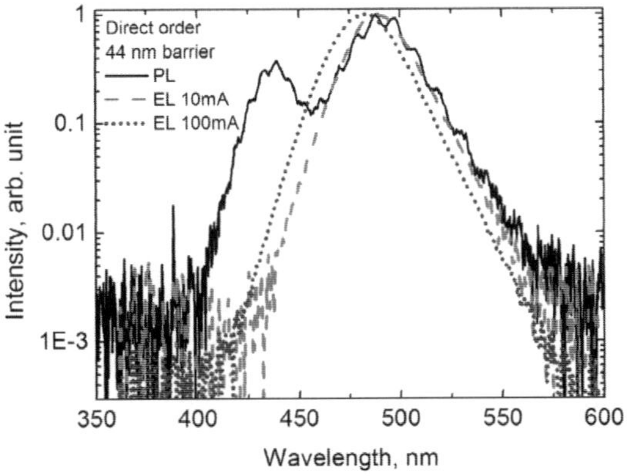

Figure 5. The PL and EL spectra of the sample with a 44 nm-thick barrier and a direct wells order.

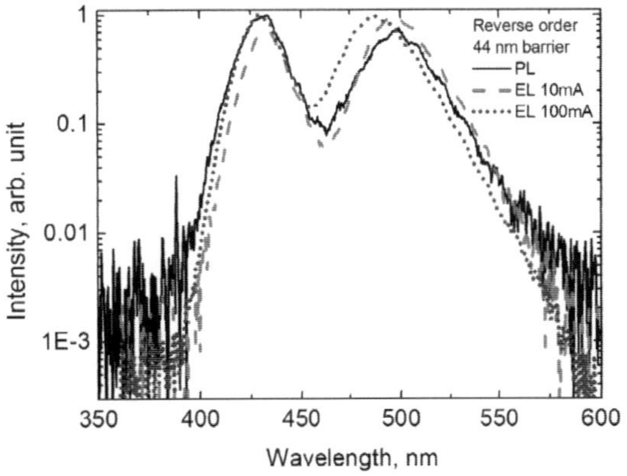

Figure 6. The PL and EL spectra of the sample with a 44 nm-thick barrier and a reverse wells order.

The design of the sample with a reverse QWs order was chosen for further optimization. Two additional samples with barrier thicknesses of 22 nm and 66 nm were grown. A comparison of figures 6, 7 and 8 shows that changing the thickness of the barrier between the quantum wells leads to a change in the ratio of the intensities of the peaks. When the barrier thickness is 22 nm, the ratio of the peaks intensities is 1:3 and it depends on the current. For the 44 nm-thick barrier, the peaks have almost the same intensity and the ratio weakly depends on the current. Increasing the barrier thickness up to 66 nm results in the change of the intensities ratio to 1.5:1. This effect of barrier thickness on the peaks intensity ratio is similar to the one observed for blue/green LEDs [4].

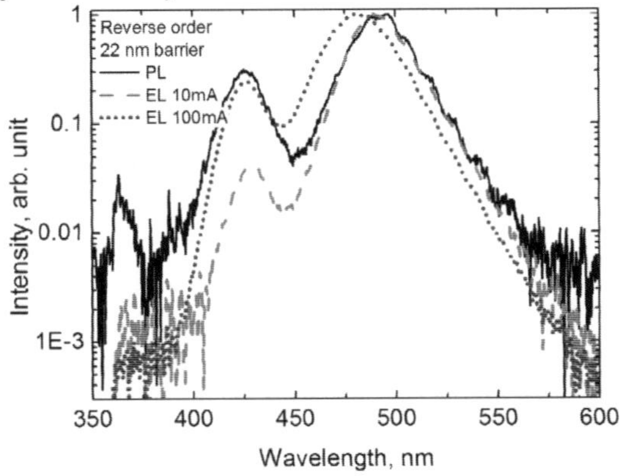

Figure 7.The PL and EL spectra of the sample with a 22 nm-thick barrier and a reverse wells order.

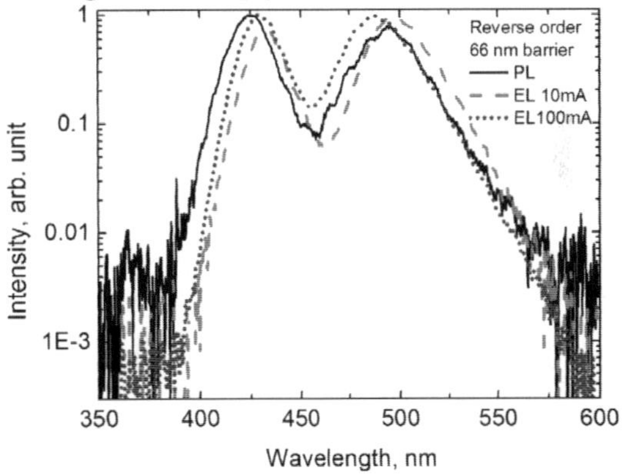

Figure 8. The PL and EL spectra of the sample with a 66 nm-thick barrier and a reverse wells order.

Unfortunately, thickening the barrier between quantum wells causes external quantum efficiency (EQE) decrease, as shown in figure 9. Apparently, it may be due to non-radiative recombination at defects in the barrier, which are generated due to the low temperature required during the growing of this layer. The EQE of the sample with a 22 nm-thick barrier is almost the same as for our standard blue LED; for the sample with a 44-nm barrier, EQE at high currents is nearly unchanged, indicating that the concentration of non-radiative recombination centres in the barrier layer is relatively low, so the most traps are occupied at operating currents of ~ 100 mA.

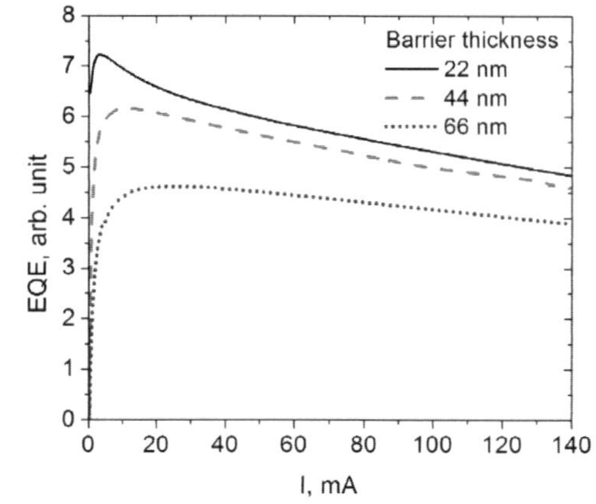

Figure 9. The external quantum efficiencies of samples with a reverse quantum wells order.

To change the peaks intensity ratio of the emission spectrum, doping of the barrier layer can be employed. The sample, similar to the previous one with a 22 nm-thick barrier, but with a barrier that is Si-doped to ~ $2{\cdot}10^{17}$ cm^{-3}, was grown. As one can see in figure 10 (blue and red lines), the EL peaks intensity ratio is about 3:1, while the sample with an unintentionally doped barrier has a ratio of 1:3. The EQE of the sample with an n-doped barrier is lower at low currents, but at high operating currents its efficiency is higher than the EQE of the sample with an undoped one. In sum, the EQE doesn't change much with the doping level of the barrier between the QWs.

To prove the proposed approach, several structures were processed into 1 mm^2 chips, and white-light LEDs were made applying a commercially available thin-film phosphor. Figure 11 shows the EL spectra recorded at 100 mA for blue and dichromatic LEDs with and without a phosphor. It can be seen that some improvement of spectral filling was obtained on the short-wavelength side of the visibility curve. CRI calculation for such a dichromatic LED with a phosphor gives a value over 97 varying from chip to chip and it is limited by the red component.

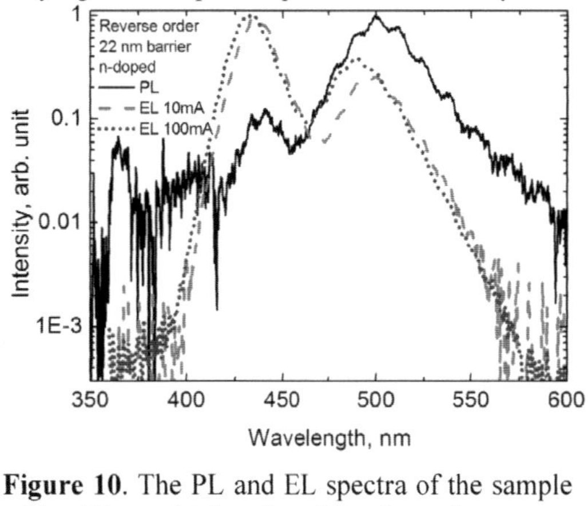

Figure 10. The PL and EL spectra of the sample with a 22 nm-thick n-doped barrier and a reverse wells order.

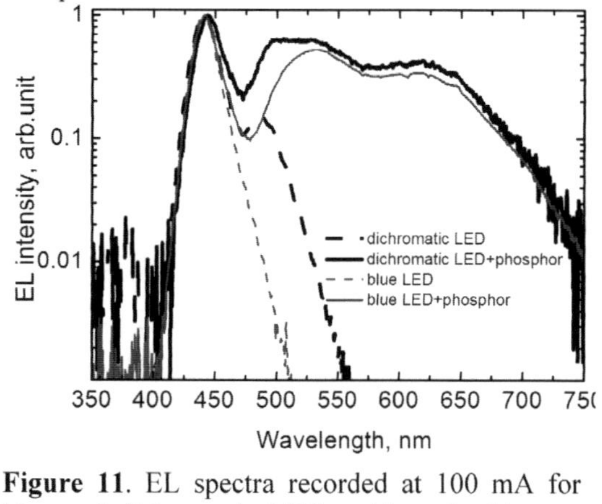

Figure 11. EL spectra recorded at 100 mA for blue and dichromatic LEDs with and without a phosphor

4. Conclusions

The influence of design of the dichromatic LEDs on their properties was investigated. Investigations by photoluminescence, electroluminescence and EQE measurement were employed to study characteristics of the samples. The analysis of the data indicates that varying different structure parameters, e.g. the quantum wells order, the thickness of the barrier between them and its doping level, allows to control the peaks intensity ratio without a significant reduction of the emission efficiency. The obtained results allowed to optimize the design of the LEDs to be used as a pumping source for phosphor-converted white high-CRI LEDs. Combination of such a dichromatic LED with a phosphor allows to create a white LED with CRI>97.

References

[1] Schubert E F 2006 *Light-Emitting Diodes Second Edition* (New York: Cambridge University Press)

[2] Cree Inc., official website http://www.cree.com

[3] Lundin W V,Sakharov A V, Tsatsulnikov A F and Ustinov V M 2011 *Semicond. Sci. Technol.* **26** 014039

[4] Karpov S Yu, Cherkashin N A, Lundin W V, Nikolaev A E, Sakharov A V, Sinitsin M A, Usov S O, Zavarin E E and Tsatsulnikov A F 2016 *Phys. Status Solidi A* **213** 19

IOP Conf. Series: Journal of Physics: Conf. Series **816** (2017) 012039 doi:10.1088/1742-6596/816/1/01

Investigation of the thermal annealing effect on electrical properties of Ni/Au, Ni/Mo/Au and Mo/Au Schottky barriers on AlGaN/GaN heterostructures

E V Sleptsov[1,2], A V Chernykh[1,2], S V Chernykh[1,2], A A Dorofeev[1], N B Gladysheva[1], M N Kondakov[1,2], A A Sleptsova[1,2], A V Panichkin[2], M P Konovalov[2] and S I Didenko[2]

[1] Joint-Stock Company "Scientific & Production Enterprise "Pulsar", Moscow, Russian Federation

[2] National University of Science and Technology "MISiS", Moscow, Russian Federation

E-mail: evgeniy.v.sleptsov@gmail.com

Abstract. Investigation of the thermal annealing effect on Schottky barrier parameters and the leakage current of Ni/Au, Ni/Mo/Au and Mo/Au Schottky barriers on AlGaN/GaN heterostructures has been performed. Improvement of Schottky barrier parameters after annealing of the investigated metallization schemes was observed. Ni/Au and Mo/Au contacts drastically degrade after annealing at the temperatures higher than 400 °C, whereas the Ni/Mo/Au contact exhibits excellent parameters after 500 °C annealing ($q\phi_b$ = 1.00 eV, n = 1.13 и I_{leak} = 5 μA).

1. Introduction

Gallium nitride (GaN) and its alloys, especially aluminium gallium nitride (AlGaN), are widely used in optoelectronics, microwave electronics and high power electronics due to the high breakdown field, temperature stability and radiation hardness. Schottky barriers on GaN and AlGaN/GaN are being successfully used in high power and microwave diodes and as gates in high electron mobility transistors (HEMT). At the same time, the optimum quality Schottky contact fabrication is still a problem that must be solved. The most important Schottky barrier issues in GaN-electronics are the high leakage current and barrier height inhomogeneity across the contact area.

The most widely used metallization scheme for a Schottky barrier on GaN and AlGaN/GaN is Ni/Au, however, many investigations of various metallization schemes were performed. It is known that the Schottky barrier parameters (the ideality factor, the Schottky barrier height) and the leakage current are determined by crystal perfection [1, 2], crystal polarity (Ga-faced or N-faced [3, 4]), epitaxial layers' growth (MOCVD, MBE, HVPE [5]), pre-deposition surface treatment [6, 7], a metallization deposition technique, passivation presence [8, 9] and post-annealing [10, 11]. In addition, compared to conventional semiconductors (Si, Ge), the influence of metal work function on the Schottky barrier height is higher due to the ionic nature of GaN [12].

Many studies report that annealing can be used in order to improve the contact parameters and to reduce the leakage current. However, high-temperature annealing (at temperatures higher than 500 °C) usually leads to contact degradation and loss of a rectifying behavior due to metal and semiconductor intermixing. The temperature behavior of Ni Schottky barriers on AlGaN/GaN heterostructures is

IOP Conf. Series: Journal of Physics: Conf. Series 816 (2017) 012039 doi:10.1088/1742-6596/816/1/01

similar to the one on GaN: the contact can be stable at the temperatures up to 450–500 °C [13-15]. It is demonstrated that during annealing of Ni-based contacts at temperatures higher than 600 °C, Ga diffusion in the Ni film and Ga-Ni alloys formation occur [16, 17]. Presence of gold in the metallization scheme (Ni/Au) can negatively affect the temperature stability: metal penetration into GaN occurs at 800 °C in case of a Ni contact and at 700 °C in case of a Ni/Au one [18]. During annealing in the presence of oxygen at 500 °C, contact oxidizing and Ni and Au intermixing take place due to NiO formation [19].

An additional metal layer between Ni and Au can be inserted to improve the contact quality and stability [20, 21]. However, it was found that the annealed contact properties are determined by the inserted metal: the Ni/Pt/Au or Ni/Ir/Au metallization scheme can be applied to achieve the low leakage current, whereas the Ni/Mo/Au one exhibits a leaky behavior. Moreover, Ni plays a significant role in achieving better contact qualities, as Ni-free contacts demonstrated worse properties. Ramesh et al [22] report that the Mo Schottky contact is relatively stable upon prolonged annealing at 400 °C and also exhibits a rectifying behavior after 600 °C annealing. The Mo/Pt/Au-gate HEMT [23] demonstrated a better breakdown voltage and transconductance compared to the Ni/Pt/Au-gate, however, the latter showed a slightly higher cutoff and maximum oscillation frequencies.

Ni-based contacts are widely used as Schottky barrier contacts, but there are certain problems related to temperature stability. As an alternative for Ni, a refractory metal-based contact such as Mo can be used, as there are some evidences that a Mo contact has better temperature stability. At the same time, contradictory investigations on Mo, Ni/Mo/Au and Mo/Pt/Au Schottky contacts were reported. In present paper, the effect of thermal annealing on Ni/Au, Ni/Mo/Au and Mo/Au Schottky barrier parameters has been investigated.

2. Experiment

AlGaN/GaN epitaxial layers were grown on a 400 μm thick 2" sapphire substrate by MOCVD. The epi-structure consists of 18 nm thick $Al_{0.27}Ga_{0.73}N$ barrier layer, a 0.7 nm thick AlN interlayer and a 2.5 μm thick unintentionally doped GaN buffer layer (figure 1). The eddy-current testing showed a sheet resistance of 320 Ohm per square. The Hall-effect measurement conducted at room temperature showed a two-dimensional electron gas (2DEG) sheet density of $1.1 \cdot 10^{13}$ cm^{-2} and a carrier mobility of 1760 $cm^2 \cdot V^{-1} \cdot s^{-1}$.

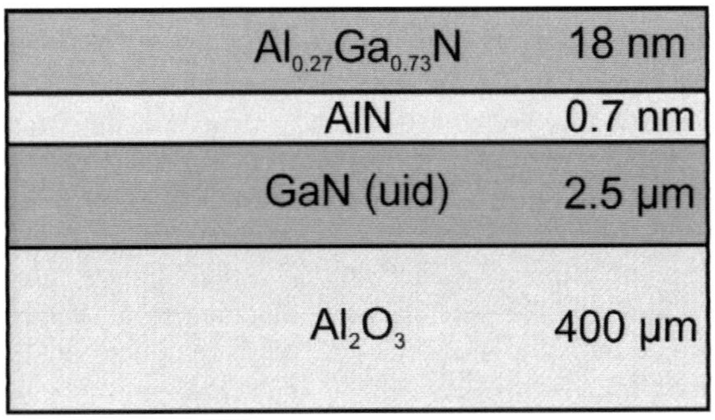

Figure 1. Schematic diagram of a AlGaN/GaN heterostructure on sapphire.

The Mo/Al/Mo/Au metallization scheme was used for ohmic contacts formation [24]. The substrates were dipped in $HCl:H_2O$ (1:1) solution for 60 s in order to remove the native oxide. The standard "lift-off" technique was used for ohmic contact patterning. Ohmic contacts were annealed at 800 °C in N_2 ambient for 30 s in rapid thermal annealing. Prior to Schottky barrier deposition, the samples were treated in $(NH_4)_2S$ (20%):H_2O (1:2) solution for 30 s. Ni/Au (30 nm/300 nm), Ni/Mo/Au (30 nm/20 nm/300 nm) and Mo/Au (30 nm/300 nm) metallization schemes deposited by e-beam evaporation have been analyzed. The fabricated test structure is shown in figure 2. The distance between ohmic and Schottky contacts was 10 and 5 μm for 300, 200, 100 and 80, 60, 40, 20 μm diameters, respectively. The Schottky barriers were annealed in N_2 ambient at 300, 350, 400, 450 and 500 °C for 10 minutes.

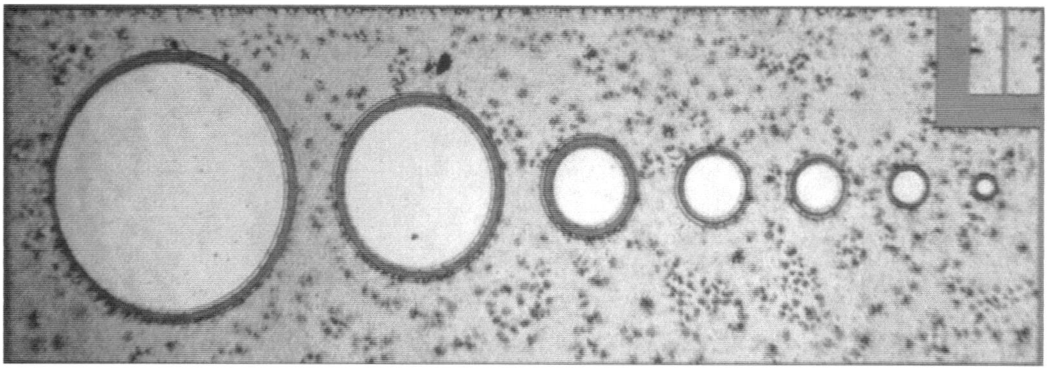

Figure 2. Fabricated Schottky barrier test (for diameters of 300, 200, 100, 80, 60, 40, 20 μm).

Current-voltage (I - V) characterization was performed with an Agilent B1500A parameters analyzer at temperatures ranging from 25 to 125 °C. The forward I - V characteristics were measured from 0 to 2 V with a 5 mV step. The reverse I - V characteristics were measured from -10 to 0 V with a 10 mV step. Prior to reverse measurements, the samples were exposed for 1 minute at -10 V for leakage current stabilization.

The Schottky barrier parameters extraction was carried out using a standard semi-logarithmic I - V plot and an activation energy I - T plot, taking into account the thermionic emission theory with series resistance. The Schottky barrier height was determined from a I - V plot using the known values of the Schottky contact area and the effective Richardson constant (calculated according to the Vegard's law for $Al_{0.27}Ga_{0.73}N$).

3. Results and discussion

Figure 3 shows the forward and reverse I - V curves for as-deposited Schottky barriers (of 100 μm in diameter). The Schottky barrier parameters were the same for different diameters, and the forward I - V curves were proportional to the contact area so all measurements are presented for the 100 μm diameter Schottky barriers. It can be seen that the forward I - V plots linearize poorly due to the high leakage current, which is confirmed by the reverse I - V plots. Using the thermionic emission expression for the forward current, the ideality factors for Ni/Au, Ni/Mo/Au and Mo/Au Schottky barriers were determined to be 4.40, 1.73 and 1.98, respectively. The determined Schottky barrier heights were equal to 0.57, 0.97 and 0.67 eV, respectively.

Figure 3 demonstrates the forward and reverse I - V plots for every metallization scheme after annealing. Figure 4 shows the variation of the Schottky parameters as a function of annealing temperature. As it can be seen, the best Schottky barrier parameters and leakage current for Ni/Au, Ni/Mo/Au and Mo/Au contacts are achieved after 350 °C, 500 °C, and 400 °C annealing, respectively. Increase of annealing temperature leads to gradual degradation of the Ni/Au contact, which is followed by reduction of the Schottky barrier height and rise of the ideality factor and leakage current. The Mo/Au contact also degrades after high-temperature annealing, but its parameters are better than Ni/Au contact parameters at the same temperature. The Ni/Mo/Au contact demonstrates excellent parameters after 500 °C annealing, though annealing at 350 and 400 °C has caused a slight leakage current increase.

Temperature dependences of forward I - V characteristics were measured for the best samples. By the activation energy plot, the Schottky barrier height and the effective Richardson constant were derived. The obtained results are summarized in table 1. Figure 5 shows the temperature dependence of forward I-V characteristics of the Ni/Mo/Au Schottky barrier.

It should be noted that the Schottky barrier height determined by the activation energy plot is independent of the Schottky contact area and the effective Richardson constant, which means that it correctly characterizes the barrier height (figure 5).

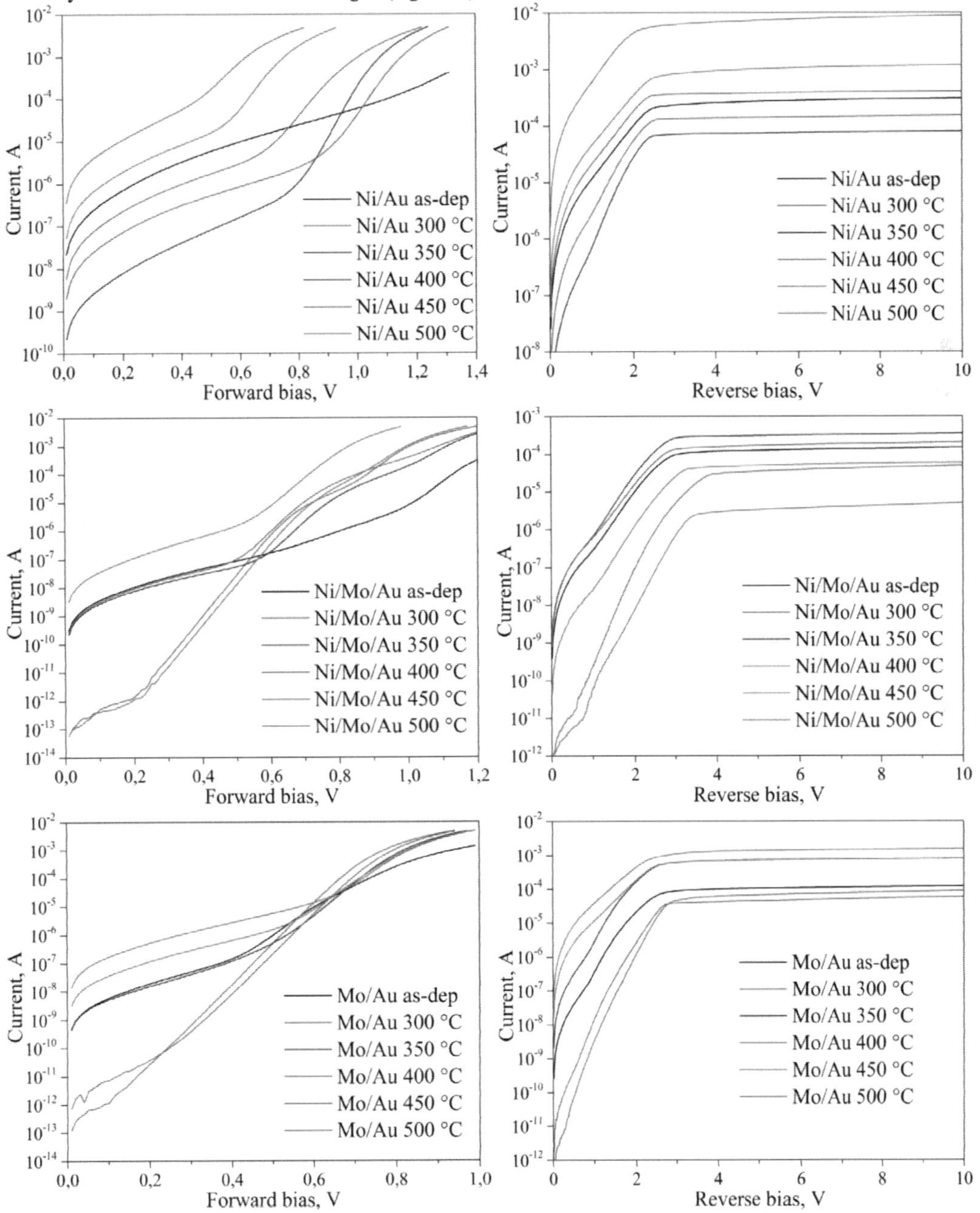

Figure 3. Forward and reverse I - V plots for every metallization scheme after annealing in comparison with as-deposited Schottky barriers.

Table 1 shows that the effective Richardson constant for Ni-based contacts is several orders of magnitude lower than the theoretical value, whereas the discrepancy for the Mo/Au contact is not so great. Such discrepancy can be explained in terms of Schottky barrier height inhomogeneity [25] or due to the presence of an interfacial insulation layer between a metal and a semiconductor [26, 27].

Table 1. Schottky barrier parameters obtained by the activation energy plot.

Metallization scheme	Annealing temperature (°C)	I_{leak}[a] (μA)	n	$q\phi_{b(I\text{-}V)}$ (eV)	$q\phi_{b(I\text{-}T)}$ (eV)	A^{**} (A·K^{-2}·cm^{-2})
Ni/Au	350	80	1.32	1.07	0.75	0.007
Ni/Mo/Au	500	5	1.12	0.99	0.75	0.006
Mo/Au	400	60	1.17	0.90	0.87	20

[a] Leakage current at -10 V.

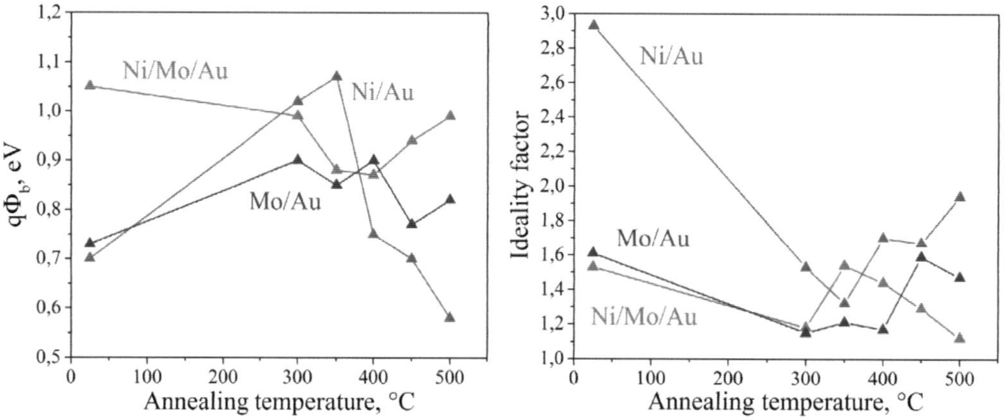

Figure 4. Variation of the Schottky parameters as a function of annealing temperature.

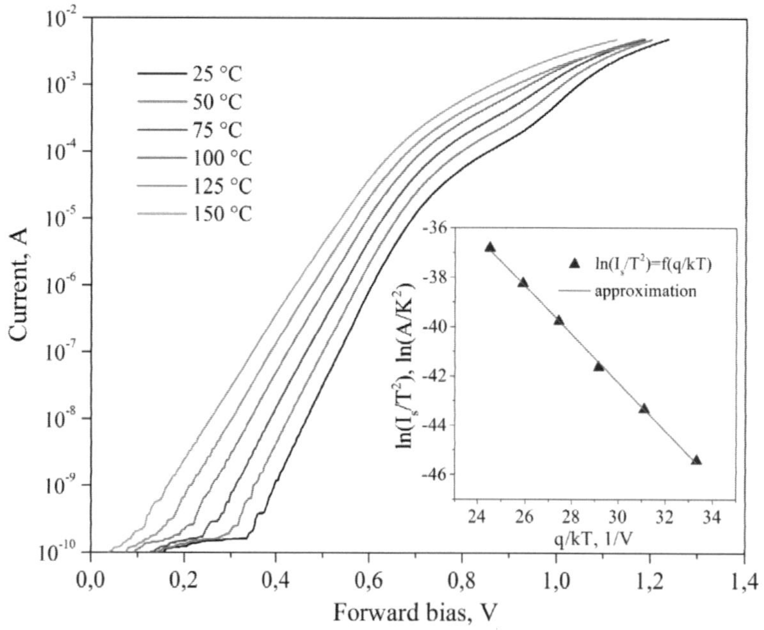

Figure 5. Activation energy plot of the Ni/Mo/Au contact annealed at 500 °C.

4. Conclusion

The investigation of the thermal annealing effect on the barrier height, ideality factor and leakage current of Ni/Au, Mo/Au and Ni/Mo/Au Schottky barriers on $Al_{0.27}Ga_{0.73}N$/GaN heterostructures has been performed. Increase of the barrier height and reduction of the ideality factor and leakage current after annealing were shown. The best values of $q\phi_b$ (0.75 eV), n (1.32) and I_{leak} (80 μA) for the Ni/Au contact can be obtained after annealing at 350 °C. Further increase of annealing temperature causes contact degradation and exhibition of a leaky behavior. Insertion of a Mo interlayer makes a contact more stable. The Ni/Mo/Au contact annealed at 500 °C demonstrates the excellent parameters: $q\phi_b = 0.75$ eV, $n = 1.12$ and $I_{leak} = 5$ μA. It was observed that the Mo/Au contact has better temperature stability as compared to the standard Ni/Au one, also the best Schottky barrier parameters can be achieved after 400 °C annealing ($q\phi_b = 0.87$ eV, $n = 1.17$, $I_{leak} = 60$ μA). It can be concluded that the Ni/Mo/Au metallization scheme has the best temperature stability and Schottky barrier parameters after thermal annealing at 500 °C, if compared to the Ni/Au and Mo/Au ones.

References

[1] Shiojima K, Suemitsu T and Ogura M 1991 *Appl. Phys. Lett.* **78** 3636–8
[2] Suresh S, Balaji M, Ganesh V and Baskar K 2010 *J. Alloys Compd.* **506** 615–9
[3] Fang Z-Q, Look D C, Visconti P, Wang D-F, Lu C-Z, Yun F, Morkoc H, Park S S and Lee K Y 2001 *Appl. Phys. Lett.* **78** 2178–80
[4] Jang H W, Lee J-H and Lee J-L 2002 *Appl. Phys. Lett.* **80** 3955–7
[5] Hellman E S 1998 *MRS Internet J. Nitride Semicond. Res.* **3** e11
[6] Gao Z, Hou B, Liu Y and Ma X 2016 *Microelectron. Eng.* **154** 22–5
[7] Kumar A, Singh T, Kumar M and Singh R 2014 *Curr. Appl. Phys.* **14** 491–5
[8] Broas M, Graff A, Simon-Najasek M, Poppitz D, Altmann F, Jung H and Blanck H 2016 *Microelectron. Reliab.* (*Preprint MR-12073*)
[9] Kaushik J K, Balakrishnan V R et al. 2016 *Thin Solid Films* **612** 147–52
[10] Akkaya A, Esmer L, Kantar B B, Cetin H and Ayyildiz E 2014 *Microelectron. Eng.* **130** 62–8
[11] Akkaya A and Ayyildiz E 2016 *Materials Today: Proceedings* **3** 1248–54
[12] Foresi J S and Moustakas T D 1993 *Appl. Phys. Lett.* **62** 2859–61
[13] Liu Q Z, Yu L S, Deng F, Lau S S and Redwing J M 1998 *J. Appl. Phys.* **84** 881–6
[14] Arulkumaran S, Egawa T, Ishikawa H et al. 2001 *IEEE Trans. Electron Devices* **48** 573–80
[15] Zhao J, Lin Z, Chen Q, Yang M, Cui P, Lv Y and Feng Z 2015 *Appl. Phys.* A **121** 1271–6
[16] Venugopalan H S, Mohney S E, Luther B P et al. 1997 *J. Appl. Phys.* **82** 650–4
[17] Grodzicki M, Mazur P, Zuber S, Pers J et al. 2014 *Appl. Surf. Sci.* **304** 24–8
[18] Venugopalan H S, Mohney S E, DeLucca J M, Luther B P and Bulman G E 1998 *J. Vac. Sci. & Technol.* A **16** 607–10
[19] Chen L-C, Chen F R, Kai J-J, Chang L, Ho J-K, Jong C-S, Chiu C C, Huang C-N, Chen C-Y and Shih K-K 1999 *J. Appl. Phys.* **86** 3826–32
[20] Monroy E, Calle F, Ranchal R, Palacios T, Verdu M, Sanchez F J, Montojo M T, Eickhoff M, Omnes F, Bougrioua Z and Moerman I 2002 *Semicond. Sci. Technol.* **17** L47–54
[21] Miura N, Nanjo T, Suita M, Oishi T, Abe Y, Ozeki T, Ishikawa H, Egawa T and Jimbo T 2004 *Solid-State Electron.* **48** 689–95
[22] Ramesh C K, Reddy V R and Choi C-J 2004 *Mater. Sci. & Eng.* B **112** 30–3
[23] Endoh A, Yamashita Y, Ikeda K, Higashiwaki M, Hikosaka K, Matsui T, Hiyamizu S and Mimura T 2004 *Japan. J. Appl. Phys.* **43** 2255–8
[24] Kondakov M N et al. 2016 *Russian Microelectronics* **24** L16
[25] Werner J H and Guttler H H 1991 *J. Appl. Phys.* **69** 1522–33
[26] Hacke P, Detchprohm T, Hiramatsu K and Sawaki N 1993 *Appl. Phys. Lett.* **63** 2676–8
[27] Guo J D, Feng M S, Guo R J, Pan F M and Chang C Y 1995 *Appl. Phys. Lett.* **67** 2657–9

Optical leakage versus luminescent coupling in a multijunction solar cell: how to recognize a source of additional photocurrent in subcells

S A Kozhukhovskaia, E D Filimonov and M Z Shvarts

Ioffe Institute, St. Petersburg, 194021, Russia

e-mail: kozhukhovskaya@mail.ioffe.ru

Abstract. A procedure for estimating an optical leakage of incoming radiation through a wide-bandgap subcell in a GaInP/GaInAs/Ge multijunction solar cell to photoactive layers of narrow-bandgap ones is proposed. A sequence for experimental determination of currents induced in a Ge subcell owing to an optical leakage or via a luminescent coupling is described.

1. Introduction

In a number of cases, experimental study of multijunction solar cell (MJ SC) photovoltaic characteristics requires to ensure selective light exposure on separate p-n junctions (subcells) by choosing sources of narrow-bandgap radiation (lasers, light diodes), or interference filters for broad-band emitters (sunlight simulators). In forming the light flux spectral composition, it is important to prevent uncontrollable radiation penetration into a narrow-bandgap subcell (being a part of a MJ SC and located under an adjacent wide-bandgap one) by searching the emitter selective band within the spectral sensitivity of a corresponding subcell. An example of an overlap of spectral ranges is presented in figure 1.

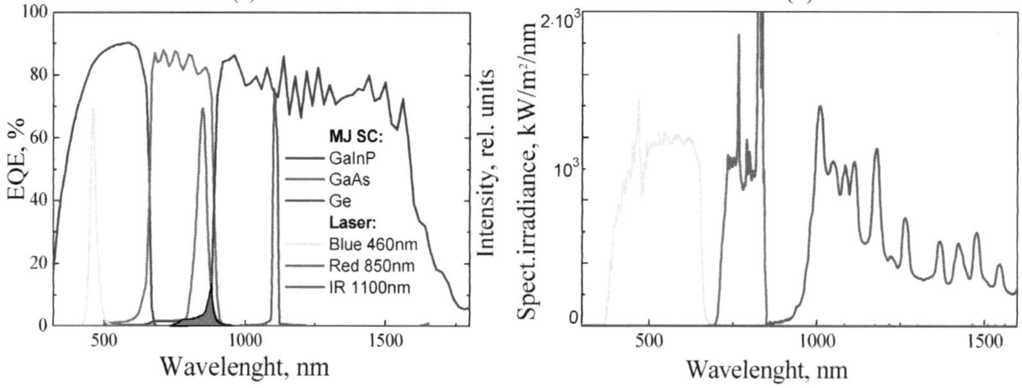

Figure 1. Spectral characteristics of: (a) – external quantum efficiency of MJ SCs and of narrow-band radiation sources (red – the overlap range of the narrow-band source spectral line with the GaAs and Ge photosensitivity areas); (b) – wide-bandgap radiation sources.

IOP Conf. Series: Journal of Physics: Conf. Series **816** (2017) 012040 doi:10.1088/1742-6596/816/1/012040

It is known that appearance of a subcell photoresponse in high-efficient MJ SCs outside a range of its photosensitivity may be associated with the luminescent coupling mechanism, when the luminescent radiation flux from a wide-bandgap subcell is absorbed in an adjacent narrowband one inducing an additional photocurrent [1-4].

In the experimental routine, it is often impossible to separate and estimate properly the influence of these two processes (light penetration and luminescent reemission) by studying only the emitter spectral composition. For this reason, in choosing emitters for selective spectral effect on a subcell, it is important to know the cause of the arising photocurrent to be either due to optical leakage (OL) of external radiation into a corresponding subcell through thin semiconductor layers, or a result of luminescent coupling (LC) between the wide- and narrow-bandgap subcells.

In the present work, a procedure for evaluating the values of photocurrents (arising due to optical leakages or luminescent reemission) has been proposed.

2. Model

Let us consider the effect of different conditions of external illumination on the photocurrent formation dynamics in a GaInP/GaAs/Ge MJ SC with a strong luminescent coupling. There could be following practical cases:

1. Irradiation from a light source within the sensitivity range of GaInP (E^{TOP}): a cascade of luminescent processes is activated. In such a mode, the GaInP subcell photocurrent J^{TOP} increases proportionally with the irradiation level, whereas in the adjacent GaAs narrow-bandgap one – J_{LC}^{MID} is induced by luminescence light at the internal current J_i^{pn} through the GaInP p-n junction (figure 2(a)).

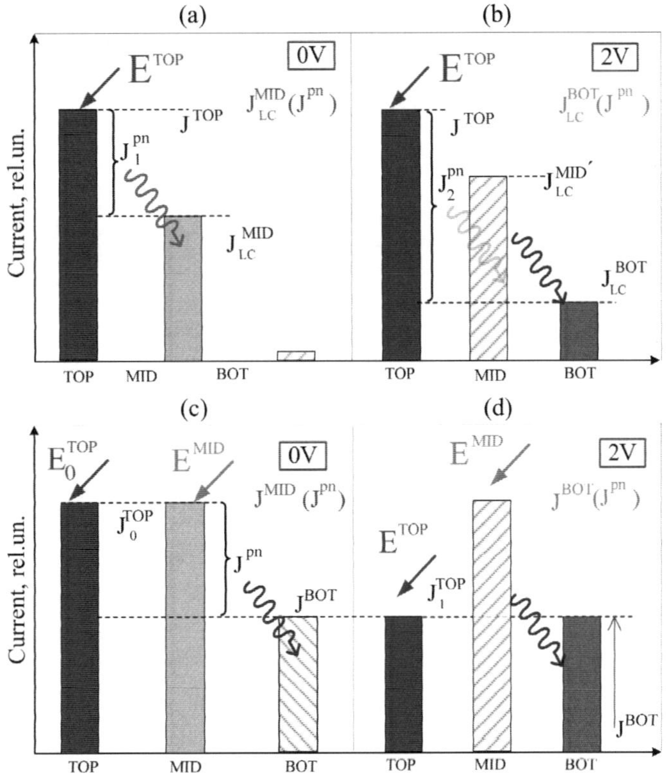

Figure 2. Bar charts of photocurrents in conditions of illumination from an external radiation source focused on the sensitivity range of: (a), (b) – GaInP; (c), (d) – GaInP и GaAs.

In case of a sharp reverse branch of the Ge subcell I-V characteristic, no current limitation takes place from the bottom Ge subcell side at $V = 0$ V and the relation $J_1^{pn} = J^{TOP} - J_{LC}^{MID}$ is valid. In applying bias voltage of about $V \approx 2V$, that corresponds to the short circuit current mode for the Ge subcell, the induced current J_{LC}^{BOT} is recorded, while $J_2^{pn} = J^{TOP} - J_{LC}^{BOT}$ and $J_{LC}^{MID} < J_{LC}^{MID\prime}$. Due to cascade emission/absorption of the luminescent reemission in a MJ SC in tuning illumination level for the top subcell, the dependencies presented in figure 3 could be formed, every point on which corresponds to its own conditions of a current balance (dependencies $J_{LC}^{MID}(J^{pn})$ and $J_{LC}^{BOT}(J^{pn})$ in figure 3). See [5] for more detailed description of the photocurrent formation processes.

2. Irradiation from two light sources in the sensitivity ranges of GaInP (E^{TOP}) and GaAs (E^{MID}) subcells for creating controllable current balance of pairs of subcells. Suppose, at the initial time, that there is a constant level of illumination of the GaInP subcell (E_0^{TOP} = const), and photocurrents corresponding to this irradiation are J_{LC0}^{MID} and J_{LC0}^{BOT} (points 1, 1' in figure 3).

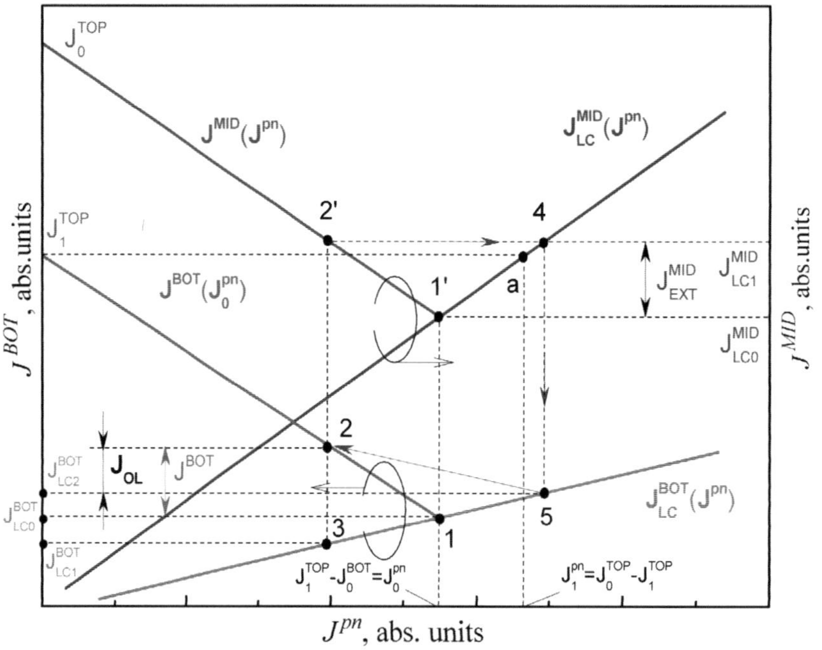

Figure 3. Dynamic plane of photocurrents of GaAs (J^{MID}) and Ge (J^{BOT}) subcells at different illumination conditions. Points 1, 1' on the plane correspond to the initial measurement conditions. Points 2, 2' – describe different current states at external illumination of the GaAs. Points 4, 5 are the states equivalent to the 2', 2 ones, but in presence of the luminescence reemission in pairs GaInP-GaAs and GaAs-Ge. Point 3 represents the value of Ge current induced by LC in presence of OL through GaAs.

The increase in illumination of GaAs results in the rise of its photocurrent up to the value J_0^{TOP}, being registered at 0 V (figure 2(c)). Hence, the dependence $J^{MID}(J^{pn})$ is formed (violet line in figure 3). To determine the photocurrent J^{BOT} induced in the Ge subcell, it is necessary to create different illumination conditions for subcells. Note that the value J^{BOT} cannot be determined by simple moving along the I-V characteristic towards the $V \approx 2$ V point [5]. Then, illumination of GaInP should be decreased until $J^{BOT} = J_1^{TOP}$ (figure 2(d)). In such a case, the dependence $J^{BOT}(J^{pn})$ is formed (green line in figure 3). Both lines $J^{MID}(J^{pn})$ and $J^{BOT}(J^{pn})$ can be determined graphically by pairs of points

(J^{pn}; J_{LC0}^{BOT}+J^{pn}) and (J^{pn}; J_{LC0}^{MID}+J^{pn}). In such a way, a dynamic plane reflecting the response of each subcell to external illumination variation is arranged.

To determine the causes of the induced current arising in Ge, it is required to find equivalent current conditions $J^{MID} = J_{LC1}^{MID}$ for GaAs at presence of external illumination, when the latter either penetrates into Ge ($E^{MID} \neq 0$) or does not ($E^{MID} = 0$). On the dynamic plane, which links all the currents mentioned above and their interrelations, this is equivalent to the transition from the state 1 to the state 2 for Ge (the total current in Ge is equal to $J_0^{BOT} + J^{BOT}$) and, correspondingly, for the GaAs subcell – from the state 2' (total current in GaAs is determined by a sum J_{EXT}^{MID} and J_{LC0}^{MID} , where J_{EXT}^{MID} is the current induced by external radiation E^{MID}) to the state 4 (the current in GaAs is determined only by LC nature). In the state 4, the internal current through the emitting GaAs p-n junction would induce J_{LC2}^{BOT} in Ge (point 5), and $J^{BOT} + J_{LC0}^{BOT} = J_{LC2}^{BOT} + J_{OL}$ would be valid, where J_{OL} – is a photocurrent originated as a result of penetration of external radiation E^{MID} into the Ge subcell. If during graphical passway the Y-component of the points 4 and 5 ($J_{OL} = 0$), then J^{BOT} is a response only to the LC processes in a MJ SC. If the Y-coordinate of the point 2 is above a corresponding coordinate of the point 5, then an optical leakage takes place ($J_{OL} \neq 0$), and the induced photocurrent J_{LC1}^{BOT} (point 3) represents the impact of luminescence coupling.

3. Conclusion

The causes of generation of additional photocurrent in MJ SC narrow-bandgap subcells at selective light exposure in certain wavelengths ranges are considered. Principles for plotting of the current plane linking the lines of currents induced by luminescence and external radiation in subcells of MJ SCs are described. A procedure for determining the values of the current induced by luminescence and of that resulting from transmission of short wavelength radiation through the semiconductor layers into the narrow-bandgap subcells has been proposed.

Acknowledgments
This work was supported by the Russian Science Foundation (Grant № 14-29-00178).

References

[1] Yoon H, King R R, Kinsey G S, Kurtz S and Krut D D 2003 *Proc. 3rd World Conf. Photovoltaic Energy Convers* **1** 745-748
[2] Baur C, Hermle M, Dimroth F and Bett A W 2005 *Appl. Phys. Lett.* **90** 192109
[3] Steiner M, Geisz J, Moriarty T, France R, McMahon W, Olson J, Kurtz S and Friedman D 2013 *Journal of Photovoltaics, IEEE* **3(2)** 879 – 887
[4] Allen C R, Lim S H, Li J.-J and Zhang Y.-H. 2011 *Proc. 37th IEEE Photovoltaic Spec. Conf.*, 000452 - 000453
[5] Shvarts M Z, Emelyanov V M, Evstropov V V, Mintairov M A, Filimonov E D and Kozhukhovskaia S A 2016 *AIP Conference Proceedings* **1766** 060005

IOP Conf. Series: Journal of Physics: Conf. Series 816 (2017) 012041 doi:10.1088/1742-6596/816

Contactless nondestructive method for determination of the carrier diffusion length in semiconductors and dielectrics

V V Manukhov[1] and A B Fedortsov[2]

[1]Saint Petersburg State University, Universitetskaya nab. 7/9, St. Petersburg, 199034, Russia
[2]Saint Petersburg Mining University, Vasil'evskii Ostrov, 21-liniya 2, St. Petersburg, 199106, Russia

E-mail: borisovitch-f@yandex.ru

Abstract. We propose the contactless nondestructive method for determination of the carrier diffusion length in semiconductors and dielectrics. The method is based on optical generation of non-equilibrium carriers at one point of the studied sample and the laser interference measurement of their concentration at another point. When changing the distance between these points, a decrease in the carrier concentration is observed. It depends on the carrier diffusion length, which is determined by comparing the experimental and theoretical dependences of the probe signal on the divergence of the injector and probe beams. We have studied silicon samples protected by an insulator layer and without any covering. The method can be used in scientific research and the electronics industry.

1. Introduction

The carrier diffusion length is an important physical characteristic of a material and it governs operation of a great number of electronic devices. At present, a number of methods for measuring the diffusion length are known [1, 2]. The guided probe method is often used. However, using this and many other methods requires an electrical contact between the sample and the probe. It is impossible to investigate dielectric samples as well as semiconductor samples with a dielectric coating by using contact methods. We propose a contactless laser method for the investigation of the carrier diffusion length. This method allows one to avoid destruction of material and not to disturb the electron system. The proposed method is based on optical generation of non-equilibrium charge carriers in the sample using one laser beam and nonintrusive optical data reading with the help of another laser probe beam with the photon energy less than the energy gap of the material to be investigated.

2. Interference modulation mechanism

The data on the excess carrier diffusion length are obtained through the interference modulation of the monochromatic long-wave IR radiation for which the material being tested is transparent ($\lambda > hc/E_g$). Here, h is the Planck's constant, c is the speed of light, E_g is the semiconductor gap width. This radiation passes through the sample as an optical probe. The samples are plane-parallel wafers with parallel edges forming a Fabry-Perot interferometer for the probe beam. The passing probe beam intensity depends on the sample refractive index and its absorption coefficient. Simultaneously, the sample is exposed to the optical injector which is the source of radiation with the photon energy sufficient for the intrinsic photoeffect ($\lambda < hc/E$), generating excess carrier pairs. The excess concentration of electrons (ΔN) and that of holes (ΔP) depend on the fluid density of the injector photons absorbed in the sample. The changing carrier concentration results in the change of the refractive index $\Delta\eta$ and that of the absorption coefficient $\Delta\alpha_p$ for the probe laser wavelength. According to this, the intensity of probing light passing through the sample is changed by the value of ΔI and thus the modulation coefficient $M = \Delta I/I$ depends on the excess carrier concentration. This interference mechanism was proposed in [3] and used in [4-7] for investigation of different parameters of semiconductors such as the lifetimes of the excess electrons and holes, the surface recombination velocity, the recombination centers concentration, the capture coefficients for electrons and holes, etc.

3. Experimental set-up

To test this method for the diffusion length determination, we have developed an experimental apparatus which is shown in figure 1. A long wave infrared laser LG-GNIK-1 was used as a source of a probing beam (1). The wavelength of this laser is 3.39 μm, the power is 6 mW, the noise/signal ratio is < 0.001. The beam of this long wave IR laser (1) transmits through the studied sample (2) and is measured by a photodetector FE-724-1 (3). The intensity I of the probing light passing through the sample is changed by the value of ΔI and depends on the excess carrier concentration at the point to be tested.

We use a semiconductor laser module KLM-H980-200-5 as an injecting radiation source (4). The wavelength of this radiation is 0.980 μm, the power is 200 mW. The modulation frequency is 1400 Hz. Due to their diffusion, non-equilibrium charge carriers, generated by this radiation at one point of the material surface, change the carrier concentration at another (tested) point located at a distance r. This distance may be regulated by a special precise stage (5). By changing the distance r between the carrier generating and probing points, it is possible to decrease the carrier concentration at the probing point.

So, the carrier concentration measured at the probing point is a function of a distance r and a carrier diffusion length L. The carrier concentration results in the change of the refractive index $\Delta\eta$ and the absorption coefficient $\Delta\alpha_p$ for the probe laser wavelength. As a result, the photodetector signal and the carrier concentration depend equally on the distance r.

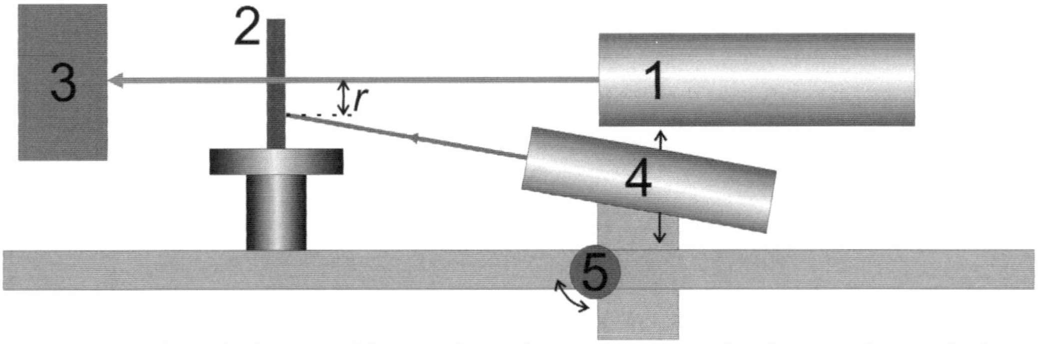

Figure 1. Schematic diagram of the experimental setup: 1 - source of probe beam, 2 - sample, 3 - photodetector, 4 - injecting radiation source, 5 - precise stage.

4. Measurements

The series of experiments has been carried out based on the method presented here. The samples made of Si crystal of a resistivity of 160 Ω·m were investigated. They were divided into two lots. Samples of one lot had a clean surface; samples of another lot were covered with an insulating layer of silicon dioxide. Such samples are used in electron device manufacturing. They are parallel enough to be used for interference measurements due to a long wavelength of the probe beam. The dependences of the photodetector signal on the distance r between the carrier generating and probing points $U(r)$ were measured. Each sample was tested several times at different points of generation.

Similarly to the contact methods [1,2], to determine the carrier diffusion length L we compared the experimental curve $U(r)$ with the dependences calculated theoretically for different values of L. Figure 2 shows that theoretical and experimental data correspond well only for large values of the distance r between the points of carrier generating and probing. This is because the diameter of the generation area is not small enough to neglect it. So, the experimental values of the carrier diffusion length L were determined in the range of large r values. These values agree well for different samples of each lot and for different points on each sample. However, they differ significantly for the samples covered with a dielectric (L = 530 μm) and uncovered ones (L = 400 μm). The carrier diffusion length L is much longer in covered samples due to low surface recombination. The statistical error of the experiments is 5 to 7 %.

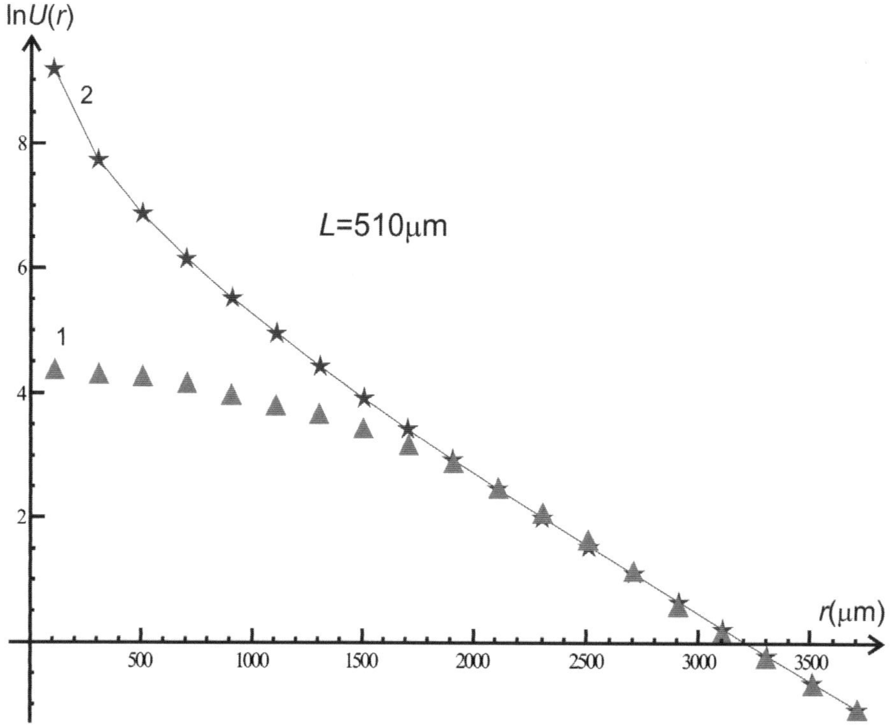

Figure2. Typical superposition of the experimental (1) and calculated (2) curves lnU(r).

5. Conclusion

The efficiency of the proposed contactless nondestructive method for determination of the carrier diffusion length in semiconductor and dielectric samples was experimentally demonstrated in a series of experiments. The method does not require special sample preparation and can be used not only for scientific and technological research but also in the production of electron devices.

References

[1] Pavlov L P 1987 *Measurements on the Parameters of Semiconductor Materials* (Moscow: Vyshaya Shkola).
[2] Batavin V V, Kontsevoi Yu A and Fedorovich Yu V 1985 *Measurements of Parameters of Semiconductor Materials and Structures* (Moscow: Radio i Svyaz').
[3] Fedortsov A B and Churkin Yu V 1988 *Sov. Tech. Phys. Lett.* **14**, 142.

[4] Areshkin A G, Vorob'ev L E, Ivanov A S, Komarovskikh K E, Letenko D G, Fedortsov A B and Churkin Yu V 1992 *Izv. Akad. Nauk, Ser. Fiz.* **56** (12), 121.
[5] Letenko D G, Fedortsov A B, Sawateyev V N and Churkin Yu V 1993 *J. Mater. Sci.: Mater. Electron.* **3**, 203.
[6] Fedortsov A B, Letenko D G, Churkin Yu V, Tsentsiper L V and Vedde I 1996 *Proc. Electrochem. Soc.* **96**, 481.
[7] Voronkov V B, Ivanov A S, Komarovskikh K F, Letenko D G, Fedortsov A B and Churkin Yu V 1991 *Sov. Tech. Phys.* **36**, 184.

The development of a bistable microdrive for the micromechanical gyroscopes

Y B Enns, R V Kleimanov, A N Kazakin, E N Pyatishev and A V Korshunov

Peter the Great St.Petersburg Polytechnic University, Russia, 195251, St.Petersburg, Polytechnicheskaya, 29

E-mail: pen.mems@gmail.com

Abstract. In present work, the model and methodology for designing of arcuate arc-shallow MEMS with buckling was developed and tested. The efficiency of the model is confirmed experimentally. The result of the present survey is an increase of the efficiency of the exciting electrostatic comb drive MMG of high precision. The result of this research allows to reduce the control voltage and noise of a microgyroscope.

1. Introduction

Microelectromechanical systems (MEMS) are an evolution of microelectronics and are based on main microelectronics technological processes but it needs some kind of specific technological solutions and processes for development and manufacturing of MEMS devices. Semiconductor materials and technological processes help to produce micromechanical devices (such as gyroscopes) with improved parameters in smaller scales. Combination of mechanical elements and electronic components based on the same technologies makes it possible to improve the parameters and workability of a device.

Nowadays, there is a discussion in microsystems technology (MST) about usage of postprocess with the addition of supplementary active functional elements to avoid limitations of deep plasma-chemical etching and formation of the electrostatic pectinate structure with a high profile and narrow gaps [1]. These methods suppose the usage of additional active elements for the displacement and fixation of a comb drive. These active bistable systems are described in [2]. Supplementary elements significantly complicate the construction and reduce the reliability of MEMS due to a large number of active elements. In this paper, we propose to use a bistable actuator based on the buckling [3]. This solution allows to form the structure for displacement of the comb drive without adding active elements that will significantly improve the device parameters. The main limitation of the construction of the comb drive structures is deep plasma-chemical etching.

2. Technology of post-process for micromechanical gyroscopes

A micromechanical gyroscope (MMG) is a device that is used for measuring the angular velocity by generating the Coriolis force in the mechanical system. The mechanical structure is formed by the resonator exciting primary oscillations and the measuring resonator of secondary oscillations caused by the Coriolis force. There are known designs of vibratory MMG differing by the number of oscillating masses, the direction of oscillation, the structure and construction of an actuator and a measuring unit, manufacturing technology and materials [4]. A micromechanical gyroscope is a measuring device for navigation, it has high requirements to the sensitivity, a level of noise, stability under mechanical loads, etc.

Figure 1 shows the frame structures of MMG, where 1 and 4 are the resonators of the primary and secondary oscillations, 2 and 3 are the exciting and measuring sense combs consisting of a system of movable and stationary electrodes formed in the body of the resonators. Increase of the profile depth can result in a significant increase of the ratio of the flexible suspension stiffness of the resonators and an increase of the comb drive area. The comb drive area increasing while maintaining the gap between the electrodes will significantly increase the excited electrostatic force or reduce the drive voltage for excitation of oscillations. The presence of parasitic capacitive links in the structure of MMG leads to mutual influence of the exciting and measuring electrodes. The reduction of the Drive voltage will reduce the influence of the excitation signal in the measuring electric circuit.

During the creation of the volumetric structure, the main technological operation is Deep Reactive Ion Etching (DRIE) of silicon. The limitation for this process is a depth of etching (DRIE) in the form of the aspect ratio. Nowadays, the achievable ratio is approximately 1:30 and during the formation of structures with a high profile, a large gap is necessary. For an electrostatic actuator, increasing of the gap will lead to a significant reduction of the excited force.

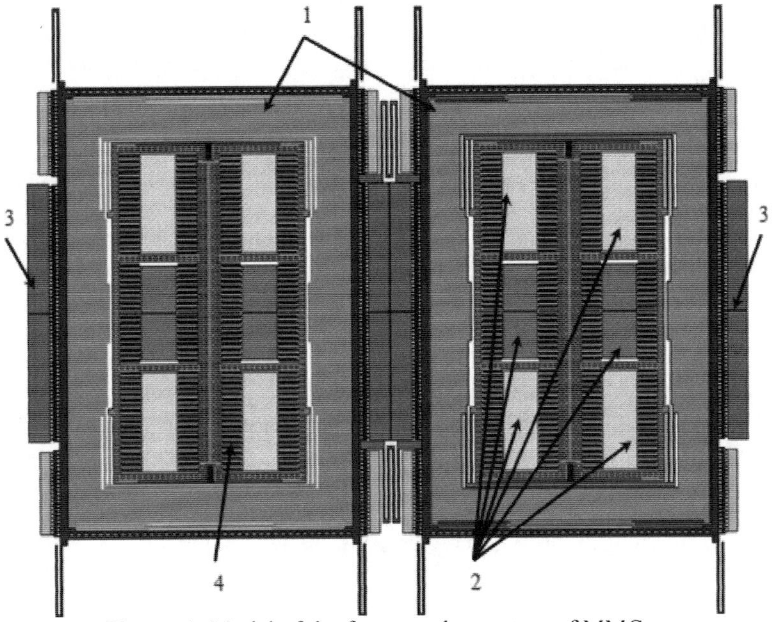

Figure 1. Model of the framework structure of MMG

There are known solutions for reduction of the gap, using the postprocess and setting [3, 4]. In case of the active adjustment, the additional mechanical structure with its own actuator (electrostatic or thermal) is formed, removing and holding the structure in the position with a reduced gap. The postprocess also uses a supplementary mechanical structure for changing the gap but involves the fixation of this structure after the movement. These methods allow to increase the capacity, however, require additional mechanical components that increase the size of the device and reduce reliability. The present study proposes the usage of a bistable system with buckling for displacement of the comb drive and the creation of pectinate-like structures with a high electrostatic force.

3. Bistable structures for postprocess

The buckling of mechanical elements which occurs under the influence of an axial load causing an abrupt changing of the deflection curve, is one of the commonly considered types of instability in microsystems technology. For the majority of devices, buckling leads to uncontrolled variance of the device parameters. For the bistable system, the presence of bifurcation points and two energy minima is characteristic, which allows the system to stay in one of two stable positions without energy consumption. Such systems need energy only for the conversion from one stable condition to another.

There are known solutions for the usage of flexible elements with buckling for designing of electrostatic and optical switches, band pass filters, valves, etc. There are two options of formation of such flexible elements: due to internal tension in thin films [2] or etching (plasma-chemical etching) of an arc-shallow suspension [5, 6]. In the first case, the parameters of the bistable element have a low repeatability and are used in surface MEMS. In this article, we consider the production of an arc-shallow suspension using the second method, due to its predictability, repeatability, and compatibility with the fabrication technology of the MMG.

An arc-shallow flexible element is a silicon beam formed by DRIE. The shape of its frontal profile is described by a function of the height of the deflection ω:

$$\omega(x) = \frac{h}{2[1 - \cos\left(2\pi \frac{x}{l}\right)]},$$

where l is the distance between the clamped tips, x is the lengthwise coordinate, h is the amplitude of the displacement of the beam midpoint ($x = l/2$). One condition must be met for this structure: $t << b$, where t is the thickness and b is the height of the profile (arc) [5]. The structure of the bistable device including alike beam is also called snap- through.

The behavior of this structure is well described by the dependence of the displacement of the point $\omega(l/2)$ on the external load, as shown in figure 2. Initially, the system stays in equilibrium without an applied load, which corresponds to the point A. With application of an external force F, high axial loads arise because of high stiffness against stretching and compression. Thus, the displacement of $\omega(l/2)$ in the direction to points A and B is minimal. Near the point C, the response of the system increases, the stiffness of the beam decreases because of the bending deformations. When the force reaches the value corresponding to the point C, there is a buckling and abrupt transition to the point D. In the same way as for the segment AB, a further increase of the external load leads to a minimal movement due to the high stiffness against stretching. In the absence of active forces, the system will not return to the point A, but will remain on the segment DE. The stationary position is determined by the presence of a second energy minimum and stability of this position depends on the depth of the potential well. The limit stable value corresponds to the point E (negative forces), where the system loses the equilibrium position and an abrupt transition to the point B occurs. When the load is removed, the system returns to the original state corresponding to the point A. Therefore, the points C and E are the points of transition (bifurcation), and their mutual arrangement determines the presence of two stable positions and their stability.

In the present survey, it is proposed to use the structure hung on a system of arcuate arc-shallow suspensions instead of the stationary part of the excitation actuator (figure 1). In the initial position, the pectinate drive stays in a separated condition, that is, the area of overlap tends to zero. One part of the actuator is formed on the body of the resonator of the primary oscillations, and the second one, on the arcuate arc-shallow suspension. The manufacturing of a gap for etching satisfies the condition of the aspect ratio, and the structure with a high profile is formed. However, the gap value exceeds any technological capabilities while working. The transition to the working position occurs by supplying a voltage (of about one hundred volts) between the pectinate drives. In this case, buckling and abrupt transition to the stationary comb occur. A high value of stiffness against stretching (much greater than the stiffness of the resonator of the primary oscillation) gives the opportunity to suppose that the actuator hung on the arcuate arc-shallow suspension is conventionally static. The resulting structure allows to increase the efficiency of the comb drive due to a smaller gap while maintaining the reliability.

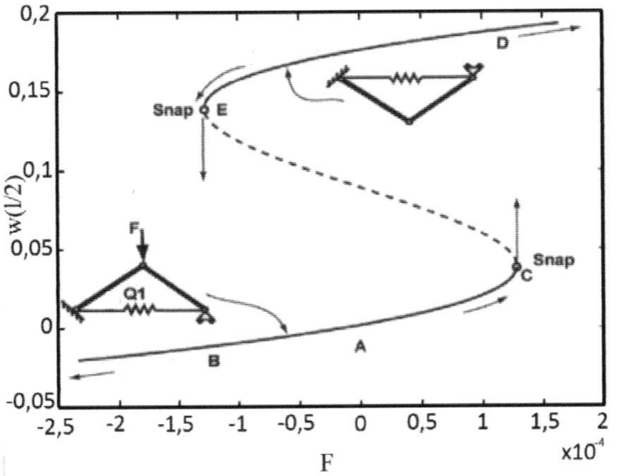

Figure 2. The dependence of the displacement of the center point of an arcuate suspension on the attached external loads [6]

4. The prototype of the bistable structures

The problem of modeling an arcuate arc-shallow suspension relates to the field of nonlinear systems with two potential minima. The position of the bifurcation points corresponding to figure 4 can be found analytically. The approximate values will be determined as:

$$F_C = 8\pi^4\frac{EIh}{l^3},\ F_E = 4\pi^4\frac{EIh}{l^3},\ dx_{CE} = 1.33h,\ dx_C = \frac{8t}{3Q},\ dx_E = 2h - \frac{8t}{3Q},\ dx_{ED} = 2h - \frac{4t}{3Q},$$

where: F is the force, E is the Young's modulus, Q is the ratio of the thickness of the beam to the height of deflection.

However, this approach cannot give a complete picture of the system behavior over the entire range of displacement.

For a buckling research, modeling in the software package COMSOL was used. Based on the theoretical dependence in figure 2, it shows that there is a region of uncertainty leading to a non-unique solution. The considered system acquired the first and third modes of the middle line (figure 3) during buckling. The aim was to avoid the second (rotational) mode of buckling.

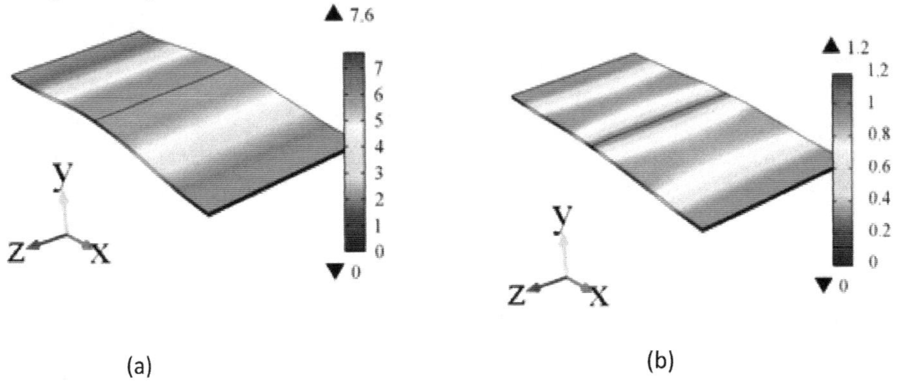

Figure 3. The map of displacements in the first (a) and third (b) buckling modes

In the course of the calculation, the dependence was obtained that described the behavior of the buckling in the "force-displacement" coordinates (figure 4). As shown, the results are in good agreement with the data of the analytical calculation.

To validate the modeling results, the test samples were fabricated (figure 5), in which the external force was formed by a double-sided electrostatic pectinate drive. The values of electrical tension were obtained for the bifurcation points C and E, and the discrepancy with the results of buckling calculation is within 8%.

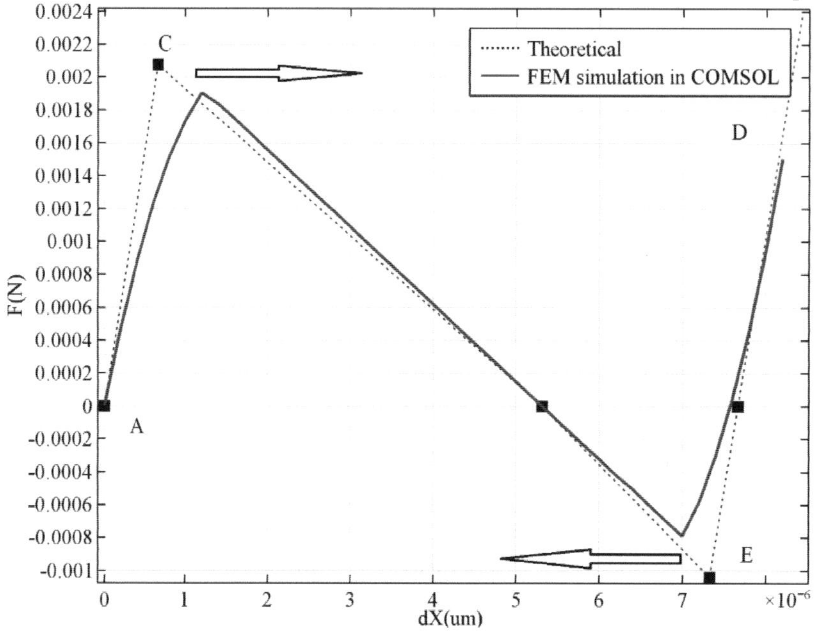

Figure 4. The dependence of the force on displacement

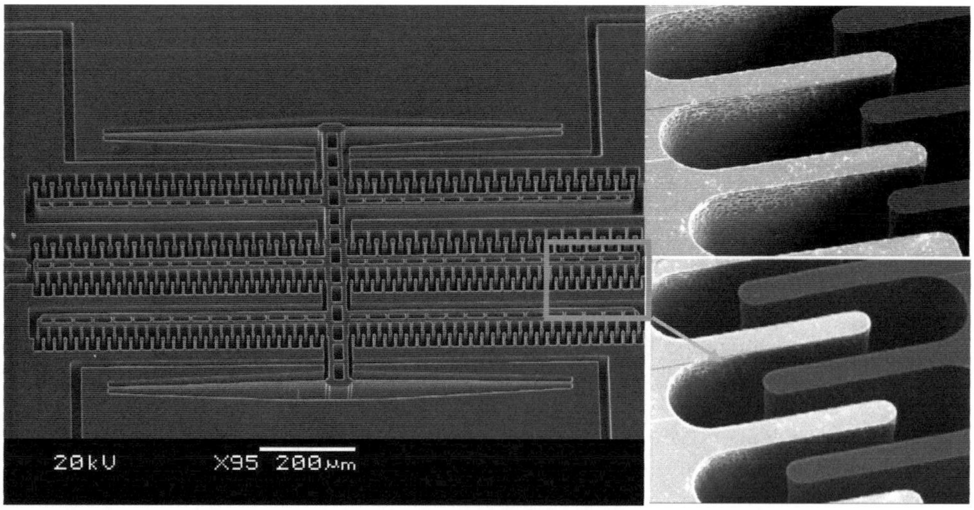

Figure 5. The pectinate actuator of the bistable suspension

After fabrication of micromechanical gyroscopes, the bistable actuator is transferred to the second stable condition. The system must be stable and able to withstand significant external loads. The usage of a structure with bistable beams of a variable thickness was proposed for this. From figure 6 it follows that the height of the potential barrier near the point E is substantially higher. This means that the system of beams of a variable thickness is more stable. During the analysis of the shape of transition and internal tension, it was decided to use a construction with a non-uniform beam profile across the thickness. An arcuate arc-shallow beam can be represented as a V–shaped beam with strong shoulders and joints at the point of fastening and in the center.

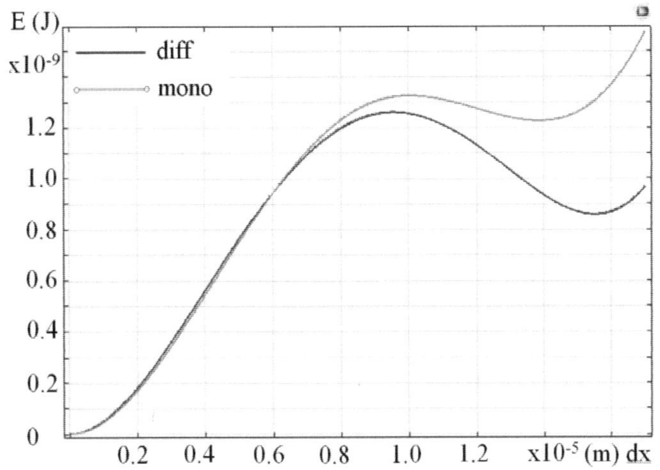

Figure 6. The dependence of potential energy on displacement of bistable beams with a constant (mono) and variable thickness (diff)

5. Conclusion

The present survey considers an actual issue of designing bistable MST devices. The model and methodology for designing arcuate arc-shallow MEMS with buckling was developed and tested in this research. The efficiency of the model is confirmed experimentally. The developed model will significantly simplify and improve designing of bistable systems, and can be used for designing microwave MEMS switches, optical switches, valves and elements of the nonvolatile memory. The main result of the present survey is an increase of the efficiency of the exciting electrostatic comb drive MMG of high precision. The result of this research allows to reduce the control voltage and noise of a microgyroscope.

Acknowledgments
This work was performed with the support of the Federal target program "Research and development on priority directions of development of scientific-technological complex of Russia for 2014 – 2020" 14.578.21.0123 on the topic: Development of design and manufacturing technology of micromechanical sensing elements for navigation systems of high accuracy

References

[1] Acar C and Shkel A M 2005 *J. Micromech. Microeng.* **15** 1092–1101
[2] Farmer S K and Carr W 1998 *Conf. IEEE MEMS* 154-159
[3] Yang Y and Kim C J 1995 *Conf. Solid- State Sensors Actuators.* **2** 337–340
[4] Acar C and Shkel A M 2009 *Springer* 262
[5] Qiu J, Lang J H and Slocum A H 2004 *J. Micromech. Microeng.* **15** 353–356
[6] Zhang Y, Yisong W, Zhihong L, Yubo H and Dachao L 2007 *J. Micromech. Microeng.* **16** 684